Biogas Plants

Wiley Series
in
Renewable Resources

Series Editor:

***Christian V. Stevens**, Faculty of Bioscience Engineering, Ghent University, Belgium*

Titles in the Series:

Wood Modification: Chemical, Thermal and Other Processes
Callum A. S. Hill
Renewables-Based Technology: Sustainability Assessment
Jo Dewulf, Herman Van Langenhove
Biofuels
Wim Soetaert, Erik Vandamme
Handbook of Natural Colorants
Thomas Bechtold, Rita Mussak
Surfactants from Renewable Resources
Mikael Kjellin, Ingegärd Johansson
Industrial Applications of Natural Fibres: Structure, Properties and Technical Applications
Jörg Müssig
Thermochemical Processing of Biomass: Conversion into Fuels, Chemicals and Power
Robert C. Brown
Biorefinery Co-Products: Phytochemicals, Primary Metabolites and Value-Added Biomass Processing
Chantal Bergeron, Danielle Julie Carrier, Shri Ramaswamy
Aqueous Pretreatment of Plant Biomass for Biological and Chemical Conversion to Fuels and Chemicals
Charles E. Wyman
Bio-Based Plastics: Materials and Applications
Stephan Kabasci
Introduction to Wood and Natural Fiber Composites
Douglas D. Stokke, Qinglin Wu, Guangping Han
Cellulosic Energy Cropping Systems
Douglas L. Karlen
Introduction to Chemicals from Biomass, 2nd Edition
James H. Clark, Fabien Deswarte
Lignin and Lignans as Renewable Raw Materials: Chemistry, Technology and Applications
Francisco G. Calvo-Flores, Jose A. Dobado, Joaquín Isac-García, Francisco J. Martín-Martínez
Sustainability Assessment of Renewables-Based Products: Methods and Case Studies
Jo Dewulf, Steven De Meester, Rodrigo A. F. Alvarenga
Cellulose Nanocrystals: Properties, Production and Applications
Wadood Hamad
Fuels, Chemicals and Materials from the Oceans and Aquatic Sources
Francesca M. Kerton, Ning Yan
Bio-Based Solvents
François Jérôme and Rafael Luque
Nanoporous Catalysts for Biomass Conversion
Feng-Shou Xiao and Liang Wang
Thermochemical Processing of Biomass: Conversion into Fuels, Chemicals and Power, 2nd Edition
Robert Brown
Chitin and Chitosan: Properties and Applications
Lambertus A.M. van den Broek and Carmen G. Boeriu
The Chemical Biology of Plant Biostimulants
Danny Geelen, Lin Xu
Biorefinery of Inorganics: Recovering Mineral Nutrients from Biomass and Organic Waste
Erik Meers, Evi Michels, René Rietra, Gerard Velthof
Process Systems Engineering for Biofuels Development
Adrián Bonilla-Petriciolet, Gade P. Rangaiah
Waste Valorisation: Waste Streams in a Circular Economy
Carol Sze Ki Lin, Chong Li, Guneet Kaur, Xiaofeng Yang
High-Performance Materials from Bio-based Feedstocks
Andrew J. Hunt, Nontipa Supanchaiyamat, Kaewta Jetsrisuparb, Jesper T. Knijnenburg
Handbook of Natural Colorants, 2nd Edition
Thomas Bechtold, Avinash P. Manian and Tung Pham
Biogas Plants: Waste Management, Energy Production and Carbon Footprint Reduction
Wojciech Czekała

Biogas Plants

Waste Management, Energy Production and Carbon Footprint Reduction

Edited by

WOJCIECH CZEKAŁA

Poznań University of Life Sciences, Poland

WILEY

This edition first published 2024

Registered Offices
John Wiley & Sons, Inc., 111 River Street, Hoboken, NJ 07030, USA
John Wiley & Sons Ltd, The Atrium, Southern Gate, Chichester, West Sussex, PO19 8SQ, UK

For details of our global editorial offices, customer services, and more information about Wiley products visit us at www.wiley.com.

Wiley also publishes its books in a variety of electronic formats and by print-on-demand. Some content that appears in standard print versions of this book may not be available in other formats.

Library of Congress Cataloging-in-Publication Data

Names: Czekała, Wojciech, editor. | Stevens, Christian V., editor.
Title: Biogas plants : waste management, energy production and carbon footprint reduction / edited by Wojciech Czekała, Christian V Stevens.
Description: Hoboken, NJ : Wiley, 2024. | Series: Wiley series in renewable resources | Includes index.
Identifiers: LCCN 2023046450 (print) | LCCN 2023046451 (ebook) | ISBN 9781119863786 (hardback) | ISBN 9781119863779 (adobe pdf) | ISBN 9781119863922 (epub) | ISBN 9781119863946 (oBook)
Subjects: LCSH: Biogas. | Renewable energy sources.
Classification: LCC TP359.B48 B537 2024 (print) | LCC TP359.B48 (ebook) | DDC 665.7/76–dc23/eng/20231107
LC record available at https://lccn.loc.gov/2023046450
LC ebook record available at https://lccn.loc.gov/2023046451

Cover Design: Wiley
Cover Image: © Lulub/Shutterstock

Set in 10/12pt TimesLTStd by Straive, Chennai, India
Printed and bound by CPI Group (UK) Ltd, Croydon, CR0 4YY
C9781119863786_291123

To my mother and father, who never stopped believing in me.
To my wife for understanding me better than everyone.
To my sons, who fill my heart with joy each and every day.

Contents

List of Contributors

Mohamed Abdallah Department of Civil and Environmental Engineering, University of Sharjah, Sharjah, United Arab Emirates

Muhammad Arslan Department of Energy Systems Engineering, University of Agriculture, Faisalabad, Pakistan

Raul Carlsson Certification Development Unit, RISE Research Institutes of Sweden, Jönköping, Sweden

Yelizaveta Chernysh Faculty of Tropical AgriSciences, Department of Sustainable Technologies, Czech University of Life Sciences Prague, Suchdol, Czechia

Department of Ecology and Environmental Protection Technologies, Faculty of Technical Systems and Energy Efficient Technologies, Sumy State University, Sumy, Ukraine

Viktoriia Chubur Faculty of Tropical AgriSciences, Department of Sustainable Technologies, Czech University of Life Sciences Prague, Suchdol, Czechia

Department of Ecology and Environmental Protection Technologies, Faculty of Technical Systems and Energy Efficient Technologies, Sumy State University, Sumy, Ukraine

Wojciech Czekała Department of Biosystems Engineering, Poznań University of Life Sciences, Poznań, Poland

Yanjun Dai Energy and Environmental Sustainability for Megacities (E2S2) Phase II, Campus for Research Excellence and Technological Enterprise (CREATE), Singapore

School of Mechanical Engineering, Shanghai Jiao Tong University, Shanghai, China

Renjie Dong College of Engineering, China Agricultural University, Beijing, China

Nalok Dutta Department of Biochemical Engineering, University College London, London, UK

Bioproducts Sciences and Engineering Laboratory, Washington State University, USA

Amal Elfeky Department of Civil and Environmental Engineering, University of Sharjah, Sharjah, United Arab Emirates

Kazi Fattah Department of Civil, Environmental, and Architectural Engineering, University of Kansas, Kansas, United States of America

Likui Feng State Key Laboratory of Urban Water Resources and Environment (SKLUWRE), School of Environment, Harbin Institute of Technology, Harbin, China

Jan Jasiński Department of Biosystems Engineering, Poznań University of Life Sciences, Poznań, Poland

Tomasz Jasiński Tomasz Jasiński Biogas Consulting, Nowe, Poland

Mengmeng Jiang College of Engineering, China Agricultural University, Beijing, China

Piotr Jurga Department of Bioeconomy and Systems Analysis, Institute of Soil Science and Plant Cultivation, Pulawy, Poland

Muhammad U. Khan Department of Energy Systems Engineering, University of Agriculture, Faisalabad, Pakistan

Martyna Kulińska Department of Biosystems Engineering, Poznań University of Life Sciences, Poznań, Poland

Jonathan T. E. Lee NUS Environmental Research Institute, National University of Singapore, Singapore

Energy and Environmental Sustainability for Megacities (E2S2) Phase II, Campus for Research Excellence and Technological Enterprise (CREATE), Singapore

Jianju Li State Key Laboratory of Urban Water Resources and Environment (SKLUWRE), School of Environment, Harbin Institute of Technology, Harbin, China

Ee Y. Lim Department of Chemical and Biomolecular Engineering, National University of Singapore, Singapore

Yu Liu State Key Laboratory of Urban Water Resources and Environment (SKLUWRE), School of Environment, Harbin Institute of Technology, Harbin, China

Kai-Chee Loh Energy and Environmental Sustainability for Megacities (E2S2) Phase II, Campus for Research Excellence and Technological Enterprise (CREATE), Singapore

Department of Chemical and Biomolecular Engineering, National University of Singapore, Singapore

Aleksandra Łukomska Department of Biosystems Engineering, Poznań University of Life Sciences, Poznań, Poland

Tatiana Nevzorova Certification Development Unit, RISE Research Institutes of Sweden, Stockholm, Sweden

Wei Qiao College of Engineering, China Agricultural University, Beijing, China

Hynek Roubík Faculty of Tropical AgriSciences, Department of Sustainable Technologies, Czech University of Life Sciences Prague, Suchdol, Czechia

Stelios Rozakis BiBELab, Department of Chemical and Environmental Engineering, Technical University of Crete, Chania, Greece

Abid Sarwar Department of Irrigation and Drainage, University of Agriculture, Faisalabad, Pakistan

Yapeng Song College of Engineering, China Agricultural University, Beijing, China

Yen Wah Tong NUS Environmental Research Institute, National University of Singapore, Singapore

Energy and Environmental Sustainability for Megacities (E2S2) Phase II, Campus for Research Excellence and Technological Enterprise (CREATE), Singapore

Department of Chemical and Biomolecular Engineering, National University of Singapore, Singapore

Katerina Troullaki BiBELab, Department of Chemical and Environmental Engineering, Technical University of Crete, Chania, Greece

To-Hung Tsui NUS Environmental Research Institute, National University of Singapore, Singapore

Energy and Environmental Sustainability for Megacities (E2S2) Phase II, Campus for Research Excellence and Technological Enterprise (CREATE), Singapore

Department of Engineering Science, University of Oxford, Oxford, UK

Simon M. Wandera Department of Civil, Construction & Environmental Engineering, Jomo Kenyatta University of Agriculture & Technology, Nairobi, Kenya

Agnieszka Wawrzyniak Department of Biosystems Engineering, Poznań University of Life Sciences, Poznań, Poland

Liangliang Wei State Key Laboratory of Urban Water Resources and Environment (SKLUWRE), School of Environment, Harbin Institute of Technology, Harbin, China

Xinhui Xia State Key Laboratory of Urban Water Resources and Environment (SKLUWRE), School of Environment, Harbin Institute of Technology, Harbin, China

Hang Yu State Key Laboratory of Urban Water Resources and Environment (SKLUWRE), School of Environment, Harbin Institute of Technology, Harbin, China

Jingxin Zhang Energy and Environmental Sustainability for Megacities (E2S2) Phase II, Campus for Research Excellence and Technological Enterprise (CREATE), Singapore

China-UK Low Carbon College, Shanghai Jiao Tong University, Shanghai, China

Le Zhang NUS Environmental Research Institute, National University of Singapore, Singapore

Energy and Environmental Sustainability for Megacities (E2S2) Phase II, Campus for Research Excellence and Technological Enterprise (CREATE), Singapore

Department of Resources and Environment, School of Agriculture and Biology, Shanghai Jiao Tong University, Shanghai, China

Weixin Zhao State Key Laboratory of Urban Water Resources and Environment (SKLUWRE), School of Environment, Harbin Institute of Technology, Harbin, China

Series Preface

Renewable resources, their use and modification, are involved in a multitude of important processes with a major influence on our everyday lives. Applications can be found in the energy sector, paints and coatings, and the chemical, pharmaceutical, and textile industries, to name but a few.

The area interconnects several scientific disciplines (agriculture, biochemistry, chemistry, technology, environmental sciences, forestry, etc.), which makes it very difficult to have an expert view on the complicated interactions. Therefore, the idea to create a series of scientific books, focusing on specific topics concerning renewable resources, has been very opportune and can help to clarify some of the underlying connections in this area.

In a very fast-changing world, trends are not only characteristic of fashion and political standpoints; science too is not free from hypes and buzzwords. The use of renewable resources is again more important nowadays; however, it is not part of a hype or a fashion. As the lively discussions among scientists continue about how many years we will still be able to use fossil fuels – opinions ranging from 50 to 500 years – they do agree that the reserve is limited and that it is essential not only to search for new energy carriers but also for new material sources.

In this respect, the field of renewable resources is a crucial area in the search for alternatives for fossil-based raw materials and energy. In the field of energy supply, biomass- and renewables-based resources will be part of the solution alongside other alternatives such as solar energy, wind energy, hydraulic power, hydrogen technology, and nuclear energy. In the field of material sciences, the impact of renewable resources will probably be even bigger. Integral utilization of crops and the use of waste streams in certain industries will grow in importance, leading to a more sustainable way of producing materials. Although our society was much more (almost exclusively) based on renewable resources centuries ago, this disappeared in the Western world in the nineteenth century. Now it is time to focus again on this field of research. However, it should not mean a "retour à la nature", but should be a multidisciplinary effort on a highly technological level to perform research towards new opportunities, and to develop new crops and products from renewable resources. This will be essential to guarantee an acceptable level of comfort for the growing number of people living on our planet. It is "the" challenge for the coming generations of scientists to develop more sustainable ways to create prosperity and to fight poverty and hunger in the world. A global approach is certainly favored.

This challenge can only be dealt with if scientists are attracted to this area and are recognized for their efforts in this interdisciplinary field. It is, therefore, also essential that consumers recognize the fate of renewable resources in a number of products. Furthermore, scientists do need to communicate and discuss the relevance of their work. The use and modification of renewable resources may not follow the path of the genetic engineering concept in view of consumer acceptance in Europe. Related to this aspect, the series will certainly help to increase the visibility of the importance of renewable resources. Being convinced of the value of the renewables approach for the industrial world, as well as for developing countries, I was myself delighted to collaborate on this series of books focusing on the different aspects of renewable resources. I hope that readers become aware of the complexity, the interaction, and interconnections, and the challenges of this field, and that they will help to communicate on the importance of renewable resources.

I certainly want to thank the people of Wiley's Chichester office, especially David Hughes, Jenny Cossham, and Lyn Roberts, in seeing the need for such a series of books on renewable resources, for initiating and supporting it, and for helping to carry the project to the end.

Last, but not least, I want to thank my family, especially my wife Hilde and children Paulien and Pieter-Jan, for their patience, and for giving me the time to work on the series when other activities seemed to be more inviting.

Christian V. Stevens
Faculty of Bioscience Engineering, Ghent University, Belgium
Series Editor, "Renewable Resources"
June 2005

1

Anaerobic Digestion Process and Biogas Production

Liangliang Wei, Weixin Zhao, Likui Feng, Jianju Li, Xinhui Xia, Hang Yu, and Yu Liu

State Key Laboratory of Urban Water Resources and Environment (SKLUWRE), School of Environment, Harbin Institute of Technology, Harbin, China

1.1 Introduction

The increasing amount of organic wastes worldwide has become problematic for most countries due to the continuous deterioration of land and water conditions, which poses serious risks to the safety of our community [1]. Moreover, the improper treatment of these organic wastes might lead to the undesired release of huge greenhouse gases (GHGs) into the atmosphere [2, 3]. It was estimated by the Intergovernmental Panel on Climate Change (IPCC) and US Environmental Protection Agency (US EPA) that the global anthropogenic methane emission from municipal solid wastes (MSWs) reached 1077 million metric ton of CO_2 equivalent in 2020 and is expected to increase by 17% in the year 2030. Mitigation practices have forced global action to adopt a technology that can address anthropogenic methane emissions [4]. Numerous available mitigation opportunities currently include the treatment of the organic portion of MSW in a controlled facility and recovering methane as a fuel for on-site or off-site electricity generation [5].

Energy generation from the MSW and the other alternative sources will benefit climate change mitigation and minimize the alarms posed to the environment [6]. There has been a high uptake of renewable energy technologies (RETs) worldwide to deal with the detrimental effects paused by fossil-related energy generation technologies. For a purpose of increasing the energy accessibility while simultaneously restricting the

Biogas Plants: Waste Management, Energy Production and Carbon Footprint Reduction,
First Edition. Edited by Wojciech Czekała

worldwide temperature increased within 2 °C before 2050, adoption of RETs should be highly encouraged and raised significantly. This growing impetus for alternative avenues for renewable energy demands the consideration of different feedstocks, exploring of novel techniques, and improvements of existing technologies.

Bioenergy has been regarded as the most substantial renewable energy source due to its cost-effective advantages and great potential for substituting nonrenewable fuels. Bioenergy derived from biomass materials, such as biological organic matter obtained from plants or animals, is renewable and green. Generally, those biomass energy sources include but are not limited to terrestrial plants, aquatic plants, timber processing residues, MSWs, animal dung, sewage sludge, agricultural crop residues, and forestry residues. Undoubtedly, bioenergy is one of the most versatile renewable energies because it can be made available in solid, liquid, and/or gaseous forms. Different avenues can be explored to harvest energy from biomass materials. Biomethane has a high heating value ranging between 50 and 55 MJ m^{-3} and a low heating value ranging between 30 and 35 MJ m^{-3} [7].

Anaerobic digestion (AD) is practiced extensively for the treatment of biodegradable waste for biomethane generation [8]. This technology has the capability of managing the typical organic wastes such as food waste, lignocellulosic biomass and residues, energy crops, and the organic fraction of municipal solid waste (OFMSW) [9], and its environmentally sound features attracted worldwide attention for biogas production. AD is a microbe-driven, multiphase, and complex biochemical process, and four typical biochemical phases such as hydrolysis, acidogenesis, acetogenesis, and methanogenesis are involved in its whole process. Organic matter could be efficiently metabolized by bacteria and archaea and finally converted into methane and carbon dioxide [10, 11]. However, AD processes are always limited by three main factors: (i) hydrolysis of substrates is the rate-limiting factor for the bioconversion phase; (ii) inefficient utilization of key intermediates such as propionic and butyric acid; (iii) slow growth of anaerobes of methanogenesis [12], and finally lead to a low biomethane recovery rate during their practical operation [13]. Thus, the advancements in the AD process are largely aimed toward one goal: improving biogas production and recovery.

There is currently considerable potential for biogas technology to be developed as a RET that addresses energy and environmental issues. Biogas is a critical technology that provides renewable energy from processing a variety of digestible biomass types. Substrates such as straw, forestry residues, animal and poultry manure, and other organic wastes can be treated within AD systems. The purified biomethane can be integrated into conventional fossil energy supply systems and guarantee the AD technology in energy transformation and ecological civilization construction. However, the biogas industry faces many challenges, including low gas productivity, short biogas tank life, high deterioration rates of digesters, difficulty in digestion residue utilization, and limited economic benefits [14, 15]. To improve the biogas and highlight its role in energy and environmental problem-solving, it is necessary to develop new approaches for the purpose of extending the industrial chain and further exploring new models that can promote the commercialization.

1.2 Basic Knowledges of AD Processes and Operations

1.2.1 Fundamental Mechanisms and Typical Processes of AD

AD, full microbiological degradation process under anaerobic conditions, represents one of the most promising processes to convert diverse organic substrates (animal manure,

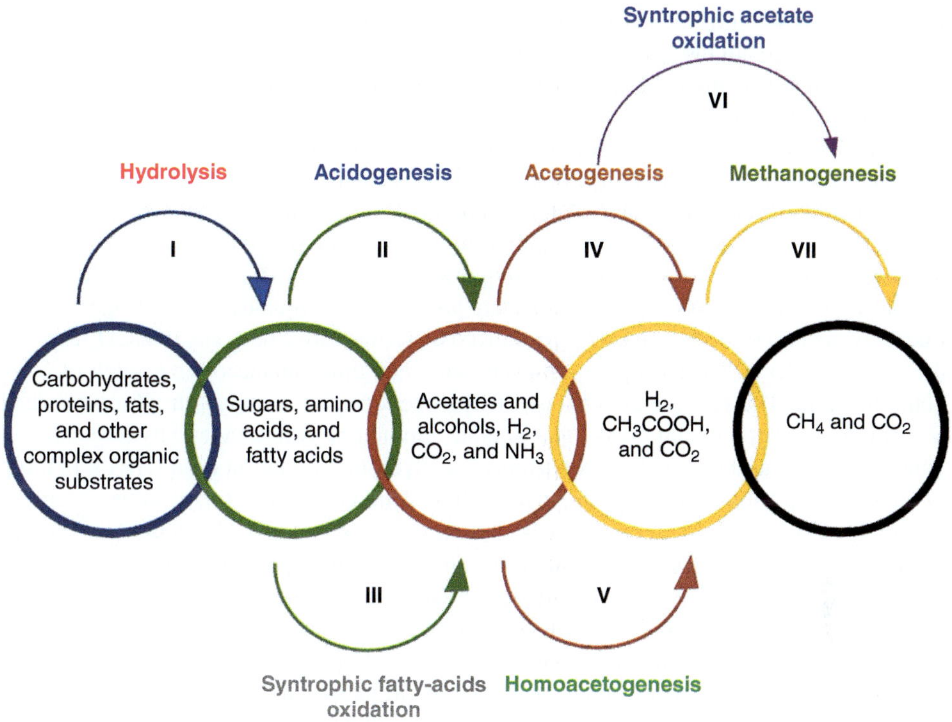

Figure 1.1 *General biochemical process involved in anaerobic digestion. Source: D'Silva et al. [17]/with permission of Elsevier.*

food waste, MSW, and lignocellulosic biomass as agricultural waste) into energy carriers (produced biogas mainly 55–75% CH_4 and 25–45% CO_2) [16].

Microbial ecology in anaerobic digesters is quite complex, and different bacterial and archaeal communities are involved in the digestion process. The AD process is composed of four main steps, namely hydrolysis, acidogenesis, acetogenesis, and methanogenesis (Figure 1.1). The hydrolysis process is the primary step (stage I) in AD where organic polymers (i.e. cellulose, lipids, carbohydrates, polysaccharides, proteins, and nucleic acids) are hydrolyzed into monomers, simple sugars, saccharides, peptides, glycerol, amino acids, and other higher fatty acids, which could be summarized in Eq. (1.1):

$$(C_6H_{10}O_5)_n + nH_2O \rightarrow n(C_6H_{12}O_6) \quad \Delta G^{0'} = -215.67 - 357.87\,\text{kJ} \tag{1.1}$$

Hydrolytic bacteria, also known as primary fermenting bacteria, are facultative anaerobes that hydrolyze the substrate with extracellular enzymes. A wide range of enzymes, i.e. cellulases, hemicellulases, proteases, amylases, and lipases, were generated in this stage and played a great role in the substrate degradation [18]. Undoubtedly, the generation of the aforementioned enzymes enhanced the whole hydrolysis. By contrast, the lack of the suitable enzymes would negatively affect the biogas generation, for instance, the hydrolyzation of lignocellulosic substrates becomes the rate-limiting step of the AD process [18]. During acidogenesis (stage II), primary fermentative bacteria convert hydrolysis products into volatile fatty acids (VFAs), including acetate, propionate, butyrate, valerate, and other acids (i.e. lactate, succinate, and alcohols). Acidogenic bacteria are able to metabolize organic compounds at a very low pH around 4. Methanogenic microorganisms cannot directly use all products from the acidogenic step. Except for acetate, H_2 and CO_2 and some other micromolecular organic acids were abundantly generated during the so-called acetogenic phase (stage III) by secondary fermenting bacteria, also called obligate hydrogen-producing bacteria (OHPB). However, the thermodynamics of these reactions are unfavorable, and these microorganisms can only live in syntrophy with end-product users, i.e. methanogens.

The methanogenic step (stage IV) corresponds to the final conversion of acetate, carbon dioxide (CO_2), and hydrogen (H_2) into biogas, and the obligate anaerobic archaea of hydrogenotrophic and acetoclastic methanogens abundantly exist in the digesters and could transform the mixture of CO_2/H_2 and acetate into methane. Specifically, hydrogenotrophic microorganisms convert H_2 and CO_2, produced by fermentative bacteria, into CH_4 and keep the reactor under a low hydrogen partial pressure and thus enhanced the growth of acetogenic bacteria. The relative abundance of hydrogenotrophic and acetotrophic is variable according to environmental factors (i.e. acetate, ammonia, hydrogen, and hydrogen sulfide concentrations), and operational conditions (i.e. hydraulic retention time [HRT], pH, type of substrate, and source of inoculum), as well as solid contents [19]. It has been reported that the hydrogenotrophic methanogens (i.e. *Methanoculleus* and *Methanobacterium*) are predominated during the start-up of anaerobic digesters and lead to a subsequent decline of the H_2 concentration; Then, a shift of the methanogens into the acetoclastic methanogens (i.e. *Methanosarcina* and *Methanosaeta*) were observed after the stabilization of the reactor [20]. In addition, a high concentration of ammonia of the anaerobic digester benefited for the growth of hydrogenotrophic methanogens in mesophilic anaerobic digestors [21], and approximately 65–70% of the methane generation was closely related to the degradation of acetate; otherwise, the oxidation of acetate to H_2 and CO_2 is the main pathway in the absence of acetoclastic methanogens (such as *Methanosaeta* sp.) [22].

1.2.2 Factors Affecting the AD Process of Biogas Production

1.2.2.1 Temperature

Three different temperature regimes, namely psychrophilic, mesophilic, and thermophilic conditions, with varied optimum temperature ranges for the domination of different strains of methane-forming bacteria, were traditionally used in anaerobic digesters [23]. Specifically, psychrophilic digesters usually operate at about 25 °C, whereas mesophilic ones operate at around 35 °C and thermophilic ones at around 55 °C. Generally, the metabolic activity and bioconversion rate of microorganisms at higher temperature are usually higher than that at lower temperature. However, the much more energy is required for maintaining a high temperature in the fermenter, which increases cost in practical operation [23]. For instance, a much higher degradation rate of fatty acids was observed for the digester operated under 55 °C with a 11 HRT than that operated under 38 °C condition with a 27 day HRT [23]. Similarly, an increase of 54–61% in CH_4 yield from algal remnants was observed when the temperature increased from 25 to 35 °C [24]. In addition, some of the recent works also revealed that reported that the variation of operational temperature, even under a very small range, would decline the biogas production rate of the digesters [25], and the fluctuation of the temperature even 1 °C per day would deteriorate the operation [26].

1.2.2.2 pH

Operational pH might be another main factor that would significantly affect the performance of the digesters, and the most favorable range of pH to achieve maximal biogas yield in AD is 6.8 to 7.2 [23]. Specifically, the methanogenic bacteria are extremely sensitive to pH fluctuations, and their preferred pH was around 7.0, and the growth rate of methanogens was seriously inhibited once the pH declined to <6.6 [27]. Acid-forming bacteria are less pH-sensitive, and the optimal pH for hydrolysis and acidogenesis is between 5.5 and 6.5, despite their tolerated pH ranged from 4.0 to 8.5 [26, 27]. Therefore, some designers prefer the isolation of the hydrolysis/acidification and acetogenesis/methanogenesis processes into two separate stages [27]. At the beginning of the fermentation, the significant accumulation of acids and CO_2, as a consequence of the growth of acidogens and acetogens, leads to a significant decline in the pH. Afterward, the consumption of these acids by the methane-producing bacteria would maintain the digester under a stable condition [23] Excessive fatty acids, hydrogen sulfide, and ammonia are toxic only in their nonionized forms (FA and H_2S–pH below 7, NH_3–pH above 7); thus, the proportional distribution of ionized and nonionized forms of inhibitors of methanogenesis was essential for the stable operation of the digesters.

1.2.2.3 Organic Loading Rate (OLR)

Organic loading rate (OLR), generally defined as kilograms of VS loaded per volume of digester per day, is hence considered as one of the main parameters for stable operation of AD systems [28]. The production of biogas and methane in continuous systems is highly dependent on the OLR value (related to the TS in the digester and the composition of feedstock), and the variation of the OLR would lead to significant variation of the methane yields and system stability. The recent work of Nizami and Murphy (2010) [29] stated that the optimum OLR of the anaerobic digesters ranged from 12 to 15 $kg\,VS\,m^{-3}\,d^{-1}$ for corn silage, while 8.5 $kg\,VS\,m^{-3}\,d^{-1}$ for other substrates [30] clearly demonstrated that the OLR values are highly dependent on the feedstock compositions. Practically, the accumulation of inhibitory compounds, such as VFA or ammonia, negatively affected the increasing operational OLR values of the digesters [31]. Many authors highlight the need for understanding. Thus, OLR needs to be carefully selected by simultaneously considering the feedstock characteristics, inhibitory compound existences, and co-digestion opportunities, to maximize waste treatment capacity and enhance the renewable energy productivity.

1.2.2.4 Carbon–Nitrogen Ratio

Feedstock total organic carbon (TOC), total nitrogen (TN) and their ratio are also critical for the stable operation of the AD systems. The addition of co-substrates, for the purpose of element balance, has been regarded as one of the most common practices for a purpose of achieving stable co-digestion [32], and the optimal C:N ratio of digesters was always ranged from 20 to 30 [33]. The nitrogen in the AD reactor is mainly derived from proteins, and it plays a key role in microbial growth. However, a low C:N ratio in the digesters system (high amount of nitrogen) can produce an ammonia accumulation, subsequently affecting the biogas and methane yields and eventually causing the system to deteriorate [34]. Thus, the additive paper waste or agricultural waste has been traditionally applied to increase the feedstock's carbon content [35].

1.2.2.5 Inoculum-to-Substrate Ratio (ISR)

Inoculum-to-substrate ratio (ISR), which determines the initial ratio between microbial populations, is an important parameter for starting up of anaerobic digesters [36]. The more the inoculum, the higher the number of methanogens in the anaerobic digesters and the better the buffering capacity. Raposo et al. used sunflower oil cake as the substrate to explore the effect of different inoculation rates on AD [37], and they found that the volatile acids were not accumulated under the operational conditions of ISR 1.0–3.0, whereas significant volatile acid accumulation occurs when the inoculation rate is less than 1.0. For instance, the ratio of total volatile acid to total alkalinity was much higher than other experimental groups once the inoculation ratio declined to 0.5.

1.2.2.6 Solids Concentration

The reduced water content of the organic wastes within the digesters is generally regarded as the main reason for the difficulty in the gas and liquid diffusion and the accumulation of inhibitors and in turn reduces the substrate availability and affects their metabolism [38]. A number of studies reported that an increase in the water content of substrate increases the methane yielding and also leads to an excellent homogenization of the AD systems, efficient element diffusion, and effective interaction between microorganisms and nutrients. In addition, the recent work of Le Hyaric et al. (2012) [39] reported that there was a linear increase in the specific methanogenic activity with the increase in water content, ascribing to the improvement of the homogeneity of the digestion reactors [40].

1.2.2.7 Hydraulic Retention Time (HRT)

Retention time of the digesters refers to both HRT and solid retention time (SRT) and was an another important parameter used for designing and optimization of anaerobic digesters (represented in Eqs. (1.2) and (1.3)) [41]. Specifically, HRT represents the retention time of the liquid phase, whereas SRT denotes the retention time of the microbial culture in the digester. Assuming that the feedstock and microbial mixed cultures existed in the same phase in the anaerobic digester, the HRT value of the digestion system equals to SRT. For example, in the AD systems, using food waste, kitchen waste, and MSW as the substrates, the HRT of the system is essentially SRT. In contrast, the interaction between solids and microbial cultures is biphasic for the digesters using waste-activated sludge and primary sludge as substrates and leads to quite different distribution of HRT and SRT:

$$HRT = V \times Q \tag{1.2}$$

$$SRT = \frac{V \times Q}{Q_x \times X_x} \tag{1.3}$$

where V refers to the individual reactor volume (m^3), Q is the influent flow rate ($m^3\ d^{-1}$), X presents the mixed liquid suspended solids in an individual reactor ($mg\ L^{-1}$), Q_x denotes the excess biosolids removal rate ($m^3\ d^{-1}$), and X_x is the mixed liquid suspended solids in excess biosolids flow ($mg\,L^{-1}$).

In general, the chosen HRT during the AD systems operation closely depended on the feedstock compositions, reactor volumes, operational parameters, and biomass activities. For example, those substrates with simple structure (e.g. starch and sucrose) can be easily hydrolyzed and digested, which only needed a much shorter retention time compared to the digesters using complex substrates (e.g. lignin and cellulose). In addition, a high operational temperature increases the decomposition rate of substrates and benefits the declining of the HRT that might be the main reason why majority of the thermophilic reactors are operated under a lower HRT than mesophilic reactors. Generally, a shorter HRT poses serious threat to the bacterial mobilization and consequently elevates the stress of the methanogens [42]. Therefore, the optimization of the operational HRT is usually neither too long nor too short (majority cases lie between 10 and 25 days), although a very high HRT in the order of 50–100 or more days may be needed for digesters operated in colder climates.

1.3 Current Challenges of AD Process and Biogas Production

1.3.1 Ammonia Inhibition

Due to the sensitivity of AD, the accumulation of certain substances in digesters can result in their performance inhibition or process failure. Ammonia is an essential nutrient for bacterial metabolisms, and an optimal ammonia level guarantees sufficient buffer capacity of methanogenic medium in AD systems and subsequentially improves the stability of the digestion [43]. On the other hand, it may also inhibit the methane production if the concentration of ammonia keeps at a high concentration, especially when treating the mixed substrates such as manure or MSW [44]. During the AD, the majority of the nitrogen organic compounds, principally in the form of proteins and urea, are finally converted to ammonia ion (NH_3) and free ammonia (NH_4^+-N) [45]. Therefore, NH_3 and NH_4^+-N are the two major forms of inorganic ammonia nitrogen, which can directly or indirectly cause the inhibition in AD. Particularly, free ammonia (FAN) and total ammonium nitrogen (TAN) exhibit a stronger effect of inhibition in AD when their concentrations reach the threshold value.

The knowledge of how ammonia inhibition occurs is limited, and few studies using pure culture have confirmed that ammonia may affect methanogenic bacteria in two ways [43], mainly including (i) direct suppression of methane-producing enzymes by ammonium ion and (ii) diffusion of the hydrophobic ammonia molecule into the cell and the occurrence of the proton imbalance, as well as the potassium deficiency [34, 46, 47] (Figure 1.2). Specially, partial of NH_3 will convert into NH_4^+ by absorbing protons once they enter the cells and negatively leading to a significant variation in pH. To keep balance in proton, the cells must then use a potassium (K^+) pump to maintain the intracellular pH, thereby increasing the energy consumption and further causing inhibition of the activities of the specific enzymes [46]. The previous studies suggested that the methanogens are the least tolerant and are most likely to be inhibited at higher ammonia concentrations among four types of anaerobic microorganism [34, 49]. Thus, the most recent work of the inhibition of AD process by ammonia is mainly focused on the evolution of methanogenic populations with increasing TAN concentrations.

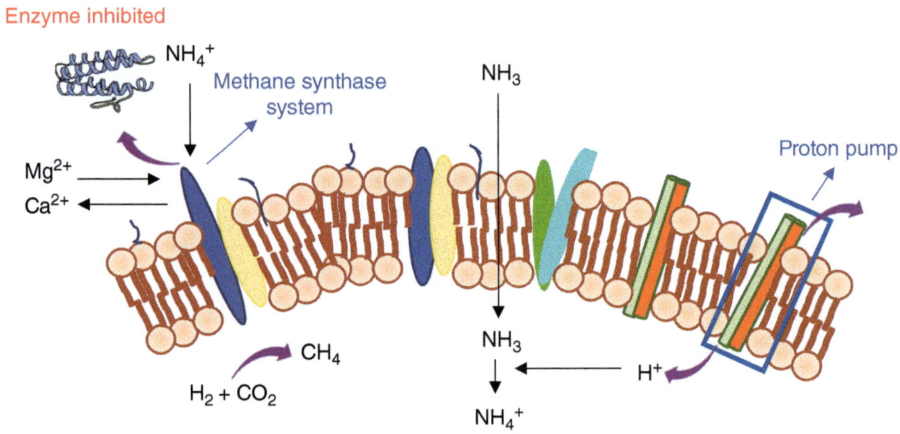

Figure 1.2 *Mechanisms of ammonia inhibition occurred in anaerobic digestion systems [43, 46, 48].*

Tremendous researches have been focused on clarifying the threshold values of ammonia inhibition with different substrates for optimizing AD performance. For example, the recent work of J. Prochazka et al. revealed that the AD bioreactors with different substrates (pig slurry, primary/excess activated sludge, and maize) exhibited a high methane productivity at TAN concentrations of 600–800 mg L^{-1}, whereas low buffering capacity and the subsequent lack of nitrogen as nutrient were observed for the reactor with a lower TAN concentration [48]. For comparison, Abouelenien et al. found that a TAN concentration of 8000–14,000 mg L^{-1} would suppress the AD system fed with chicken manure [50], while 800–1400 mg L^{-1} TAN for the AD of swine slurry [51]. In another study using piggery wastes as the substrate, the concentration of 3000 mg L^{-1} TAN would partially inhibit the digesters [44]. This discrepancy in ammonia concentrations on the inhibition of the digesters is probably caused by the differences in chemical characteristics of the substrates, inocula, and environmental conditions (temperature or pH) [45].

Notably, AD is a biochemical process with multiple phases, and its stability and efficiency closely depend on external and multiple syntrophic interactions among different taxa [52]. For instance, a stable AD reactor with synthetic acetic acid substrate was inhibited at TAN levels of >5000 mg L^{-1} (the corresponding FAN was 256 mg L^{-1}), under an operational pH of 8.0 [43]. For the substrates of 9–10% sewage sludge, as high as a TAN concentration of 2500 mg L^{-1} would cause noteworthy inhibition of the microorganisms under thermophilic conditions [50]. Moreover, a much higher tolerance of TAN concentration even under 5000 mg L^{-1} was observed for the digester operated under a long 40-day SRT, in comparison with the reactor operated under 25-day SRT [53], implying that the ammonia inhibition also correlated with the SRT of the digesters fed with sludge. The recent work of Yenigun and Demirel summarized the ammonia inhibition occurred in AD [44], and the relevant results are cited in Table 1.1.

For the purpose of alleviating ammonia inhibition, tremendous approaches including air stripping, bioaugmentation, and ammonia binding have been widely applied to counteract ammonia inhibition in AD process [60]. However, high operational costs and technical challenges associated with these approaches further hinder their full-scale practical

Table 1.1 Summarization of the threshold values of ammonia inhibition in different anaerobic digestion processes.

Substrates	Digester type	Organic loading rate	Temperature (°C)	pH	TAN	FAN	Acclimation	References
Sludge	Laboratory scale	—	30	7.2–7.4	>5000 mg L^{-1}	—	Yes	[44]
Piggery mature	Laboratory scale	—	30	7.2–7.4	>3075 mg L^{-1}	—	Yes	[44]
Cattle mature	Continuously	—	55	7.9	>4000 mg L^{-1}	900 mg L^{-1}	No	[54]
Slaughterhouse wastes	Laboratory scale	2–3 kg COD m^3 d^{-1}	38	8.1	>6000 mg L^{-1}	—	Yes	[55]
Cattle mature	Continuous stirred tank reactor (CSTR)	2.5 and 6.0 g N L^{-1}	40–64	7.4–7.9	—	>700 mg L^{-1}	Yes	[56]
Slaughterhouse wastes	Semi-batch mode	4.2 kg COD m^3 d^{-1}	38	8.0	>6000 mg L^{-1}	—	Yes	[57]
Cattle manure	CSTR	—	45	7.4–7.9	6000 mg L^{-1}	700 mg L^{-1}	Yes	[56]
Food waste	Anaerobic batch	—	35	7.7	>6 g L^{-1}		Yes	[58]
Piggery manure	CSTR	9.4 g COD L^{-1} d^{-1}	51	8.0	11 g L^{-1}	1450 mg L^{-1}	Yes	[59]

TAN concentration is the start of inhibition concentration.
MSW, municipal solid waste.

application [61]. Exploring low-cost input, easy maintenance, and practical method for alleviating ammonia inhibition will be still the main stream in AD field.

1.3.2 Volatile Fatty Acid Inhibition

VFAs, which mainly include acetic acid, propionic acid, butyric acid, and valeric acid, usually act as the most important intermediate products in the acidogenesis and acetogenesis steps and play a key role in the overall performance of AD systems. During these four steps of AD processes, VFA acts as an important intermediate metabolite that connects acid formers (e.g. acidogenic bacteria and acetogens) and utilizers (e.g. methanogens). As is known to us all, more than 72% of the methane production is derived from acetate, and the majority of the acetic acid in the AD reactors is converted from ethanol, propionate, and butyrate.

Despite VFAs being essential nutrients for methanogens, excessive VFA generation, especially that of propionic acid and butyric acid, under some special cases (e.g. organic overload, nutrient deficiency, toxicant exposure, or other factors) might be the essential reason for the deterioration of AD due to their inhibition of methanogen activity (Figure 1.3). From the perspective of thermodynamics, the degradation process of VFAs usually belongs to the thermodynamic nonspontaneous reaction due to the high Gibbs free energy as described in Eqs. (1.4) and (1.5):

$$CH_3CH_2COOH + 2H_2O \rightarrow CH_3COOH + CO_2 + 3H_2 \quad \Delta G = +76.1\ \text{kJ mol}^{-1} \tag{1.4}$$

$$CH_3CH_2COOH + 2H_2O \rightarrow 2CH_3COOH + 2H_2 \quad \Delta G = +48.1\ \text{kJ mol}^{-1} \tag{1.5}$$

Generally, the degradation of VFA would occur under the digestion condition with a low hydrogen partial pressure, as well as a low concentration of degradation products. However, a large amount of H_2 could not be easily consumed in a short time due to the tightness of the anaerobic fermentation tank, and the accumulated degradation products are also not easily consumed due to the complex microbial relationships; thus, those digesters should be carefully operated.

The inhibition ability of VFAs during the operation of AD reactors correlated well with the operational pH and unionized VFAs concentration. For instance, the activated sludge microorganisms are easily inhibited by unionized VFAs when pH declines to 6.0 and severely inhibited at a lower pH [62]. Furthermore, the continuous accumulation of VFAs will lead to a further declining of system pH and enhance the conservation of VFAs from ionized phase to unionized one [63]. Theoretically, the ionized VFAs could not penetrate the membrane due to the lipid-bilayer base structure of the bacterial plasma membrane; thus, the damage of those ionized VFAs to the cell is negligible. By contrast, the unionized VFA could penetrate the membrane freely due to its smaller size and nonpolar characteristics [65] and undoubtedly cause serious damage to DNA and proteins, ascribing to its lipophilic characteristics for passing through the cell membrane freely [64]. This suggests that the dissociation state of the organic acids is more decisive for microbial activity than the total concentration of VFAs. In addition, the dissociated H^+ can acidify the cytoplasm of the biomass within the digester; thus, the cell needs to export H^+ via a proton ATPase pump mechanism, which is energy demanding and may result in

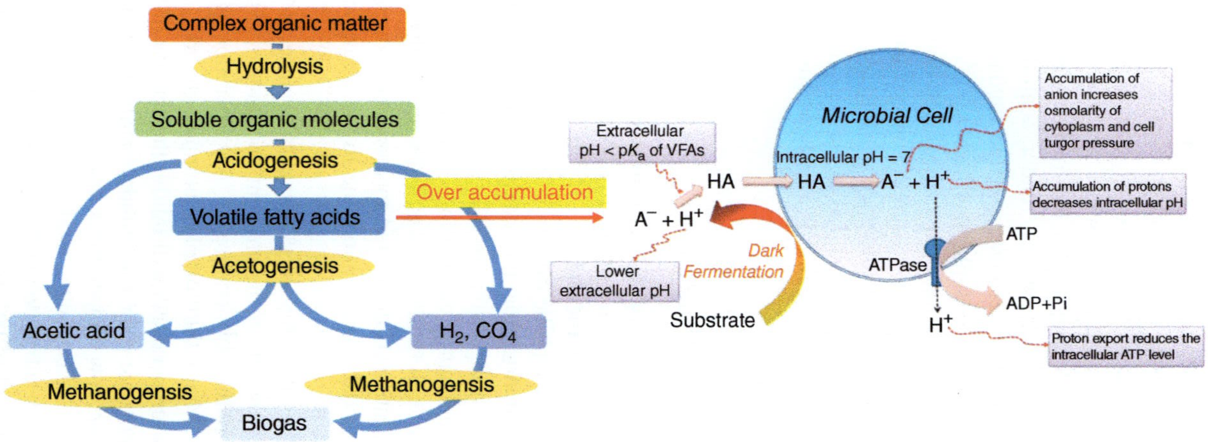

Figure 1.3 *Inhibition mechanisms of volatile fatty acids on anaerobic digestion [62–64].*

energy depletion [64]. With the gradual accumulation of H^+, the gradual decline of the intracellular pH would finally lead to the cessation of the cell growth once the pH drops to the limit value of the biomass [65] (Table 1.2).

Aforementioned experimental results revealed the inhibition mechanism of different VFAs on individual digesters. In practical AD systems, those microorganisms, such as hydrolytic acid-producing bacteria, acetic acid-producing bacteria, and methanogenic archaea, exhibit different tolerances to VFA and pH. Specifically, hydrolytic bacteria and fermentation acid-producing bacteria have a wider pH tolerance range (4.0–8.5) and a stronger tolerance to VFAs, while majority of the methanogenic archaea are susceptible to the VFAs inhibition and could grew only under neutral pH conditions (6.8–7.2). From the aforementioned analysis, it is clear that the fundamental mechanism of acid inhibition is that when the digestive system has a higher VFA concentration due to high organic loading or imbalance of substrates, the imbalance between the hydrolytic acid production (upstream of the digestive system) and the methanogenic process (downstream of digestion) occurs due to the different tolerance capacity of microorganisms, and the higher generation rate of hydrolytic acid than that of the methanogenic consumption leads to the VFA accumulation and negative feedback, which eventually leads to the collapse of the digestive system.

1.3.3 Psychrophilic Temperature Inhibition

Temperature is a critical factor affecting AD performance because of the influence of both system heating requirements and methane production. Although AD can successfully operate under psychrophilic (15–25 °C), mesophilic (35–40 °C), and thermophilic (50–60 °C) conditions, mesophilic and thermophilic digestions have been typically recommended for CH_4 generation [75]. It was reported that thermophilic process shows tremendous advantages in achieving high rates of digestion, excellent waste organics conversion, fast solid liquid separation, and insignificant accumulation of viral pathogens [44]. For comparison, mesophilic temperatures can relieve the fast FAN accumulation and are more resistant to higher FAN concentration as compared to mesophilic digestion [43]. Adversely, psychrophilic digestion attracted less attention due to the weak performance and single substrate demanding.

As for psychrophilic anaerobic digesters, the performance of CH_4 production and solid removal is closely dependent on the ambient conditions [76]. The main inhibition mechanism of the psychrophilic temperature on AD operation is the slow hydrolysis, the first step of digestion where the complex compounds are converted into soluble simple structures [77]. Low temperatures will inhibit the activity of hydrolytic microbials, especially for those methanogenic biomasses, undoubtedly a notable decline in digestion performance [78]. In addition, low temperatures also lead to a poor mixing of the sludge bed due to the considerable decreases in the biogas production [79] that might be the main reason why those cellulose-rich substrates, such as grass, were traditionally digested under the mesophilic condition instead of psychrophilic condition, for their complex chemical structures and difficult hydrolysis characteristics. Cysneiros et al. [77] pointed out that the low-temperature AD of grass could be feasible only under the condition that coupled the efficient hydrolysis to the digestion. Nowadays, more attention should be paid on how to improve the hydrolysis in AD process under psychrophilic conditions for achieving a high CH_4 production.

Table 1.2 Summary of related research on acid inhibition in wastewater anaerobic digestion.

Substrates	Reaction mode	Organic loading	Temperature (°C)	VFA concentration	Inhibition effect	References
Gin spent wash	Semicontinuous mode	32 kg COD m^{-3} d^{-1}	(36 ± 1)	14.7 g COD_{VFA} L^{-1}	Methane production rate and COD removal rate decreased	[66]
Cassava wastewater	Continuous mode	15 kg m^{-3} d^{-1}	55	350 mg L^{-1}	System pH and alkalinity sharply decreased	[67]
Cattle slaughterhouse wastewater	Semicontinuous mode	1.82 g L^{-1} d^{-1}	38	>400 mg L^{-1}	VFA accumulated rapidly and the gas production rate decreased	[68]
Ethanol wastewater	Batch mode	18 kg COD m^{-3} d^{-1}	37	100 mg L^{-1}	Biogas production rate decreased, COD of effluent increased	[69]
Glucose wastewater	Semicontinuous mode	4.67 g COD L^{-1} d^{-1}	37	1400 mg L^{-1}	Methane production rate decreased, VFA accumulated	[70]
Food wastewater	Batch mode	50 g COD L^{-1}	(28 ± 2)	7500 mg L^{-1}	Biogas production rate decreased, COD of effluent increased	[71]
Nonfat dry milk	Semicontinuous mode	2 g COD L^{-1} d^{-1}	(35 ± 2)	>1000 mg L^{-1}	pH of the effluent decreased, COD of effluent increased	[72]
Olive mill wastewaters	Continuous mode	1.87 g COD L^{-1} d^{-1}	37	60 mM	Methane yield and biogas production rate decreased	[73]
Rapeseed oil	Continuous mode	10 kg VS m^{-3} d^{-1}	50	1063 mg L^{-1}	pH decreased, VFA content increased	[74]

1.4 Proposed Strategies for Enhanced Biogas Production

1.4.1 Promoting Direct Interspecies Electron Transfer via Conductive Materials Additive

AD process is fundamentally the coupling between the oxidation and reduction of different chemical compounds, accompanying with significant electron transfer [80]. For the syntrophic methanogenesis, interspecies electron transfer is the key approach for accelerating the reaction kinetics [81, 82], in which the hydrogen and formate are electron donors and act as a shuttle for interspecies electron transfers. Thus, traditionally methanogenesis is often the rate-limiting step in the AD process for the reason that the interspecies electron transfer is driven by molecular diffusion [80]. Thus, how to enhance the electron transfer efficiency is essential and urgent for AD performance improvement.

The additive of conductive materials (e.g. biochar, magnetite, and carbon nanotube) has, recently, been proposed and recognized as one of the most promising methods for accelerating the electron transfer between methanogenic and acetogenic microbes (Figure 1.4), which could effectively improve biogas productivity [84]. Briefly, conductive material additives could not only relieve methanogenesis inhibition and recovery biogas production but also shorten the lag time and improve methane yielding (Table 1.3). As a result, those electro-active microorganisms would be efficiently enriched via the continuous additive of conductive materials, and the potential VFA metabolism may shift from a thermodynamically unfavorable interspecies hydrogen transfer (IHT) pathway to direct interspecies electron transfer (DIET) pathway [95, 96]. Usually, the DIET process exhibits a more rapid electron transferring rate (in e cp^{-1} s^{-1}) than that of IHT (4.49×10^4 versus 5.24×10^3), which is the main reason for enhancing the organics conversion to methane [97]. Besides, the cell-to-cell electron transfer needs less energy for hydrogen/formate (as electron transfer shuttle) production (Figure 1.5), which accelerated the syntrophic conversion of various organics to methane [99].

Except for increasing electrical conductivity, the addition of conductive materials provides a solid surface for microbial colonization due to its high specific surface area and abundant porosity [100]. Sequencing analysis revealed that the abundance and diversity of microbial communities in biochar-added reactors were much higher than those without biochar additive and might be the main reason for the biogas production acceleration [100]. For instance, those acidification bacteria, including *Bacteroidetes*, *Synergistetes*, *Chloroflexi*, and *Planctomycetes*, have been recently reported were enriched in the biochar-amended reactors [100]. Moreover, as compared with the bacterial community, porous biochar exhibits a more significant enrichment in those methanogenesis archaea (e.g. *Methanosaeta*, *Methanosarcina*, *Methanobacterium*, *Methanolinea*, and *Methanosarcina*) [100, 101]. Besides, biochar addition also improves the activity of key enzymes of methanogenesis (e.g. dehydrogenase enzyme and coenzyme F420), which is great meaningful to the enhancement of the methane production [102].

Besides, addition of conductive materials also significantly affects the oxidation–reduction potential (ORP) distribution of the AD system. ORP, which regulates the anaerobic metabolisms, is the key parameter for AD system operation, and a negative ORP promotes methanogenesis [103]. Generally, a range of ORP from −200 to −400 mV provides the most favorable condition for methanogenesis [103]. The recent work of

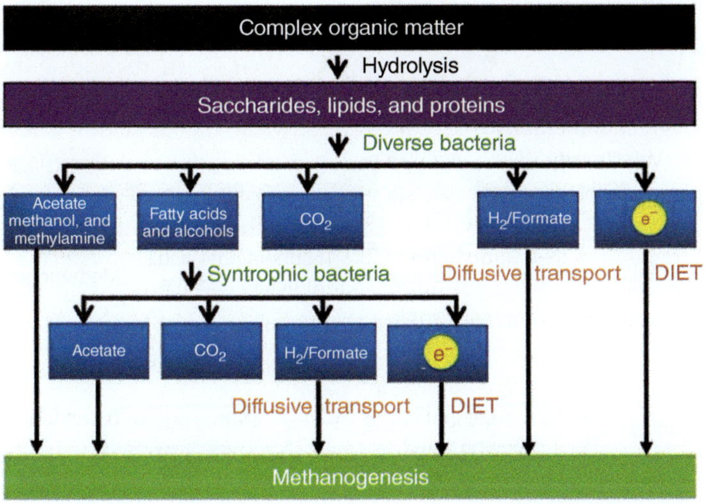

Figure 1.4 Schematic diagram of methane formation during anaerobic digestion. Source: Zhao et al. [83]/with permission of Elsevier.

Table 1.3 Performances of methane production in AD system with conductive materials additive.

Conductive materials	Dosage ($g\,L^{-1}$)	Reactor type	Substrates	Digestion time (d)	Methane production increasement (%)	References
Fe_3O_4	10	Batch	Sewage sludge and food waste	18	16	[85]
Fe_3O_4	10	Batch	Organic wastewater	11	78.3	[86]
Graphene	1	Batch	Organic wastewater	12	25	[87]
Graphene	0.03	Batch	Organic wastewater	55	13.4	[88]
Magnetite	27	Batch	Waste-activated sludge	55	7.3	[89]
Granular-activated carbon	0.5–5	Batch	Waste-activated sludge	20	17.4	[90]
Granular-activated carbon	5	Batch	Organic wastewater	6	66	[91]
Biochar	10	Batch	Food waste and sludge	30	24	[92]
Biochar	10	Batch	Oil	172	13.3	[93]
Biochar	5	Batch	Food waste	27	18	[94]
Biochar	10	Batch	Sewage sludge and food waste	18	23	[85]
Biochar	20	Batch	Food waste and sludge	22	70	[95]

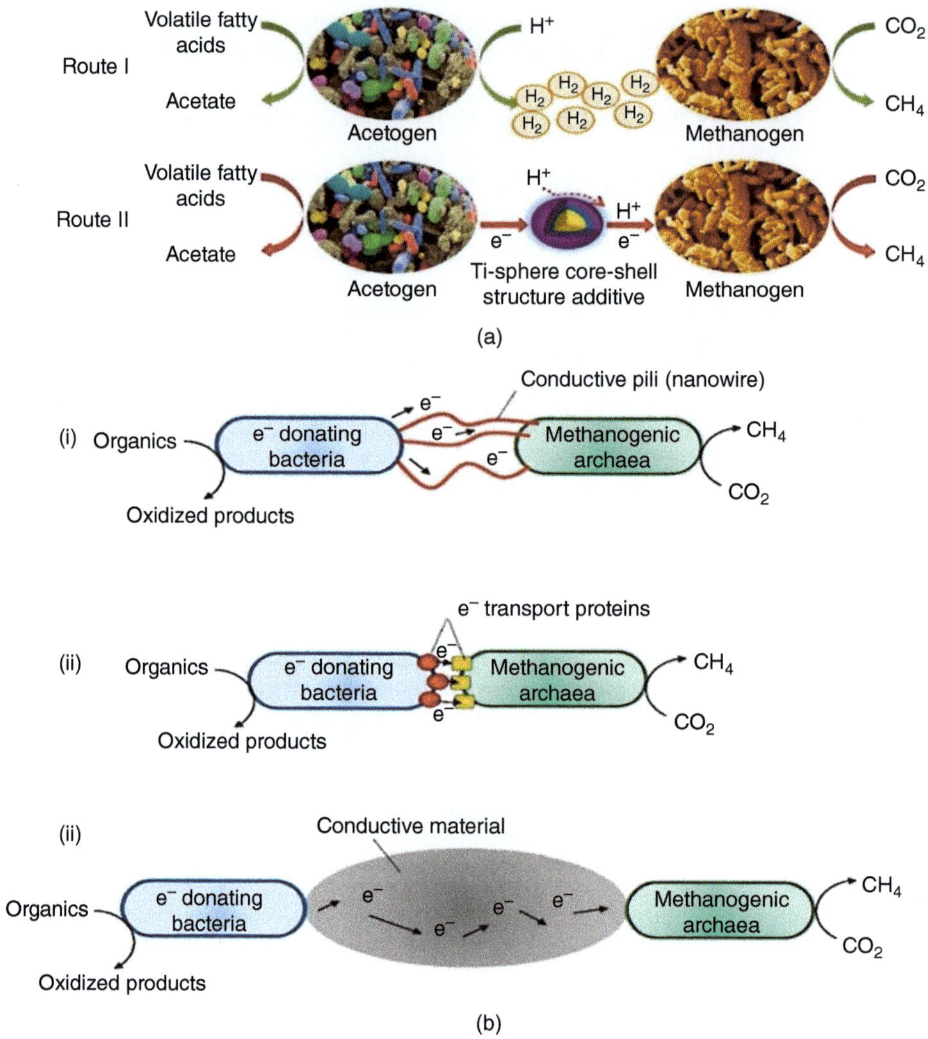

Figure 1.5 Digestion mechanism of methanogen (a) volatile fatty acid to acetate and (b) electro generation. Source: Zhang et al. [98]/with permission from Elsevier.

Salvador et al. revealed that an increase in ORP from −240 to −189 mV after carbon nanotube addition (5 g L^{-1}) promotes the growth of Methanobacterium formicicum [104].

1.4.2 Co-digestion of Different Substrates

AD of biomass wastes could be successfully operated under both individual substrates (mono-digestion) and mixture substrates (mixed-digestion or co-digestion). As compared to mono-digestion, co-digestion enhances the methane generation via the increasement of the organic loading and the availability of nutrients for microbes while reducing the inhibitory of chemical toxicity via co-substrate dilution. Mono-digestion is commonly employed for digesting sewage sludge, animal manure in smaller biogas production facilities, whereas co-digestion is frequently employed in larger facilities for various organic waste treatment (farms, residential areas, and industry).

Customarily, AD technology was meant for one feed material, but lately it has been recognized that the system turns out to be more stable when a diversity of substrates are co-digested and simultaneously enhance the biogas productivity [105–107]. The previous works revealed that a balanced C/N ratio, achieved through co-digestion of different feedstocks, would relieve the accumulation of VFA due to its higher buffering capacity, even under a high OLR [108, 109]. For example, the co-digestion of food waste with trace element-rich piggery wastewater can avoid VFA accumulation [110]. Similarly, the mixing of the sludge with organic wastes equilibrates the C/N ratio and enhances biogas production [111]. Anaerobic co-digestion benefits the synergistic interactions via nutrient balancing, trace element supplementation, toxic inhibitory compound dilution, and microbial diversity promotion. For this reason, the knowledge of the characteristics of each substrate and their digestion behavior helps scientists to identify the best "organic couples" for a better synergetic digestion (Figure 1.6).

Generally, numerous nutrients, originating from the biomaterials and organic wastes, are abundantly existed and necessary for the growth of microorganisms. The need for the nutrient is interconnected with the age, geographical origins, and species of the organic wastes. Majority of the agricultural residues and aquatic plants are enriched with high nutrients; however, their lignocellulosic recalcitrant nature renders them resistant to micro-bacterial degradation. Co-digesting these multifaceted biomaterials with animal manures and other biodegradable organic substances gives enough access to microorganisms to foster optimized degradation [113].

Briefly, a recent investigation revealed that a higher amount of biogas was generated from co-digestion of Eichhornia crassipes, poultry waste, and cow manure, as compared to the mono-digestion [114]. In addition, the advantages of immaculate digestibility, supreme manure production, excellent odor management, low operational cost, and environment-friendly behavior were also observed for the co-digestion [115]. The most commonly used wastes for sludge co-digestion are food waste, MSW, agro-industrial waste, and fatty waste. Specifically, lipids, which are widely existed in meat processing by-products, fatty wastewater, and some agro-industrial residues (such as olive and soybean residues), have been proven to exhibit the highest methane productivity potential due to its abundant existence of carbon and hydrogen atoms in their molecules. Carbohydrates in those organic wastes are easily biodegradable and well known for their rapid conversion, whereas they deliver a relatively low methane yield. Carbohydrates are widely distributed in those agricultural wastes and in the OFMSWs, especially those food wastes. Proteins are essentially found in wastes from slaughterhouses and meat processing and are suitable for co-digestion because of their high organic content. However, the high nitrogen concentration and the production/existence of ammonia of those proteins may cause a serious inhibition under a mono-digestion.

The major advantages of co-digestion are the improvement of biogas yield as well as the increase in the methane content within the biogas. Animal manures have been widely co-digested with other biodegradable materials to increase economic effectiveness while ensuring AD system stability at a commercial scale [116]. According to the recent

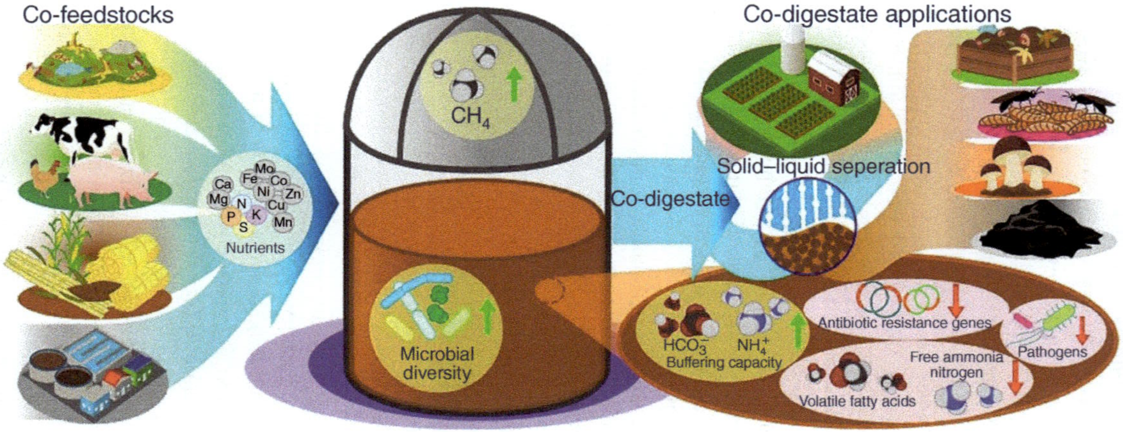

Figure 1.6 Co-digestion of different substrates to enhance methane production. Source: Karki et al. [112]/with permission of Elsevier.

survey, the majority of commercial reactors employ mono-digestion mainly due to the availability of specific substrate in large quantities within the vicinity of the digester geographic location. Other reasons for nonimplementation of anaerobic co-digestion include ignorance, insufficient co-digestion technical expertise, reluctance to adopt new technology, and avoiding the drawbacks of co-digestion. Some of the major drawbacks of co-digestion, which hamper the application of the technology with commercial reactors, include the accumulation of undigestable solids/VFAs inside the digester and high nitrogen backload [117]. The synergistic effects of the co-substrate mixture, which are brought about by the dynamics of the co-digestion process, will outcome the drawbacks. With the advancement of technology, inclusive pretreatment, regulation, and control strategies have been explored for enhancing the efficiency of the co-digestion. However, development into the co-substrate blending proportions needs to be further investigated for a wide variety of co-digestion substrates.

1.4.3 Bioaugmentation

Bioaugmentation is the practice of adding specific microorganisms to a system to achieve a desired biomass activity and provide a meaningful pathway to improve the efficiency of AD, and the previous works revealed that bioaugmentation would successfully reduce the start-up period [118], shorten the HRT [119], and decrease the recovery time of anaerobic digesters suffering from oxygen stress [72] or organic overloading [120]. Furthermore, bioaugmentation can also improve the performance of AD, including increase in methane productivity from cellulosic waste [121, 122], digested sludge (mainly proteins and polysaccharides) [123], lipid-rich wastes [124], ammonia-rich substrate [125], and long-chain fatty acids (LCFAs) [126].

Temperature is one of the key limiting factors for the worldwide application of AD technology, because fermenting microorganisms have strict temperature requirements [127]. An increasing number of reports indicate that psychrophilic AD may become an economical and easy-to-operate method of processing biomass feedstock (FS) in areas where climate conditions are not suitable [128, 129]. In terms of improving AD efficiency, the recent works found that both of the adding of propionate-utilizing cultures [120] or VFA-degrading culture [130] would reduce the propionate accumulation and improve the performance of digestion. Bioaugmentation has been proven to be an effective way to counteract ammonia inhibition, for instance, the introduction of hydrogenotrophic methanogens increases methane production at high ammonia levels [125]. However, not all bioaugmentation cases lead to a positive impact on digestion performance, for example, the addition of syntrophic acetate-oxidizing cultures enhanced the digestion performance or stability insignificantly [131]. It might be due to methanogens playing a more important role than syntrophic acetate-oxidizing culture in AD under high ammonia levels [132].

Bioaugmentation has been successfully applied in batch tests of AD processes. For continuous reactors, the major challenge for bioaugmentation is how to ensure that the introduced microorganisms are able to thrive and are not washed out from the reactor [125]. To maintain the activity of the exogenous microbial population, routine bioaugmentation for the continuous reactor has been highly recommended [133]. Some studies have suggested that temperature affects the initial hydrolysis rate of the substrate in the reactor and leads to

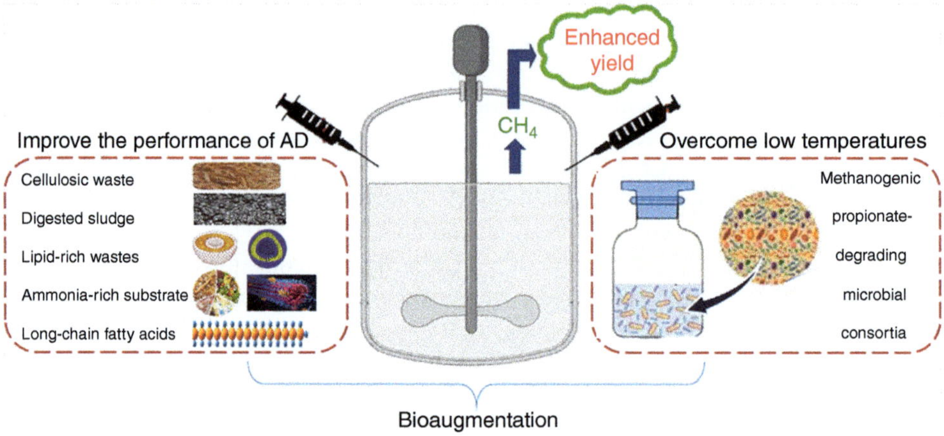

Figure 1.7 Improvement of biogas production via bioaugmentation [137, 138, 140].

longer fermentation cycles [134], thereby affecting the subsequent fermentation stage and biogas production [135, 136]. At the microbial level, a low-temperature environment has a significant impact on the performance of methanogenic and acidogenic microorganisms [127, 137], especially for those archaea, which might be the main reason for the troubles in starting up of digesters at a low temperature.

To enhance the performance of AD at low temperatures, tremendous approaches including pretreatment of substrates [138], co-digestion with other substrates, improvement of anaerobic fermentation reactors [139], introduction of physical or chemical additives [140], and acclimation of inoculums [137] have been widely explored. Among the aforementioned processes, bioaugmentation is a more targeted process (Figure 1.7), which is a direct way to improve the fermentation performance by adding microorganisms with specific biodegradability [141]. Previous studies have shown that adding methanogenic propionate-degrading microbial consortia can accelerate VFAs degradation and improve anaerobic fermentation performance [142, 143]. At present, bioaugmentation mainly focuses on the mesophilic and thermophilic AD systems.

1.4.4 Bioelectrochemical System-Assisted AD

Bioelectrochemical treatment can increase biogas production during the operation of the AD (biomass retention on electrodes rather than electrical current enhances stability in AD), and main reason could be summarized as: (i) improvement of substrate degradation rate, especially those refractory and complex substrates; (ii) excellent redox potential between electrodes benefit for the enrichment of functional microorganisms (extracellular electron transfer from cathode to microbes application for biofuel production); and (iii) accelerated oxidation of intermediate VFAs that increases the acid breakdown threshold [144].

Recently, the system of microbial electrolysis cell-assisted anaerobic digestion (MEC-AD) has been proven to be an effective system for methane production from organic waste, which enhances electron transfer and accelerates methane production from biomass hydrolysates [145]. In MEC-AD system, MEC can effectively increase

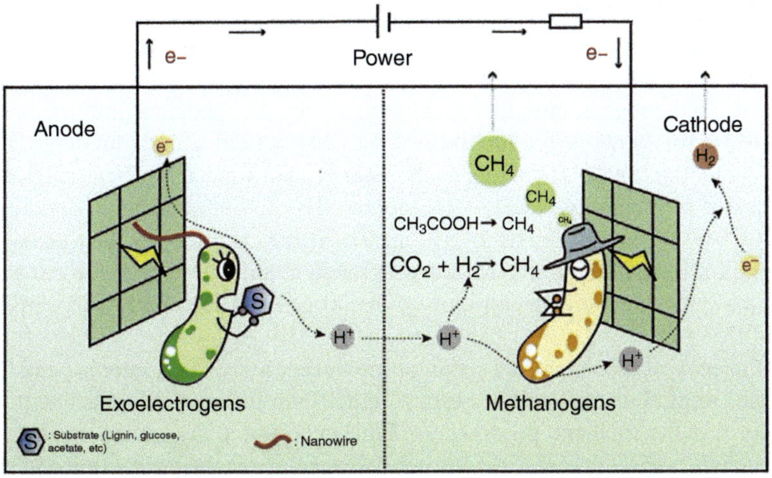

Figure 1.8 Schematic diagram of methane formation during bioelectrochemical system-assisted AD. Source: Yu et al. [146]/with permission of Elsevier.

the degradation rate of the recalcitrant and complex substrates and increase biogas production by changing the microbial community involved in the MEC-AD process (Figure 1.8) [146].

Maintaining the voltage at 0.8 V, the MEC-AD system yields much higher methane productivity of raw waste-activated sludge and heat-pretreated waste-activated sludge, which were respectively 7.8 and 2.1 times higher than those without voltage supply. Once the applied voltage was reduced to 0 V, the enhanced observation of heat-pretreated waste-activated sludge almost disappeared, whereas that of raw waste-activated sludge was still 6.2 times of that without voltage supply. The application of voltage in raw waste-activated sludge is not only beneficial to the enrichment of electricity-producing bacteria and methanogens but also essential to the enrichment of fermentation bacteria and acetotrophic bacteria in the two electrode biofilms might be the main reason for the observation of the insignificant variation of the digestion system, before and after voltage supply, using raw waste-activated sludge as substrates. Undoubtedly, the enhanced hydrolysis fermentation and synergistic effect of acetyl-producing bacteria and hydrogen-producing methanogenic bacteria may be the main reason for the efficient methanogenesis in MEC-AD [147].

Compared to the voltage-free condition, a 0.6 V voltage applied in the system can not only increase the methane production rate by nearly twice under medium and low organic loading (bovine albumin [BSA] concentrations of 500 mg L^{-1} and 4 g L^{-1}) but also a 225.4% increase in methane productivity under high organics loading (BSA concentration of 20 g L^{-1}). The mechanism exploration showed that the applied voltage significantly enhanced the acid and methane production in digestion of protein-related substrates. Microbial community analyses showed that applied voltage increased the abundance of fermentation bacteria in anode by 46.7%, while the abundance of methanobacteria in cathode increased from 10.4% to 84.3%, indicating that the methanogenic pathway changed from acetic acid decomposition to hydrogen nutrition. External circuit electron transfer calculations show that only 10% of methane productivity was attributed to the DIET. From a thermodynamic point of view, the applied external voltage reduced the cathodic potential to −0.9 V, which was meaningful to the increased methane productivity by methanogens enriched in hydrogen via the mediated interspecies electron transfer [145]. Total energy production in MEC-AD system using cow dung as substrate is higher than that in AD system (bulk energy recovery efficiency reached 324%); meanwhile, the productivity of $CH_4 + H_2$ increased by 137.9% during the steady operation of the MEC-AD system, whereas 14.5% increased for chemical oxygen demand (COD) conversion [148]. In addition, the MEC-AD system could also reduce the content of CO_2 in biogas and increase the concentration of useful gases (CH_4 and H_2) to a maximum of 95.8% [149].

The methane production rate and stabilization time of MEC-AD system, using food waste as substrate, were about 1.7 and 4.0 times higher than those of AD reactor. It should be pointed out that the methane production of the two systems is similar to the theoretical maximum methane production. Therefore, MEC accelerates the production and stability of methane through bioelectrochemical reactions but may not increase the methane production to exceed the theoretical value [150]. It is worth to note that AD system should not be combined with microbial fuel cells for a purpose of guarantying the methanogenic efficiency, because microbial fuel cells would inhibit the activity of methanogenic bacteria and enrich the electricity-producing bacteria [151].

1.5 Techno-Economic and Environmental Assessment of Anaerobic Digestion for Biogas Production

1.5.1 Techno-Economic Analysis

A techno-economic assessment of different AD processes for biogas production and production cost evaluation is summarized here, and for those economic indicators, only the costs associated with the operational phase were considered. Specifically, those economic indicators mainly included sludge management indicators, electricity consumption indicators, and chemical consumption indicators. Biogas is considered as a product; thus, it is computed for the calculation of revenues [152]. To address the technical efficiency, carbon neutrality, and economic cost reduction of AD and biogas production process, under the sustainability criteria, were highly concerned for proper sludge management and energy-saving requirements [153–155].

Meanwhile, AD systems also exhibited the economic benefits, such as reduction in fossil fuel consumption in waste management systems by utilizing the biogas as energy, electricity and heat sources, income generation by selling excess energy and heat, and fertilizer generation/utilization for improving soil fertility and structure [156]. However, significantly different among the substrates types and compositions, digester scale, operating conditions, government incentives, and potential use of products were for those AD systems constructed throughout the world. A farm-based AD system was built in a farm and has been to be profitable for effective organic waste disposal and sustainable development, via the full utilization of energy production, charging gate fees for accepting solid wastes, generating income from co-products such as compost/organic fertilizer, and potentially selling carbon credits obtained from offsetting GHGs [157]. Moreover, the synergistic management of AD plants and farms could also share resources such as labor and machineries, thus contributing toward positive economics of these systems.

A major challenge for practical utilization of AD systems is how to efficiently operate those digesters throughout the year, regardless of operationing temperature changes and substrates variations. Logistics of feedstocks/products are also essential for the economic and environmental feasibility judgment of the AD systems because a long distance transportation could increase the biogas production costs. Besides, the majority of AD systems, in certain regions such as Europe, obtained abundant subsidies from governmental agencies and various state incentive programs. Thus, the economic feasibility of AD systems needs to be re-assessed for considering the variability of all aforementioned specific aspects.

Capital and feedstock costs were one of the predominant contributions for the total energy generation cost of the AD systems, depending on the feedstock sources [158]. For instance, feedstocks such as animal manure have negligible feedstock cost, whereas they require a large digester size for their low biogas yielding, thus leading to a high capital cost [159]. The relevant publications on the key findings of the economic feasibility of AD systems, using various feedstocks, were summarized and analyzed. Regarding the capital cost, as high as US$ 100,000 (ton COD d)$^{-1}$ was reported during the practical operation of the AD-MFC system [160, 161]. By contrast, the operational costs of another AD system were estimated at about US$ 0.048 (kg COD)$^{-1}$ via lab-scale simulation [162]. The revenue values for biogas in practical AD system operation were about US$ 0.03 kWh^{-1}. Generally, the capital cost is the main contributor to the production cost for AD systems.

Operating cost of the AD systems, varied from US$ 20 to US$ 110 t^{-1} of feedstock handled, depends on AD plant size. A recent case study, via the evaluation of 38 AD systems in the United States, demonstrated that the investment on the electrical generation equipment costs approximately 36% of the total capital cost [163]. The cost of electricity production, for each kWh of electricity generated, varies from US$ 0.06 to US$ 0.23 during the AD system's operation. The electricity production cost depends on the type of AD plant used in the system as well as the feedstock used [158]. For instance, organic fraction of MSW is more economical than that of animal manure [164], ascribing to the higher gate fee and higher biogas yielding ability of the MSW. Electricity generation cost is usually lower for the digesters with a large plant size due to the effect of economies of scale [158, 165].

In addition, chosen strategies of bioenergy conversion pathways of the different AD systems are examined, following the life cycle analysis, by considering, on the one hand, the electricity production costs and, on the other hand, the profitability of the pathways (expressed as gain or loss by operating a specific bioenergy plant). Specifically, the calculation of the profitability of the AD technical pathways is based on the annuity method (a dynamic investment calculation), where the net present value of an investment is equally distributed over the lifetime of an investment project (Figure 1.9). The obtained annuity, regular, and equal payments per period over the considered lifetime were calculated, and a positive result represents that the investment was feasible (the invested cost plus the interest rate paid could be received back), whereas a negative value implies that the project is not feasible [158]. For the bioenergy generation, it means the average yearly gain or loss of operating a bioenergy plant, e.g. for electricity (and heat) provision. The electricity production costs are determined by dividing the sum of total annual costs (in €) into the investment (capital-related), feedstock (consumption-related), maintenance (operating-related),

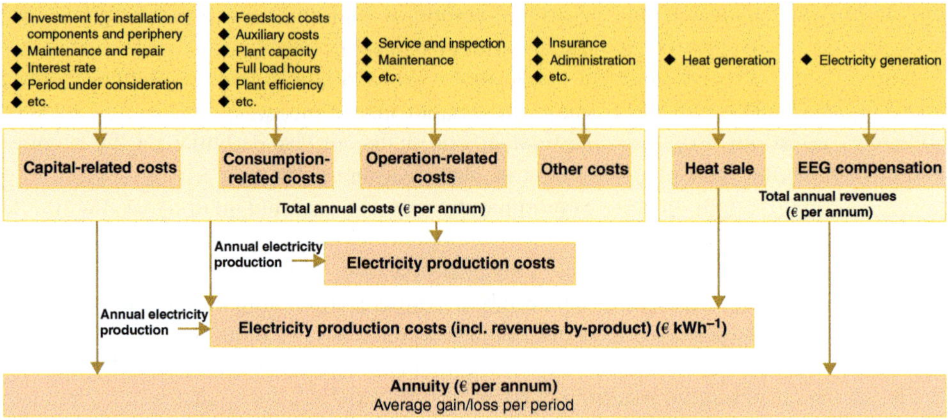

Figure 1.9 Social, economic, and environmental impacts of renewable energy resources.

and other costs by the annual electricity production (in kWh). Assuming that the total annual revenues consist of the compensation payment within the Renewable Energy Sources Act (EEG) of 2009, and the incoming from heat sales cogenerated, the results revealed that the level of the EEG 2009 compensation payments mainly depends on the size of the plant, the technical routes applied, the fuel used, and the quality of cogenerated heat produced.

Economic benefits would be achieved in the AD systems via the following approaches [156]: (i) further utilization of composting products to reduce fertilizer, chemical herbicide, and pesticide demands; (ii) high efficient utilization of energy produced (biogas/electricity/heat) in the whole treatment process to reduce fossil energy demands; (iii) acquisition of new income via composting products and energy sales (biogas/electricity/heat); (iv) improving soil/agricultural productivity through long-term utilization of compost fertilizer; (v) recovery of material to be recycled for the utilization and improvement of their economic prospects; (vi) reduction of landfill space and consequently land costs. With the further development of capital expenditure, operational expenditure, electricity prices, gas costs, and efficiencies, an economic production of synthetic methane via the operation of the AD system for the year 2030, especially 2050, is feasible.

1.5.2 Environmental Feasibility and Benefit Assessment

AD is a promising management option yielding a high potential GHG savings, nutrients recovery, and nitrogen availability in fertilizer for plants. Life cycle assessment of the environmental and economic feasibility of the AD systems is very helpful for the optimization of the digestion processes [166]. The main environmental issues of AD systems are the potential methane emission, which is negative to the GHGs controlling target achieving. Therefore, it is necessary to set up a combustion system to burn off the excess gases to ensure the safety of the projects, as well as reduce the negative impact on the environment. And using methane produced by AD as energy will reduce the emission of 27.7 kg CO_2eq t^{-1} relative to fossil fuels [167]. 2.4 MWh and 658.0 kg of N-manure were estimated to be generated as a consequence of the AD utilization, potentially offsetting 0.13 ± 0.01 kg of CO_2eq kg lwg^{-1} or 95% (±45%) of total direct emissions from the manure management. Furthermore, by replacing fossil fuel sources, i.e. diesel oil, this offset could be increased to 169% (±47%) [168]. Under the conditions of appropriate incentives (about 0.28 € kWh^{-1} of power supply revenue), by supplying power to the grid and selling unpacked digestate in the price of 20 € t^{-1}, the use of AD to treat energy crops can obtain a return on investment of 5% and reduce the environmental impact by about half compared with fossil fuels [169].

The challenges involved in ADs include the lack of short-term process optimization strategies and proper designing, which often lead to a low biogas production and instable operation. Because most countries have not implemented a central separation system for solid waste treatment and disposal, the development of separation technology for impurities in organic wastes should be promoted [170].

Environmental life cycle assessment and an exergy levelized cost evaluation of the treatment scheme for MSW in Germany show that the compost after AD is most in line with the economic and environmental requirements. Furthermore, AD of organic MSW can only be achieved through source-isolated waste collection; thus, maintaining or expanding a separate waste collection system is not only reasonable but also meaningful. The effect of organic fraction on the environment during the MSW incineration is similar to that of AD, but the levelized costs of exergy are higher. When the waste is pre-dried before incineration, the environmental or economic balance cannot be improved, which reflects the superiority of AD in environmental and economic aspects [171]. Therefore, for MSW with more biodegradable components, AD is better than incineration [172]. In 2019, a total of 39.04 million tons of sludge (80% moisture content) were produced in China, in which the majority of the sludge treatment methods are land use, incineration, sanitary landfills, and construction material utilization, which shows great environmental and economic optimization space [173].

Considering the co-digestion, how to achieve the purpose for a simultaneous biomethane production and organic waste treatment has been widely explored. The recent work revealed that food waste can yield 47% of biomethane, and hence for Australia, an estimated total of 0.07–1.54 million m^3 of biogas can be produced and up to 414,898 tons of GHG emissions reduced annually. The work also revealed that another 52.36 GW electricity and 554.4 TJ heat, per year, would be generated via the effective utilization of the biogas instead of fossil fuels. Environmental assessments show that full conversion of food waste can reduce GHG emissions by up to 507,434 tons per year in Australia and about US$ 5,238,000,000 income from electricity production. Therefore, the conversion of food waste into biofuel through AD systems would play an important role in replacing fossil energy for electricity generation and income generation [174].

Landfill gas without recycling energy has a high global warming potential, and the conversion of landfill gas into energy technology can reduce the global warming potential by 71.5%. Among them, the utilization of AD technology for landfill upgrading can reduce the global warming potential by 92.7% [175]. Life cycle assessment results from a factory in Austria showed that in the impact categories of climate change, acidification, and eutrophication, direct life cycle emissions from AD systems are much higher than from combustion and gasification due to differences in the nature of the co-products (digestate or ash), in which the AD will be the best choice if energy crops are selected as substrates [176].

The study found that in different stages, the main factors of AD impact on the environment are different, resulting in phased differences in the impact on the environment [177]. Focusing on the intermediate process of AD can tap more potential valuable by-products such as biopolymers, medium-chain fatty acids, and biohydrogen [178]. The results of life cycle assessment of the AD systems closely depend on the characteristics of MSWs, classification level, recycling, and utilization technologies. Therefore, a separate evaluation must be carried out according to local conditions [172]. Different scenarios (the type of substrate, the scale, product demand, and policies) have different constraints and consequently different solutions. The trade-offs between cost and environmental performance should be a future extension of this work [179].

References

1. Jain, M. (2018). Anaerobic membrane bioreactor as highly efficient and reliable technology for wastewater treatment—a review. *Advances in Chemical Engineering and Science* 08 (2): 82–100.
2. Demirel, B., Yenigun, O., and Onay, T.T. (2005). Anaerobic treatment of dairy wastewaters: a review. *Process Biochemistry* 40 (8): 2583–2595.
3. Watts, N., Amann, M., Arnell, N. et al. (2018). The 2018 report of the lancet countdown on health and climate change: shaping the health of nations for centuries to come. *The Lancet* 392 (10163): 2479–2514.
4. Christensen, T.R., Arora, V.K., Gauss, M. et al. (2019). Tracing the climate signal: mitigation of anthropogenic methane emissions can outweigh a large Arctic natural emission increase. *Scientific Reports* 9 (1): 1146.
5. Barrera, E.L., Rosa, E., Spanjers, H. et al. (2016). A comparative assessment of anaerobic digestion power plants as alternative to lagoons for vinasse treatment: life cycle assessment and exergy analysis. *Journal of Cleaner Production* 113: 459–471.
6. Kang, J., Wei, Y.M., Liu, L.C. et al. (2020). Energy systems for climate change mitigation: a systematic review. *Applied Energy* 263: 114602.
7. Paritosh, K., Balan, V., Vijay, V.K., and Vivekanand, V. (2020). Simultaneous alkaline treatment of pearl millet straw for enhanced solid state anaerobic digestion: experimental investigation and energy analysis. *Journal of Cleaner Production* 252: 119798.
8. Caposciutti, G., Baccioli, A., Ferrari, L., and Desideri, U. (2020). Biogas from anaerobic digestion: power generation or biomethane production? *Energies* 13 (3): 743.
9. Ardolino, F., Parrillo, F., and Arena, U. (2018). Biowaste-to-biomethane or biowaste-to-energy? An LCA study on anaerobic digestion of organic waste. *Journal of Cleaner Production* 174: 462–476.
10. Han, S., Liu, Y., Zhang, S., and Luo, G. (2016). Reactor performances and microbial communities of biogas reactors: effects of inoculum sources. *Applied Microbiology and Biotechnology* 100 (2): 987–995.
11. Maaz, M., Yasin, M., Aslam, M. et al. (2019). Anaerobic membrane bioreactors for wastewater treatment: novel configurations, fouling control and energy considerations. *Bioresource Technology* 283: 358–372.
12. Cayetano, R.D.A., Kim, G.B., Park, J.H. et al. (2021). Anaerobic digestion of waste activated sludge using dynamic membrane at varying substrate concentration reveals new insight towards methanogenic pathway and biofilm formation. *Chemical Engineering Journal* 423: 130249.
13. The Scientific World (2017). Retracted: Microbial ecology of anaerobic digesters: the key players of anaerobiosis. *Scientific World Journal* 2017: 3852369.
14. Chen, Q. and Liu, T. (2017). Biogas system in rural China: upgrading from decentralized to centralized? *Renewable and Sustainable Energy Reviews* 78: 933–944.

15. Zhang, T., Yang, Y., and Xie, D. (2015). Insights into the production potential and trends of China's rural biogas. *International Journal of Energy Research* 39 (8): 1068–1082.
16. Tayibi, S., Monlau, F., Bargaz, A. et al. (2021). Synergy of anaerobic digestion and pyrolysis processes for sustainable waste management: a critical review and future perspectives. *Renewable and Sustainable Energy Reviews* 152: 111603.
17. D'Silva, T.C., Isha, A., Chandra, R. et al. (2021). Enhancing methane production in anaerobic digestion through hydrogen assisted pathways – a state-of-the-art review. *Renewable and Sustainable Energy Reviews* 151: 111536.
18. Taherzadeh, M.J. and Karimi, K. (2008). Pretreatment of lignocellulosic wastes to improve ethanol and biogas production: a review. *International Journal of Molecular Sciences* 9 (9): 1621–1651.
19. Abbassi-Guendouz, A., Trably, E., Hamelin, J. et al. (2013). Microbial community signature of high-solid content methanogenic ecosystems. *Bioresource Technology* 133: 256–262.
20. Illmer, P., Reitschuler, C., Wagner, A.O. et al. (2014). Microbial succession during thermophilic digestion: the potential of *Methanosarcina* sp. *PLoS One* 9 (2): 86967.
21. Kampmann, K., Ratering, S., Baumann, R. et al. (2012). Hydrogenotrophic methanogens dominate in biogas reactors fed with defined substrates. *Systematic and Applied Microbiology* 35 (6): 404–413.
22. Karakashev, D., Batstone, D.J., Trably, E., and Angelidaki, I. (2006). Acetate oxidation is the dominant methanogenic pathway from acetate in the absence of Methanosaetaceae. *Applied and Environmental Microbiology* 72 (7): 5138–5141.
23. Kwietniewska, E. and Tys, J. (2014). Process characteristics, inhibition factors and methane yields of anaerobic digestion process, with particular focus on microalgal biomass fermentation. *Renewable and Sustainable Energy Reviews* 34: 491–500.
24. Ehimen, E.A., Sun, Z.F., Carrington, C.G. et al. (2011). Anaerobic digestion of microalgae residues resulting from the biodiesel production process. *Applied Energy* 88 (10): 3454–3463.
25. Chae, K.J., Jang, A.M., Yim, S.K., and Kim, I.S. (2008). The effects of digestion temperature and temperature shock on the biogas yields from the mesophilic anaerobic digestion of swine manure. *Bioresource Technology* 99 (1): 1–6.
26. Appels, L., Baeyens, J., Degrève, J., and Dewil, R. (2008). Principles and potential of the anaerobic digestion of waste-activated sludge. *Progress in Energy and Combustion Science* 34 (6): 755–781.
27. Ward, A.J., Hobbs, P.J., Holliman, P.J., and Jones, D.L. (2008). Optimisation of the anaerobic digestion of agricultural resources. *Bioresource Technology* 99 (17): 7928–7940.
28. Fagbohungbe, M.O., Dodd, I.C., Herbert, B.M. et al. (2015). High solid anaerobic digestion: operational challenges and possibilities. *Environmental Technology and Innovation* 4: 268–284.
29. Nizami, A. and Murphy, J.D. (2010). What type of digester configurations should be employed to produce biomethane from grass silage? *Renewable and Sustainable Energy Reviews* 14 (6): 1558–1568.
30. Duan, N., Dong, B., Wu, B., and Dai, X. (2012). High-solid anaerobic digestion of sewage sludge under mesophilic conditions: feasibility study. *Bioresource Technology* 104: 150–156.
31. Fernández-Rodríguez, J., Pérez, M., and Romero, L.I. (2014). Dry thermophilic anaerobic digestion of the organic fraction of municipal solid wastes: solid retention time optimization. *Chemical Engineering Journal* 251: 435–440.
32. Karthikeyan, O.P. and Visvanathan, C. (2013). Bio-energy recovery from high-solid organic substrates by dry anaerobic bio-conversion processes: a review. *Reviews in Environmental Science and Bio/Technology* 12: 257–284.
33. Bouallagui, H., Lahdheb, H., Romdan, E.B. et al. (2009). Improvement of fruit and vegetable waste anaerobic digestion performance and stability with co-substrates addition. *Journal of Environmental Management* 90 (5): 1844–1849.
34. Chen, Y., Cheng, J.J., and Creamer, K.S. (2008). Inhibition of anaerobic digestion process: a review. *Bioresource Technology* 99 (10): 4044–4064.

35. Li, Y., Park, S.Y., and Zhu, J. (2011). Solid-state anaerobic digestion for methane production from organic waste. *Renewable and Sustainable Energy Reviews* 15 (1): 821–826.
36. Brown, D. and Li, Y. (2013). Solid state anaerobic co-digestion of yard waste and food waste for biogas production. *Bioresource Technology* 127: 275–280.
37. Raposo, F., Borja, R., Martín, M.A. et al. (2009). Influence of inoculum–substrate ratio on the anaerobic digestion of sunflower oil cake in batch mode: process stability and kinetic evaluation. *Chemical Engineering Journal* 149 (1): 70–77.
38. Ge, X., Xu, F., and Li, Y. (2016). Solid-state anaerobic digestion of lignocellulosic biomass: recent progress and perspectives. *Bioresource Technology* 205: 239–249.
39. Le Hyaric, R., Chardin, C., Benbelkacem, H. et al. (2011). Influence of substrate concentration and moisture content on the specific methanogenic activity of dry mesophilic municipal solid waste digestate spiked with propionate. *Bioresource Technology* 102 (2): 822–827.
40. Forster-Carneiro, T., Pérez, M., and Romero, L.I. (2008). Influence of total solid and inoculum contents on performance of anaerobic reactors treating food waste. *Bioresource Technology* 99 (15): 6994–7002.
41. Yoon, S., Kim, H., and Yeom, I. (2004). The optimum operational condition of membrane bioreactor (MBR): cost estimation of aeration and sludge treatment. *Water Research* 38 (1): 37–46.
42. Sarker, S., Lamb, J.J., Hjelme, D.R., and Lien, K.M. (2019). A review of the role of critical parameters in the design and operation of biogas production plants. *Applied Sciences* 9 (9): 1915.
43. Rajagopal, R., Masse, D.I., and Singh, G. (2013). A critical review on inhibition of anaerobic digestion process by excess ammonia. *Bioresource Technology* 143: 632–641.
44. Yenigun, O. and Demirel, B. (2013). Ammonia inhibition in anaerobic digestion: a review. *Process Biochemistry* 48 (5–6): 901–911.
45. Yuan, H. and Zhu, N. (2016). Progress in inhibition mechanisms and process control of intermediates and by-products in sewage sludge anaerobic digestion. *Renewable and Sustainable Energy Reviews* 58: 429–438.
46. Wittmann, C., Zeng, A.P., and Deckwer, W.D. (1995). Growth inhibition by ammonia and use of a pH controlled feeding strategy for the effective cultivation of *Mycobacterium chlorophenolicum*. *Applied Microbiology and Biotechnology* 44 (3–4): 519–525.
47. Gallert, C. and Winter, J. (1997). Mesophilic and thermophilic anaerobic digestion of source-sorted organic wastes: effect of ammonia on glucose degradation and methane production. *Applied Microbiology and Biotechnology* 48 (3): 405–410.
48. Prochazka, J., Dolejš, P., Máca, J., and Dohányos, M. (2012). Stability and inhibition of anaerobic processes caused by insufficiency or excess of ammonia nitrogen. *Applied Microbiology and Biotechnology* 93 (1): 439–447.
49. Kayhanian, M. (1994). Performance of a high-solids anaerobic-digestion process under various ammonia concentrations. *Journal of Chemical Technology and Biotechnology* 59 (4): 349–352.
50. Takashima, M. and Yaguchi, J. (2021). High-solids thermophilic anaerobic digestion of sewage sludge: effect of ammonia concentration (September, 2020). *Journal of Material Cycles and Waste Management* 23 (1): 423–423. https://doi.org/10.1007/s10163-020-01117-z.
51. Gonzalez-Fernandez, C. and Garcia-Encina, P.A. (2009). Impact of substrate to inoculum ratio in anaerobic digestion of swine slurry. *Biomass & Bioenergy* 33 (8): 1065–1069.
52. Li, L., Peng, X., Wang, X., and Wu, D. (2018). Anaerobic digestion of food waste: a review focusing on process stability. *Bioresource Technology* 248 (A): 20–28.
53. Bhattacharya, S.K. and Parkin, G.F. (1989). The effect of ammonia on methane fermentation processes. *Journal Water Pollution Control Federation* 61 (1): 55–59.
54. Angelidaki, I. and Ahring, B.K. (1993). Thermophilic anaerobic digestion of livestock waste: the effect of ammonia. *Applied Microbiology and Biotechnology* 38 (4): 560–564.
55. Lauterböck, B., Ortner, M., Haider, R., and Fuchs, W. (2012). Counteracting ammonia inhibition in anaerobic digestion by removal with a hollow fiber membrane contactor. *Water Research* 46 (15): 4861–4869.

56. Angelidaki, I. and Ahring, B.K. (1994). Anaerobic thermophilic digestion of manure at different ammonia loads: effect of temperature. *Water Research* 28 (3): 727–731.
57. Lauterböck, B., Nikolausz, M., Lv, Z. et al. (2014). Improvement of anaerobic digestion performance by continuous nitrogen removal with a membrane contactor treating a substrate rich in ammonia and sulfide. *Bioresource Technology* 158: 209–216.
58. Pan, J., Chen, X., Sheng, K. et al. (2013). Effect of ammonia on biohydrogen production from food waste via anaerobic fermentation. *International Journal of Hydrogen Energy* 38 (29): 12747–12754.
59. Nakakubo, R., Møller, H.B., Nielsen, A.M., and Matsuda, J. (2008). Ammonia inhibition of methanogenesis and identification of process indicators during anaerobic digestion. *Environmental Engineering Science* 25 (10): 1487–1496.
60. Nie, H., Jacobi, H.F., Strach, K. et al. (2015). Mono-fermentation of chicken manure: ammonia inhibition and recirculation of the digestate. *Bioresource Technology* 178: 238–246.
61. Mahdy, A., Fotidis, I.A., Mancini, E. et al. (2017). Ammonia tolerant inocula provide a good base for anaerobic digestion of microalgae in third generation biogas process. *Bioresource Technology* 225: 272–278.
62. Keefer, C.E. and Meisel, J. (1951). Activated sludge studies: III. Effect of pH of sewage on the activated sludge process. *Sewage and Industrial Wastes* 23 (8): 982–991.
63. Salsali, H.R., Parker, W.J., and Sattar, S.A. (2006). Impact of concentration, temperature, and pH on inactivation of *Salmonella* spp. by volatile fatty acids in anaerobic digestion. *Canadian Journal of Microbiology* 52 (4): 279–286.
64. Puchajda, B. and Oleszkiewicz, J. (2006). Extended acid digestion for inactivation of fecal coliforms. *Water Environment Research* 78 (12): 2389–2396.
65. Russell, J.B. and Diez-Gonzalez, F. (1997). The effects of fermentation acids on bacterial growth. In: *Advances in Microbial Physiology* (ed. R.K. Poole), 205–234. Elsevier.
66. Montes, J.A., Leivas, R., Martínez-Prieto, D., and Rico, C. (2019). Biogas production from the liquid waste of distilled gin production: optimization of UASB reactor performance with increasing organic loading rate for co-digestion with swine wastewater. *Bioresource Technology* 274: 43–47.
67. Intanoo, P., Rangsanvigit, P., Malakul, P., and Chavadej, S. (2014). Optimization of separate hydrogen and methane production from cassava wastewater using two-stage upflow anaerobic sludge blanket reactor (UASB) system under thermophilic operation. *Bioresource Technology* 173: 256–265.
68. Schmidt, T., McCabe, B.K., Harris, P.W., and Lee, S. (2018). Effect of trace element addition and increasing organic loading rates on the anaerobic digestion of cattle slaughterhouse wastewater. *Bioresource Technology* 264: 51–57.
69. Intanoo, P., Watcharanurak, T., and Chavadej, S. (2020). Evolution of methane and hydrogen from ethanol wastewater with maximization of energy yield by three-stage anaerobic sequencing batch reactor system. *International Journal of Hydrogen Energy* 45 (16): 9469–9483.
70. Zhao, W., Huang, J.J., Hua, B. et al. (2020). A new strategy to recover from volatile fatty acid inhibition in anaerobic digestion by photosynthetic bacteria. *Bioresource Technology* 311: 123501.
71. Goud, R.K., Sarkar, O., Chiranjeevi, P., and Mohan, S.V. (2014). Bioaugmentation of potent acidogenic isolates: a strategy for enhancing biohydrogen production at elevated organic load. *Bioresource Technology* 165: 223–232.
72. Schauer-Gimenez, A.E., Zitomer, D.H., Maki, J.S., and Struble, C.A. (2010). Bioaugmentation for improved recovery of anaerobic digesters after toxicant exposure. *Water Research* 44 (12): 3555–3564.
73. Mechichi, T. and Sayadi, S. (2005). Evaluating process imbalance of anaerobic digestion of olive mill wastewaters. *Process Biochemistry* 40 (1): 139–145.
74. Kleyböcker, A., Liebrich, M., Verstraete, W. et al. (2012). Early warning indicators for process failure due to organic overloading by rapeseed oil in one-stage continuously stirred tank reactor, sewage sludge and waste digesters. *Bioresource Technology* 123: 534–541.
75. Arikan, O.A., Mulbry, W., and Lansing, S. (2015). Effect of temperature on methane production from field-scale anaerobic digesters treating dairy manure. *Waste Management* 43: 108–113.

76. C. A. Khalil, L. Ibrahim, E. Ibrahim, and S. Ghanimeh (2016). Clean energy generation through psychrophilic anaerobic digestion of food and landscape wastes. *2016 3rd International Conference on Renewable Energies for Developing Countries (Redec)*, Zouk Mosbeh, Lebanon (13–15 July 2016). IEEE.
77. Cysneiros, D., Thuillier, A., Villemont, R. et al. (2011). Temperature effects on the trophic stages of perennial rye grass anaerobic digestion. *Water Science and Technology* 64 (1): 70–76.
78. Chernicharo, C.A.L., Van Lier, J.B., Noyola, A., and Bressani Ribeiro, T. (2015). Anaerobic sewage treatment: state of the art, constraints and challenges. *Reviews in Environmental Science and Bio/Technology* 14: 649–679.
79. Leon, E.S., Vargas-Machuca, J.A.P., Corona, E.L. et al. (2018). Anaerobic digestion of municipal sewage under psychrophilic conditions. *Journal of Cleaner Production* 198: 931–939.
80. Nguyen, L.N., Vu, M.T., Johir, M.A.H. et al. (2021). Promotion of direct interspecies electron transfer and potential impact of conductive materials in anaerobic digestion and its downstream processing – a critical review. *Bioresource Technology* 341: 125847.
81. Wu, Y., Wang, S., Liang, D., and Li, N. (2020). Conductive materials in anaerobic digestion: from mechanism to application. *Bioresource Technology* 298: 122403.
82. Park, J., Kang, H.J., Park, K.H., and Park, H.D. (2018). Direct interspecies electron transfer via conductive materials: a perspective for anaerobic digestion applications. *Bioresource Technology* 254: 300–311.
83. Zhao, Z., Li, Y., Zhang, Y., and Lovley, D.R. (2020). Sparking anaerobic digestion: promoting direct interspecies electron transfer to enhance methane production. *iScience* 23 (12): 101794.
84. Zhao, W., Yang, H., He, S. et al. (2021). A review of biochar in anaerobic digestion to improve biogas production: performances, mechanisms and economic assessments. *Bioresource Technology* 341: 125797.
85. Liang, J., Luo, L., Li, D. et al. (2021). Promoting anaerobic co-digestion of sewage sludge and food waste with different types of conductive materials: performance, stability, and underlying mechanism. *Bioresource Technology* 337: 125384.
86. Yin, Q., Yang, S., Wang, Z. et al. (2018). Clarifying electron transfer and metagenomic analysis of microbial community in the methane production process with the addition of ferroferric oxide. *Chemical Engineering Journal* 333: 216–225.
87. Lin, R., Cheng, J., Zhang, J. et al. (2017). Boosting biomethane yield and production rate with graphene: the potential of direct interspecies electron transfer in anaerobic digestion. *Bioresource Technology* 239: 345–352.
88. Tian, T., Qiao, S., Li, X. et al. (2017). Nano-graphene induced positive effects on methanogenesis in anaerobic digestion. *Bioresource Technology* 224: 41–47.
89. Peng, H., Zhang, Y., Tan, D. et al. (2018). Roles of magnetite and granular activated carbon in improvement of anaerobic sludge digestion. *Bioresource Technology* 249: 666–672.
90. Yang, Y., Zhang, Y., Li, Z. et al. (2017). Adding granular activated carbon into anaerobic sludge digestion to promote methane production and sludge decomposition. *Journal of Cleaner Production* 149: 1101–1108.
91. Xu, S., Zhang, W., Zuo, L. et al. (2020). Comparative facilitation of activated carbon and goethite on methanogenesis from volatile fatty acids. *Bioresource Technology* 302: 122801.
92. Kaur, G., Johnravindar, D., and Wong, J.W.C. (2020). Enhanced volatile fatty acid degradation and methane production efficiency by biochar addition in food waste-sludge co-digestion: a step towards increased organic loading efficiency in co-digestion. *Bioresource Technology* 308: 123250.
93. Lü, F., Liu, Y., Shao, L., and He, P. (2019). Powdered biochar doubled microbial growth in anaerobic digestion of oil. *Applied Energy* 247: 605–614.
94. Lim, E.Y., Tian, H., Chen, Y. et al. (2020). Methanogenic pathway and microbial succession during start-up and stabilization of thermophilic food waste anaerobic digestion with biochar. *Bioresource Technology* 314: 123751.

95. Wang, G., Li, Q., Yuwen, C. et al. (2020). Biochar triggers methanogenesis recovery of a severely acidified anaerobic digestion system via hydrogen-based syntrophic pathway inhibition. *International Journal of Hydrogen Energy* 46 (15): 9666–9677.
96. Wang, C., Liu, Y., Wang, C. et al. (2021). Biochar facilitates rapid restoration of methanogenesis by enhancing direct interspecies electron transfer after high organic loading shock. *Bioresource Technology* 320: 124360.
97. Wang, M., Zhao, Z., and Zhang, Y. (2021). Magnetite-contained biochar derived from Fenton sludge modulated electron transfer of microorganisms in anaerobic digestion. *Journal of Hazardous Materials* 403: 123972.
98. Zhang, M., Maneengam, A., Sajjad, M. et al. (2022). Meta-analysis of bio-based carbon materials for anaerobic digestion with direct interspecies electron transfer mechanism. *Materials Letters* 310: 131485.
99. Lovley, D.R. (2011). Live wires: direct extracellular electron exchange for bioenergy and the bioremediation of energy-related contamination. *Energy & Environmental Science* 4 (12): 4896–4906.
100. Chen, M., Liu, S., Yuan, X. et al. (2021). Methane production and characteristics of the microbial community in the co-digestion of potato pulp waste and dairy manure amended with biochar. *Renewable Energy* 163: 357–367.
101. Kumar, M., Dutta, S., You, S. et al. (2021). A critical review on biochar for enhancing biogas production from anaerobic digestion of food waste and sludge. *Journal of Cleaner Production* 305: 127143.
102. Qi, Q., Sun, C., Cristhian, C. et al. (2021). Enhancement of methanogenic performance by gasification biochar on anaerobic digestion. *Bioresource Technology* 330: 124993.
103. Hirano, S., Matsumoto, N., Morita, M. et al. (2013). Electrochemical control of redox potential affects methanogenesis of the hydrogenotrophic methanogen *Methanothermobacter thermautotrophicus*. *Letters in Applied Microbiology* 56 (5): 315–321.
104. Salvador, A.F., Martins, G., Melle-Franco, M. et al. (2017). Carbon nanotubes accelerate methane production in pure cultures of methanogens and in a syntrophic coculture. *Environmental Microbiology* 19 (7): 2727–2739.
105. Vivekanand, V., Mulat, D.G., Eijsink, V.G., and Horn, S.J. (2018). Synergistic effects of anaerobic co-digestion of whey, manure and fish ensilage. *Bioresource Technology* 249: 35–41.
106. Lee, J., Hong, J., Jeong, S. et al. (2020). Interactions between substrate characteristics and microbial communities on biogas production yield and rate. *Bioresource Technology* 303: 122934.
107. Maragkaki, A.E., Vasileiadis, I., Fountoulakis, M. et al. (2018). Improving biogas production from anaerobic co-digestion of sewage sludge with a thermal dried mixture of food waste, cheese whey and olive mill wastewater. *Waste Management* 71: 644–651.
108. Zhang, Y., Cañas, E.M.Z., Zhu, Z. et al. (2011). Robustness of archaeal populations in anaerobic co-digestion of dairy and poultry wastes. *Bioresource Technology* 102 (2): 779–785.
109. Wang, M., Sun, X., Li, P. et al. (2014). A novel alternate feeding mode for semi-continuous anaerobic co-digestion of food waste with chicken manure. *Bioresource Technology* 164: 309–314.
110. Wang, Z.Z., Jiang, Y., Wang, S. et al. (2020). Impact of total solids content on anaerobic co-digestion of pig manure and food waste: insights into shifting of the methanogenic pathway. *Waste Management* 114: 96–106.
111. Elalami, D., Carrere, H., Monlau, F. et al. (2019). Pretreatment and co-digestion of wastewater sludge for biogas production: recent research advances and trends. *Renewable and Sustainable Energy Reviews* 114: 109287.
112. Karki, R., Chuenchart, W., Surendra, K.C. et al. (2021). Anaerobic co-digestion: current status and perspectives. *Bioresource Technology* 330: 125001.
113. Kunatsa, T., Zhang, L., and Xia, X. (2020). Biogas potential determination and production optimisation through optimal substrate ratio feeding in co-digestion of water hyacinth, municipal solid waste and cow dung. *Biofuels* 13 (5): 631–641.

114. Kunatsa, T. and Xia, X. (2022). A review on anaerobic digestion with focus on the role of biomass co-digestion, modelling and optimisation on biogas production and enhancement. *Bioresource Technology* 344: 126311.
115. Yasar, A., Nazir, S., Tabinda, A.B. et al. (2017). Socio-economic, health and agriculture benefits of rural household biogas plants in energy scarce developing countries: a case study from Pakistan. *Renewable Energy* 108: 19–25.
116. Hegde, S. and Trabold, T.A. (2019). Anaerobic digestion of food waste with unconventional co-substrates for stable biogas production at high organic loading rates. *Sustainability* 11 (14): 3875.
117. Sembera, C., Macintosh, C., Astals, S., and Koch, K. (2019). Benefits and drawbacks of food and dairy waste co-digestion at a high organic loading rate: a Moosburg WWTP case study. *Waste Management* 95: 217–226.
118. Lins, P., Reitschuler, C., and Illmer, P. (2014). *Methanosarcina* spp., the key to relieve the start-up of a thermophilic anaerobic digestion suffering from high acetic acid loads. *Bioresource Technology* 152: 347–354.
119. Baek, G., Kim, J., Shin, S.G., and Lee, C. (2016). Bioaugmentation of anaerobic sludge digestion with iron-reducing bacteria: process and microbial responses to variations in hydraulic retention time. *Applied Microbiology and Biotechnology* 100 (2): 927–937.
120. Tale, V.P., Maki, J.S., and Zitomer, D.H. (2015). Bioaugmentation of overloaded anaerobic digesters restores function and archaeal community. *Water Research* 70: 138–147.
121. Čater, M., Fanedl, L., Malovrh, Š., and Logar, R.M. (2015). Biogas production from brewery spent grain enhanced by bioaugmentation with hydrolytic anaerobic bacteria. *Bioresource Technology* 186: 261–269.
122. Yu, J., Zhao, Y., Liu, B. et al. (2016). Accelerated acidification by inoculation with a microbial consortia in a complex open environment. *Bioresource Technology* 216: 294–301.
123. Lü, F., Li, T., Wang, T. et al. (2013). Improvement of sludge digestate biodegradability by thermophilic bioaugmentation. *Applied Microbiology and Biotechnology* 98 (2): 969–977.
124. Cirne, D.G., Björnsson, L., Alves, M., and Mattiasson, B. (2006). Effects of bioaugmentation by an anaerobic lipolytic bacterium on anaerobic digestion of lipid-rich waste. *Journal of Chemical Technology and Biotechnology* 81 (11): 1745–1752.
125. Fotidis, I.A., Wang, H., Fiedel, N.R. et al. (2014). Bioaugmentation as a solution to increase methane production from an ammonia-rich substrate. *Environmental Science & Technology* 48 (13): 7669–7676.
126. Cavaleiro, A.J., Sousa, D.Z., and Alves, M.M. (2010). Methane production from oleate: assessing the bioaugmentation potential of *Syntrophomonas zehnderi*. *Water Research* 44 (17): 4940–4947.
127. Rusín, J., Chamrádová, K., and Basinas, P. (2021). Two-stage psychrophilic anaerobic digestion of food waste: comparison to conventional single-stage mesophilic process. *Waste Management* 119: 172–182.
128. Massé, D.I., Masse, L., and Croteau, F. (2003). The effect of temperature fluctuations on psychrophilic anaerobic sequencing batch reactors treating swine manure. *Bioresource Technology* 89 (1): 57–62.
129. Saady, N.M.C. and Massé, D.I. (2013). Psychrophilic anaerobic digestion of lignocellulosic biomass: a characterization study. *Bioresource Technology* 142: 663–671.
130. Acharya, S.M., Kundu, K., and Sreekrishnan, T.R. (2015). Improved stability of anaerobic digestion through the use of selective acidogenic culture. *Journal of Environmental Engineering* 141 (7): 04015001.
131. Westerholm, M., Leven, L., and Schnurer, A. (2012). Bioaugmentation of syntrophic acetate-oxidizing culture in biogas reactors exposed to increasing levels of ammonia. *Applied and Environmental Microbiology* 78 (21): 7619–7625.
132. Fotidis, I.A., Karakashev, D., and Angelidaki, I. (2013). Bioaugmentation with an acetate-oxidising consortium as a tool to tackle ammonia inhibition of anaerobic digestion. *Bioresource Technology* 146: 57–62.

133. Martin-Ryals, A., Schideman, L., Li, P. et al. (2015). Improving anaerobic digestion of a cellulosic waste via routine bioaugmentation with cellulolytic microorganisms. *Bioresource Technology* 189: 62–70.
134. Yao, Y., Huang, G., An, C. et al. (2020). Anaerobic digestion of livestock manure in cold regions: technological advancements and global impacts. *Renewable & Sustainable Energy Reviews* 119: 109494.
135. Capson-Tojo, G., Torres, A., Muñoz, R. et al. (2017). Mesophilic and thermophilic anaerobic digestion of lipid-extracted microalgae *N. gaditana* for methane production. *Renewable Energy* 105: 539–546.
136. Petropoulos, E., Dolfing, J., Davenport, R.J. et al. (2017). Developing cold-adapted biomass for the anaerobic treatment of domestic wastewater at low temperatures (4, 8 and 15 °C) with inocula from cold environments. *Water Research* 112: 100–109.
137. Dev, S., Saha, S., Kurade, M.B. et al. (2019). Perspective on anaerobic digestion for biomethanation in cold environments. *Renewable & Sustainable Energy Reviews* 103: 85–95.
138. Chen, X., Huang, G., An, C. et al. (2019). Plasma-induced PAA-ZnO coated PVDF membrane for oily wastewater treatment: preparation, optimization, and characterization through Taguchi OA design and synchrotron-based X-ray analysis. *Journal of Membrane Science* 582: 70–82.
139. Zhang, P., Huang, G., An, C. et al. (2019). An integrated gravity-driven ecological bed for wastewater treatment in subtropical regions: process design, performance analysis, and greenhouse gas emissions assessment. *Journal of Cleaner Production* 212: 1143–1153.
140. Jang, H.M., Choi, Y.K., and Kan, E. (2018). Effects of dairy manure-derived biochar on psychrophilic, mesophilic and thermophilic anaerobic digestions of dairy manure. *Bioresource Technology* 250: 927–931.
141. Raper, E., Stephenson, T., Anderson, D.R. et al. (2018). Industrial wastewater treatment through bioaugmentation. *Process Safety and Environmental Protection* 118: 178–187.
142. Jiang, J., Li, L., Li, Y. et al. (2020). Bioaugmentation to enhance anaerobic digestion of food waste: dosage, frequency and economic analysis. *Bioresource Technology* 307: 123256.
143. Li, Y., Li, L., Sun, Y., and Yuan, Z. (2018). Bioaugmentation strategy for enhancing anaerobic digestion of high C/N ratio feedstock with methanogenic enrichment culture. *Bioresource Technology* 261: 188–195.
144. Cheng, S., Xing, D., Call, D.F., and Logan, B.E. (2009). Direct biological conversion of electrical current into methane by electromethanogenesis. *Environmental Science & Technology* 43 (10): 3953–3958.
145. Zhao, L., Wang, X.T., Chen, K.Y. et al. (2021). The underlying mechanism of enhanced methane production using microbial electrolysis cell assisted anaerobic digestion (MEC-AD) of proteins. *Water Research* 201: 117325.
146. Yu, Z., Leng, X., Zhao, S. et al. (2018). A review on the applications of microbial electrolysis cells in anaerobic digestion. *Bioresource Technology* 255: 340–348.
147. Wang, X., Zhao, L., Chen, C. et al. (2021). Microbial electrolysis cells (MEC) accelerated methane production from the enhanced hydrolysis and acidogenesis of raw waste activated sludge. *Chemical Engineering Journal* 413: 127472.
148. Hassanein, A., Witarsa, F., Lansing, S. et al. (2020). Bio-electrochemical enhancement of hydrogen and methane production in a combined anaerobic digester (AD) and microbial electrolysis cell (MEC) from dairy manure. *Sustainability* 12 (20): 8491.
149. Hassanein, A., Witarsa, F., Guo, X. et al. (2017). Next generation digestion: complementing anaerobic digestion (AD) with a novel microbial electrolysis cell (MEC) design. *International Journal of Hydrogen Energy* 42 (48): 28681–28689.
150. Park, J., Lee, B., Tian, D., and Jun, H. (2018). Bioelectrochemical enhancement of methane production from highly concentrated food waste in a combined anaerobic digester and microbial electrolysis cell. *Bioresource Technology* 247: 226–233.

151. Geng, Y., Yuan, L., Liu, T. et al. (2020). Thermal/alkaline pretreatment of waste activated sludge combined with a microbial fuel cell operated at alkaline pH for efficient energy recovery. *Applied Energy* 275: 115291.
152. Li, Y., Zhang, S., Zhang, W. et al. (2019). Life cycle assessment of advanced wastewater treatment processes: involving 126 pharmaceuticals and personal care products in life cycle inventory. *Journal of Environmental Management* 238: 442–450.
153. Lorenzo-Toja, Y., Vázquez-Rowe, I., Amores, M.J. et al. (2016). Benchmarking wastewater treatment plants under an eco-efficiency perspective. *Science of the Total Environment* 566–567: 468–479.
154. Resende, J.D., Nolasco, M.A., and Pacca, S.A. (2019). Life cycle assessment and costing of wastewater treatment systems coupled to constructed wetlands. *Resources, Conservation and Recycling* 148: 170–177.
155. Kamali, M., Suhas, D.P., Costa, M.E. et al. (2019). Sustainability considerations in membrane-based technologies for industrial effluents treatment. *Chemical Engineering Journal* 368: 474–494.
156. Taleghani, G. and Kia, A.S. (2005). Technical-economical analysis of the Saveh biogas power plant. *Renewable Energy* 30 (3): 441–446.
157. Menardo, S. and Balsari, P. (2012). An analysis of the energy potential of anaerobic digestion of agricultural by-products and organic waste. *Bioenergy Research* 5 (3): 759–767.
158. Hennig, C. and Gawor, M. (2012). Bioenergy production and use: comparative analysis of the economic and environmental effects. *Energy Conversion and Management* 63: 130–137.
159. Kythreotou, N., Florides, G., and Tassou, S.A. (2014). A review of simple to scientific models for anaerobic digestion. *Renewable Energy* 71: 701–714.
160. Pham, T.H., Rabaey, K., Aelterman, P. et al. (2006). Microbial fuel cells in relation to conventional anaerobic digestion technology. *Engineering in Life Sciences* 6 (3): 285–292.
161. Escapa, A., Gómez, X., Tartakovsky, B., and Morán, A. (2012). Estimating microbial electrolysis cell (MEC) investment costs in wastewater treatment plants: case study. *International Journal of Hydrogen Energy* 37 (24): 18641–18653.
162. Sleutels, T.H., Ter Heijne, A., Buisman, C.J., and Hamelers, H.V. (2012). Bioelectrochemical systems: an outlook for practical applications. *ChemSusChem* 5 (6): 1012–1019.
163. Ramirez-Islas, M.E., Güereca, L.P., Sosa-Rodriguez, F.S., and Cobos-Peralta, M.A. (2020). Environmental assessment of energy production from anaerobic digestion of pig manure at medium-scale using life cycle assessment. *Waste Management* 102: 85–96.
164. Murphy, J.D. and McCarthy, K. (2005). The optimal production of biogas for use as a transport fuel in Ireland. *Renewable Energy* 30 (14): 2111–2127.
165. Murphy, J.D. and McKeogh, E. (2004). Technical, economic and environmental analysis of energy production from municipal solid waste. *Renewable Energy* 29 (7): 1043–1057.
166. Li, H., Jin, C., Zhang, Z. et al. (2017). Environmental and economic life cycle assessment of energy recovery from sewage sludge through different anaerobic digestion pathways. *Energy (Oxford)* 126: 649–657.
167. Sefeedpari, P., Vellinga, T., Rafiee, S. et al. (2019). Technical, environmental and cost-benefit assessment of manure management chain: a case study of large scale dairy farming. *Journal of Cleaner Production* 233: 857–868.
168. Junior, C.C., Cerri, C.E., Pires, A.V., and Cerri, C.C. (2015). Net greenhouse gas emissions from manure management using anaerobic digestion technology in a beef cattle feedlot in Brazil. *Science of the Total Environment* 505: 1018–1025.
169. Torquati, B., Venanzi, S., Ciani, A. et al. (2014). Environmental sustainability and economic benefits of dairy farm biogas energy production: a case study in Umbria. *Sustainability* 6 (10): 6696–6713.
170. Chew, K.R., Leong, H.Y., Khoo, K.S. et al. (2021). Effects of anaerobic digestion of food waste on biogas production and environmental impacts: a review. *Environmental Chemistry Letters* 19 (4): 2921–2939.

171. Mayer, F., Bhandari, R., Gäth, S.A. et al. (2020). Economic and environmental life cycle assessment of organic waste treatment by means of incineration and biogasification. Is source segregation of biowaste justified in Germany? *The Science of the Total Environment* 721: 137731.
172. Chaya, W. and Gheewala, S.H. (2007). Life cycle assessment of MSW-to-energy schemes in Thailand. *Journal of Cleaner Production* 15 (15): 1463–1468.
173. Wei, L., Zhu, F., Li, Q. et al. (2020). Development, current state and future trends of sludge management in China: based on exploratory data and CO_2-equivaient emissions analysis. *Environment International* 144: 106093.
174. Mahmudul, H.M., Akbar, D., Rasul, M.G. et al. (2022). Estimation of the sustainable production of gaseous biofuels, generation of electricity, and reduction of greenhouse gas emissions using food waste in anaerobic digesters. *Fuel (Guildford)* 310: 122346.
175. Dan, C., Myat, S.H., and Aditya, P.N. (2020). Electricity generation using biogas from organic fraction of municipal solid waste generated in provinces of China: techno-economic and environmental impact analysis. *Fuel Processing Technology* 203: 106381.
176. Siegl, S., Laaber, M., and Holubar, P. (2011). Green electricity from biomass, part I: environmental impacts of direct life cycle emissions. *Waste and Biomass Valorization* 2 (3): 267–284.
177. Zhang, Z., Han, W., Chen, X. et al. (2019). The life-cycle environmental impact of recycling of restaurant food waste in Lanzhou, China. *Applied Sciences* 9 (17): 3608.
178. Khan, M.A., Ngo, H.H., Guo, W.S. et al. (2016). Comparing the value of bioproducts from different stages of anaerobic membrane bioreactors. *Bioresource Technology* 214: 816–825.
179. Fan, Y.V., Klemeš, J.J., Perry, S., and Lee, C.T. (2019). Anaerobic digestion of lignocellulosic waste: environmental impact and economic assessment. *Journal of Environmental Management* 231: 352–363.

2

Pretreatment of Lignocellulosic Materials to Enhance Biogas Recovery

Jonathan T. E. Lee[1,2,*], Nalok Dutta[3,*], To-Hung Tsui[1,2], Ee Y. Lim[4], Yanjun Dai[2,5], and Yen W. Tong[2,4]

[1]*NUS Environmental Research Institute, National University of Singapore, Singapore, Singapore*
[2]*Energy and Environmental Sustainability for Megacities (E2S2) Phase II, Campus for Research Excellence and Technological Enterprise (CREATE), Singapore, Singapore*
[3]*Department of Biochemical Engineering, University College London, London, UK*
[4]*Department of Chemical and Biomolecular Engineering, National University of Singapore, Singapore, Singapore*
[5]*School of Mechanical Engineering, Shanghai Jiao Tong University, Shanghai, China*

2.1 Introduction

With an ever-growing global population, estimated to hit 8 billion in 2030 [1], food and energy requirements are burgeoning correspondingly. The use of fossil fuels to meet the increasing demand for energy is untenable because of its status as an unrenewable energy source as well as the release of carbon in the form of CO_2 which was prehistorically locked within the ground and is emitted into the atmosphere when the fuels are burnt, potentially contributing to climate change phenomena. Hence, substitute energy sources should be renewable and carbon neutral, with bioenergy being one of the logical choices [2]. Lignocellulosic substrates constitute the largest proportion of biomass in the world and

*JTE Lee and N Dutta contributed equally to the work

Biogas Plants: Waste Management, Energy Production and Carbon Footprint Reduction,
First Edition. Edited by Wojciech Czekała

have long been recognized for their potential as a sustainable source of energy. Carbon in the form of carbon dioxide from the atmosphere is incorporated into plant biomass via photosynthesis, and hence any emission when lignocellulosic biomass is oxidized is biogenic in nature. Furthermore, with the exception of complete combustion treatment methods, any carbon in residues after the substrates are processed is trapped for much longer periods of time in the ground and not immediately released into the environment. Anaerobic digestion is one such method that treats lignocellulosic waste, which is expected to increase with rising food demand, to recover energy in the form of biogas as well as produce a nutrient-rich digestate that can be processed and utilized as an organic fertilizer for a circular economy. Biogas production stands out because it can be produced from waste material instead of requiring the cultivation of crops like rapeseed for biodiesel and sugar beet for bioethanol [3]. Additionally, anaerobic digestion can be used to process most parts of the biomass instead of specific components like lipids for biodiesel production and is versatile in its use from being able to produce electricity in a gas engine, to being upgraded to compressed or liquefied natural gas to serve as town gas or fuels for transportation. Finally, the biogas is easily separated from the reactor as opposed to requiring energy to separate biodiesel or bioethanol from the liquid phase [2].

2.1.1 Lignocellulosic Waste Material Production

While lignocellulosic substrates constitute the largest proportion of biomass globally, not all of it can be easily harvested and processed for biogas production. The more pertinent types of lignocellulosic waste have been variously described and include broadly the categories of agricultural residues, the cellulosic fraction of municipal solid waste (CF-MSW, which is specifically the lignocellulosic component of the larger subset of the organic fraction, OF-MSW), forestry resides, horticultural wastes, nonfood plants including grasses, and animal manures [4–6].

Paudel et al. [7] reviewed the annual production of agricultural residues and reported that cereals had an annual residues production of 1394–2698 Mt of which rice straw, wheat straw, and corn stover were indubitably the largest shares, fruit and vegetable residues were producing 253 Mt a year, annual grass silages from both temporary and permanent grasslands were estimated at 3367 Mt, and animal manure from cattle, swine, and poultry farms approximately produced 5995 Mt each year. Similarly, Kang et al. [8] published their calculations of wheat straw, rice straw, corn stover, and sorghum annual global yields of 589, 755, 876, and 36 Mt, respectively.

Lignocellulosic waste residue production varies from region to region and is largely influenced by the dominant agricultural produce as well as availability of substrates. For example, Bandgar et al. [9] recently reported on the annual lignocellulosic biomass production in India, for which rice straw (43.9 Mt), sugarcane bagasse (41.6 Mt), cotton stalk (29.7 Mt), and wheat straw (25.1 Mt) dominated the availability. In contrast, Cai et al. [10] described the difference between the main lignocellulosic biomass utilized in large-scale anaerobic digestion plants in China and Germany, with the former mainly treating corn stover (220 Mt), rice straw (207 Mt), and wheat straw (128 Mt), while the latter was predominantly using silage corn (77 Mt) with a minor fraction of energy grass (2 Mt).

2.1.2 Structural Insight of Lignocellulosic Materials

Notwithstanding the regional differences in lignocellulosic materials, these substrates share a common feature of a cell wall with a recalcitrant lignin-encrusted superstructure protecting holocellulose consisting of crystalline cellulose microfibrils encased in a matrix of more amorphous hemicellulose as shown in Figure 2.1. There are two main types of plant cell walls [11]; Primary plant cell walls contain cellulose, hemicelluloses, proteins, and pectin, except for grasses where it is replaced with glucuronoarabinoxylan [12], and secondary cell walls occur more frequently and are composed of cellulose, hemicellulose, and lignin [13]. Additionally, the cellulose microfilaments are interwoven with a closer alignment and spacing in primary plant cell walls [14].

Cellulose is a stable polymer of cellobiose, which itself is a disaccharide composed of two flipped D-glucose monomers linked by β-1–4-glycosidic bonds [15]. The polymer is aggregated into bundles called fibrils and includes crystalline and amorphous regions with an average molecular weight of 100 kDA [16]. The molecules of cellulose are linear and flat, resulting in hydrogen bonds and van der Waals forces between chains. These fibrillar units display high tensile strength and can fit closely enough to exclude water [14]. The lattice structure is polymorphic, depending on the form of cellulose, of which there are types I, II, III, IV, and X, although the native form is I [17]. Crystallinity affects how easily cellulose is degraded, with amorphous cellulose being hydrolyzed to cellobiose much more rapidly than crystalline cellulose, which likewise is dependent on the degree of polymerization [18]. While it is hydrophilic, the large size of the molecules negates its ability to dissolve in water [19].

Hemicellulose is the second most abundant constituent of lignocellulosic materials and is a heterogeneous mixture of polysaccharides. It comprises the matrix between cellulose fibrils, where it prevents microfibrillar aggregation and bestows flexibility to the cell wall. It is composed of four broad types of molecules: (i) xylans that contain a

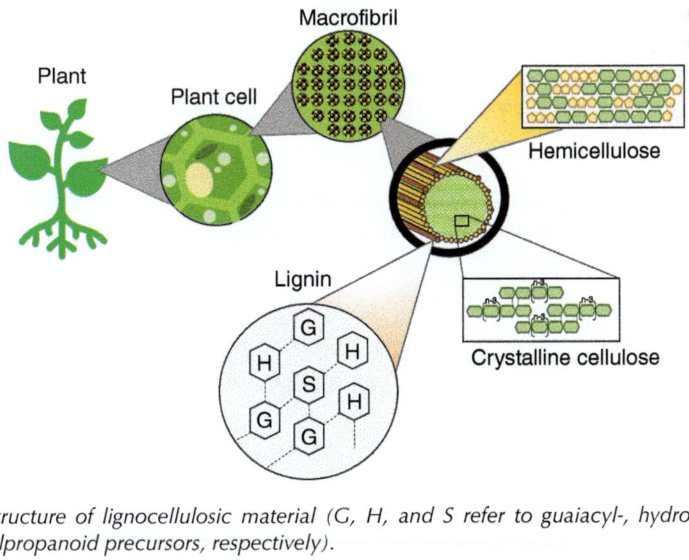

Figure 2.1 Structure of lignocellulosic material (G, H, and S refer to guaiacyl-, hydroxyphenyl-, and syringyl-phenylpropanoid precursors, respectively).

xylose backbone, (ii) mannans that instead have a mannose or mannose/glucose backbone, (iii) xyloglucans that have a glucose backbone with additional xylose branches, and (iv) mixed-linkage glucans that contain β-1–3 glycosidic bonds in addition to β-1–4 linkages [20]. Hemicellulose is randomly branched and also contains uronic acids such as glucuronic acid, methyl glucuronic acid, and galacturonic acid [21] with an average molecular weight of below 30 kDa [16].

Lignin is the biopolymer found in lignocellulosic substrates that contributes most to its recalcitrance to anaerobic digestion by acting as a protective sheath [16]. Like hemicellulose, it is quite heterogeneous and is principally composed of guaiacyl-, hydroxyphenyl-, and syringyl-phenylpropanoid precursors (or coniferyl, coumaryl, and sinapyl aromatic alcohol monomers). The lignin polymer is rigid and hydrophobic with many cross-links and is extremely resistant to degradation [13], especially under anaerobic conditions.

2.1.3 Biogas Production from Lignocellulosic Materials and the Need for Pretreatment

The production of biogas from lignocellulosic materials takes place via anaerobic digestion, whereby a complex organic substrate is degraded into simpler molecules mediated by a consortia of microorganisms. These microbes are for the most part facultative anaerobes or anaerobes; hence, the multistage process usually takes place in the absence of oxygen to produce a methane rich (50–70%) biogas with the remainder as carbon dioxide and trace amounts of contaminants [22–24]. Figure 2.2 displays the generic stages of anaerobic digestion, with the complex biopolymers like carbohydrates, proteins, and lipids first hydrolyzed into smaller substrates. In the next stage of acidogenesis, oligomers such as fatty acids are broken down into short-chain fatty acids, which in the penultimate stage of acetogenesis,

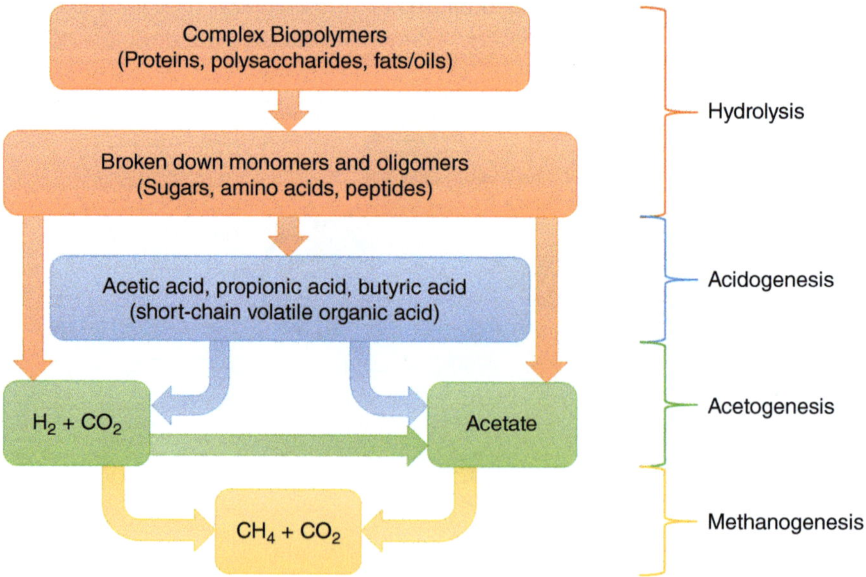

Figure 2.2 Schematic of anaerobic digestion process.

are converted into acetic acid, carbon dioxide, and hydrogen gas. Acidogenesis may not be required if the starting substrate can be hydrolyzed directly into the component monomers. Finally, methanogens convert the monomers into methane and carbon dioxide via methanogenesis [25]. Bacteria mediate the former three stages of anaerobic digestion, while only specific classes of archaea from the phylum Euryarchaeota known as methanogens possess the metabolic machinery for the latter.

For the majority of substrates processed by biogas plants utilizing anaerobic digestion, the rate-limiting step is the final stage of methanogenesis due to the slow growth rate and narrow range of optimal conditions for the methanogens. In the case of lignocellulosic materials, due to the complex structure described in the preceding section, the first step of hydrolysis is most impeded. Ergo, some form of pretreatment before the application of anaerobic digestion is required for biogas plants to maximize yield from lignocellulosic substrates.

2.2 Available Pretreatment Technologies for Lignocellulosic Materials and the Corresponding Biogas Recovery Associated

Pretreatment refers to one or more processes applied to the lignocellulosic substrates before the anaerobic digestion is carried out. Broadly, the available techniques fall into three categories based on the type of treatment utilized. In physical pretreatments, physical processes involving energy are applied, whether it is a mechanical approach, or the use of radiative or thermal energies. In chemical pretreatments, chemicals are added into the lignocellulosic materials and incubated for varying periods of time. In biological pretreatments, biological agents are employed to effect changes in the substrate before anaerobic digestion. In practice, one or more combinations of pretreatments can be performed to synergize the positive aspects of the respective methods while mitigating the negative elements and are broadly termed as hybrid pretreatment technologies. A list of the more commonly utilized pretreatments was presented in Figure 2.3.

In general, pretreatment strategies bring about the improvement in biogas yield in a variety of mechanisms. An increase in surface area to volume ratio improves the access that hydrolytic microbes and agnate enzymes have to the substrate, allowing an increase in kinetics as well as overall methane yield. Secondly, the lignin in the lignin-encrusted superstructure can be modified or removed, exposing the holocellulose within the substrates. Thirdly, the hemicellulose matrix can be loosened, allowing increased contact area with the cellulose fibers. Lastly, the crystalline regions of cellulose can be fragmented to promote degradation. The following sections introduce pretreatment technologies, explicate the mechanism of enhanced biogas recovery, and discuss the pros and cons of each technique as well as reported results on biomethane yield. A more general description will be presented for hybrid methods in Section 2.4 as the sheer number of permutations and combinations of the use of multiple methods will be challenging to list in detail.

2.2.1 Physical Pretreatment

Physical pretreatments use methods that generally reduce particle size in order to increase surface area for subsequent anaerobic digestion and methane production. Figure 2.4 presents the schematic diagram displaying how physical pretreatment technologies affect

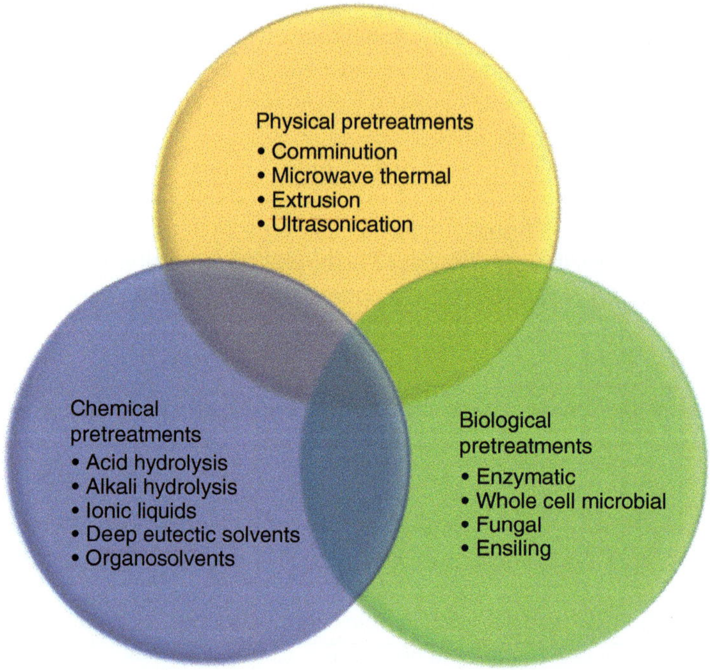

Figure 2.3 List of pretreatments according to type.

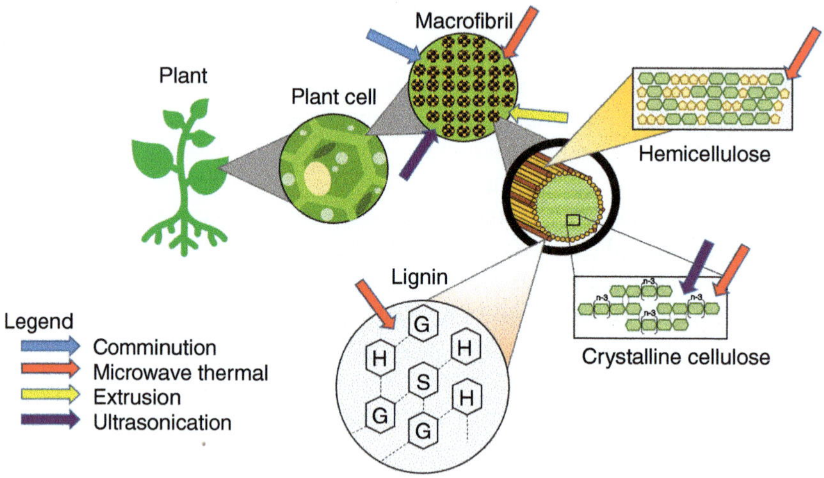

Figure 2.4 Mechanism of action of physical pretreatments (G, H, and S refer to guaiacyl-, hydroxyphenyl-, and syringyl-phenylpropanoid precursors, respectively).

the lignocellulosic structure of substrates. They are characterized by the usage of energy and are uncomplicated to integrate and operate and of low environmental impact.

2.2.1.1 Comminution

Comminution is the term that describes methods that reduce particle size of the lignocellulosic materials. Dividing substrates into smaller particles predominantly allows an increased surface area for hydrolytic microbes as well as the extracellular enzymes they excrete to digest the materials, although higher energy methods such as grinding and milling also assist to tear up the ordered structure of lignocellulose as shown in Figure 2.4. Shredding and chipping reduces particle sizes to 1–10 cm, while milling decreases the size to the millimeter scale. Comminution largely has no effect on the molecular structure of lignocellulosic materials, but with an increase in energy expended, it can reduce the crystallinity of cellulose in some cases [26].

Comminution is one of the simplest forms of pretreatment available and usually required for converting the lignocellulosic biomass into a form that is easier for movement through the processing plant. The main resource utilized is energy [27], with good returns on decreased retention times and increased biogas yield [28]. Additionally, the processes generally are fast, not complex nor hazardous to human health, nor are there any significant impacts on the environment. Smaller particle sizes are generally correlated with biogas recovery, up to a certain size, e.g. 0.4 mm, whereupon the increase in hydrolysis rate outstrips the later stages of anaerobic digestion and causes volatile fatty acid accumulation and subsequent inhibition of biogas production [29–31]. Stanley et al. [32] noted in their review that agricultural wastes, yard trimmings, grass, aspen chips, spruce, and wood chips produced a 27% increase in biogas when the respective lignocellulosic materials were reduced to sub-millimeter magnitudes. In another lab-scale batch digestion, Zieminski and Kowalska-Wentel [26] collected 29% more methane from sugar beet pulp comminuted to 2.5 mm. Sharma et al. [33] reported that wheat straw and cauliflower leaves with a particle size of 0.09 mm produced 54% and 18% more methane, respectively, while rice straw, mirabilis leaves, dhub grass, and banana peels reduced to 0.4 mm in size increased methane yield in anaerobic digestion by 52%, 18%, 66%, and 11% respectively. Conversely, Chandra et al. [34] reported an increase of 13% and 39% for rice and wheat straw with a particle size of 0.3–0.75 mm, while Dell'omo and Spena [35] disclosed a methane enhancement of 96% for the latter. Meadow grass of size 1.5 cm was reported to increase output by 22% [36], and sisal fiber ground to 2 mm particle size increased methane production by the same amount [37]. Animal manure fibers that were reduced to the same size as the latter yielded 16% more methane, while a further reduction to 0.35 mm increased the harvest to 20% [38]. Hidalgo et al. [39] pelletized vine trimming shoots to 6 mm and reported a 35% increase in methane yield. In a full-scale experiment, Monch-Tegeder et al. [40] ground horse manure before anaerobic digestion with a 80-day HRT and obtained a 27% increase in methane production.

2.2.1.2 Microwave Thermal Pretreatment

Traditional thermal pretreatment involves the heating of the lignocellulosic material through energy transfer via hot fluids with temperatures reaching 240 °C. The use of microwave irradiation avoids the formation of thermal gradients, thus allowing more precise control of the temperature, and is able to heat large quantities of the substrate easily due to its direct penetration. The elevated temperature loosens the lignocellulosic structure, aiding in lignin depolymerization and reduction in cellulose crystallinity and increasing the solubility of hemicellulose [41, 42]. Microwave pretreatment also includes nonthermal responses due to increased water molecules collision and dipoles, aiding in the solubilization of the biomass [43, 44].

Besides its effect on enhancing biogas recovery, thermal pretreatment methods have the added advantage of being able to inactivate pathogens in the waste substrates [45]. It is fast, simple, and mostly safe to operate, easily incorporated into existing anaerobic digestion plants and, for the most part, has little impact on the environment. However, the temperature and length of treatment has to be carefully optimized, as more severe conditions have been reported to result in inhibitory products for anaerobic digestion due to the modification of hemicellulose [28, 46]. The energy requirements are also among the highest of all the pretreatment techniques.

Jackowiak et al. [47] declared that the microwave pretreatment caused methane production of anaerobic digestion of rice straw to increase by 28%. Kainthola et al. [44] reported that the same substrate heated to 190 °C for four minutes by microwaves produced 41% more methane in a batch reaction, while Zhao, B. et al. [48] achieved 30% increase with hyacinth as the lignocellulosic material heated to 100 °C. Feng et al. [49] pretreated green algae and recovered the same increase as the latter, while Passos and Ferrer's [50] microwave-pretreated microalgae evidenced 59% more methane yield.

2.2.1.3 Extrusion

Contrary to the former two pretreatment methods of comminution and microwave irradiation, extrusion has much fewer variants in terms of design. In this physical pretreatment, pressure is applied on the substrate by one or two screw conveyors within an extruder barrel [51, 52]. The high shear forces between the lignocellulosic materials and the screw and barrels physically break apart the fibers, leading to a higher surface area and lower bulk density. Once extruded, the large pressure differential also causes physical damage to the cell walls and further reduces particle size [53–55].

Extrusion equipment are lower in cost and energy, simple to operate, and safe and do not have much impact on the environment. Additionally, it is easily customizable, with the option of adding heating plates around the barrel for combination with thermal pretreatment, or mixing in added chemicals for chemical pretreatment. It would require careful sorting of the lignocellulosic substrates, as contaminants such as farm implements or rocks can severely damage the screws [56, 57]. If chemicals are added, the screws and barrel would have to be made of materials that are resistant to chemical attack [58]. Finally, the viscosity of the materials has to be considered [59].

Panepinto and Genon [51] utilized a twin-screw extruder on maize silage and obtained a methane increase of 7–15%. Hjorth et al. [53] extruded grass and straw and recovered 9% and 11% more biomethane, respectively. Chen et al. [57] operated an extruder at 120 rpm for rice straw and produced 39% more methane. A mixture of 3 : 5 : 2 of rice straw silage: maize silage: triticale silage after twin screw extrusion was reported to increase methane yield by 11% [60]. Wheat straw extruded at 600 rpm was reported to yield 16% more methane than the control [61]. Hidalgo et al. [39] pretreated vine trimming shoots with a twin-screw extruder at 500 rpm and reported a 10% increase in methane yield. Khor et al. [62] extruded maize straw, grass, and sprout stem at 15 and 60 rpm and reported 26–49% increase in methane production after extrusion.

2.2.1.4 Ultrasonication

Ultrasonication is another physical irradiation method that uses ultrasound to pretreat lignocellulosic materials in solution. With a frequency of between 20 and 1000 kHz, ultrasonic equipment generate cavitation bubbles quickly that collapse and cause localized mechanical damage [63] and even the formation of free radicals (Kumari and Singh, 2018), leading to particle size reduction and microfiber shearing [64, 65]. Ultrasonication has been reported to destroy wax layers and silica bodies on the surface of lignocellulosic materials and reduced the crystallinity and degree of polymerization of cellulose [66].

Similar to most of the physical pretreatments, ultrasonication is relatively simple and safe and has little impact on the environment. It is usually used in combination with other pretreatment methods and fast but requires more cost and energy. One downside is that with high energy pretreatments, inhibitory substances for anaerobic digestion can be produced [67].

When screened cattle manure was subjected to ultrasound pretreatment, biogas yield was 121% of the control [68, 69] Zielinski et al. [69] reported 29% increased biomethane recovery when cattle manure with wheat straw was subject to full-scale anaerobic digestion. Zou et al. [70] produced 70% more methane during the co-digestion of ultrasonic-pretreated maize straw and dairy manure.

2.2.2 Chemical Pretreatment

Whereas physical pretreatments involve energy and diffuse the treatment over the whole superstructure of lignocellulosic materials, and biological pretreatments involve microbial agents and their respective specific molecules that they act on, chemical pretreatments use compounds to target the bonds between constituent monomers in a less focused manner. Chemical pretreatments are highly effective and therefore among the most commonly applied techniques, whether solo or in combination with other regimens. Figure 2.5 displays the various ways in which chemical pretreatment affects the structure of lignocellulosic biomass.

2.2.2.1 Acid Hydrolysis Pretreatment

Both inorganic and organic acids can be used in acid pretreatment, at both high concentrations and in a more dilute form. The main effect of acidic pretreatment is to solubilize hemicellulose, hydrolyzing it to release xylose and glucose. This removal of the cementitious material exposes cellulose microfibrils, increasing the effective surface area for subsequent anaerobic digestion [71, 72]. Depending on the severity of the acid pretreatment, lignin can also be affected to a certain degree [7]. However, higher severity (higher concentrations, inorganic rather than organic acids, temperature, and retention

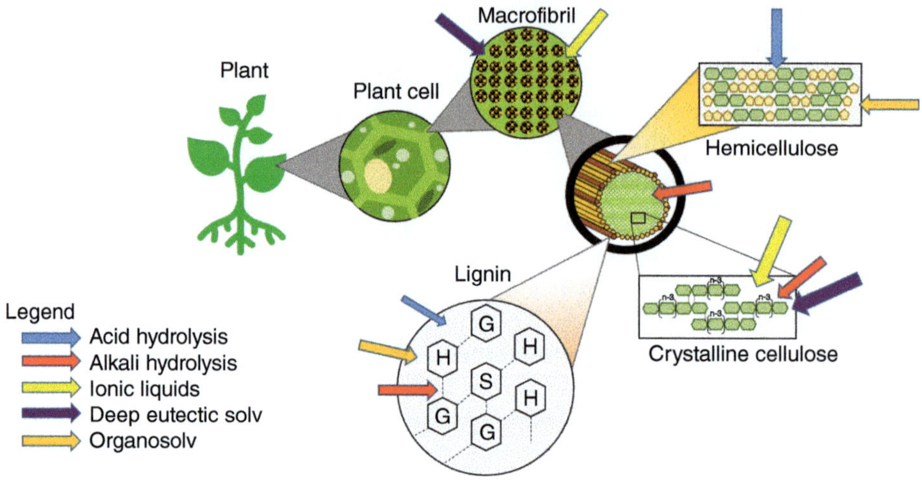

Figure 2.5 *Mechanism of action of chemical pretreatments (G, H, and S refer to guaiacyl-, hydroxyphenyl-, and syringyl-phenylpropanoid precursors, respectively).*

time) also leads to more inhibitory products due to the nonspecific reactions as well as consumption of the desired monomers [73], such that in most cases a lower severity chemical pretreatment is selected instead.

Acid pretreatments are quite effective in their function and fast; however, there is a higher cost of consumables and a high impact on the environment due to the acidic reject water [6]. The pretreated biomass can also affect the subsequent anaerobic digestion due to inhibitory compounds and a low pH. Worker safety and equipment maintenance is of paramount concern as well due to the corrosive nature of the pretreatment [74].

Fu et al. [75] pretreated *Miscanthus floridulus* with hydrochloric acid and reported a 16% increase in methane yield. Sunflower stalks were pretreated with 4% of the same acid and resulted in 21% methane yield increase [76]. Cow manure pretreated at half the concentration presented the same enhanced effect [77]. With the same amount and type of acid, corn straw was reported to produce 62% more methane [78]. The researchers also utilized sulfuric acid at the same concentration instead and declared an improvement of 75% for the substrate. One percent concentration of sulfuric acid was used to pretreat wheat, and an increase of 16% methane yield was reported by Taherdanak et al. [79]. Quadruple the amount was utilized in the pretreatment of *Salvinia molesta*, resulting in an enhancement of 82% [80]. With 5% of the same acid, Sarto et al. [81] declared that water hyacinth could produce 40% more methane. Nair et al. [82] used 1.2% phosphoric acid to treat wheat straw and gathered 1.5 times more methane. Organic acids such as acetic and citric acid have also been studied, with Saha et al. [83] producing 10% more methane from fruit waste and Peng et al. [84] treating rice straw with 2% of the former to obtain an enhancement of 24%. The latter was used by Pellera and Gidarakos [85] at 0.5 mmol g VS^{-1} to pretreat cotton waste and resulted in 57% increased biomethane. However, care must be taken to remove the methane yield contribution from the organic acids themselves.

2.2.2.2 Alkali Hydrolysis Pretreatment

Alkaline methods require less severe conditions, e.g. temperature and pressure, and have the added benefit of mitigating pH drop in the anaerobic digestion process compared to acid pretreatments (Section 2.2.2.1). The basic chemicals undergo saponification and solvation, disrupting lignin–carbohydrate bonds and degrading lignin itself. They also cause distension and increase the surface area available for subsequent enzyme hydrolysis [86, 87]. The hydroxyl ion has also been reported to break bonds between hemicellulose and cellulose and reducing cellulose crystallinity [10, 88], as well as for hemicellulose solubilization [71]. Many basic chemicals such as hydroxides, oxides, urea, and ammonia have been utilized, with factors such as cost, solubility, and ease of use contributing to their selection. As with acid pretreatment, an increased severity has been shown to produce inhibitory products [74]. Additionally, if urea or ammonia are used, it could impede the downstream anaerobic digestion process due to ammonia toxicity or an increased C:N ratio [89].

Alkali pretreatments are similar to acid pretreatments with a few exceptions. There is a little less impact to the environment, but there is the possibility of more different inhibitors present including the solubilized phenolic compounds originating from lignin [90]. Additional care has to be undertaken when optimizing pretreatment using the nitrogen-containing bases, if the substrates themselves have a higher nitrogen content, e.g. manures and some agricultural residues. It should also be noted that the use of potassium hydroxide can be potentially beneficial for the residue to be utilized as fertilizer after the digestion process [8].

Sabeeh et al. [91] reported a 71% increase in methane yield when 1.5% sodium hydroxide was used to pretreat rice straw, while the same substrate pretreated at 1.6% concentration increased production by 21% [92], and 1% concentration resulted in 34% increase in yield [93]. Manure fibers pretreated with 6% of the same alkali increased yield by 26% [94], while the same chemical pretreatment at ambient temperature increased biogas production from *Miscanthus floridulus* by 19% [75], asparagus stem by 38% [57], and corn stover by 49% [95]. The same substrate was pretreated with 1.2% and 2% concentration with enhanced methane production of 59% [96] and 73% more [97]. Corn stalk pretreated with the same concentration evidenced a methane yield enhancement of 5% [98]. Dairy manure at 10% concentration instead produced 24% more methane [77], while sorghum forage pretreated with the same concentration increased methane yield by 29% [99]. Grass silage pretreated at 7.5% was reported to produce 28% more methane [100], teff straw at 4% yielded 40% more biogas [101], poplar processing residues at 5% enhanced methane collected by 14% [102], and maize straw at 6% produced 23% more [103]. Pellar and Gidarakos [85], Mancini et al. [104], and Chandra et al. [105], respectively, pretreated wheat straw with 1.2%, 1.6%, and 4% of sodium hydroxide, and a biomethane increase of 15%, 15%, and 112% was reported. Giant reed feedstock was pretreated at 20% and presented an increase of 28% methane yield, but when the chemical pretreatment was changed to 20% calcium hydroxide instead, the increase was moderated to 23% [74]. Calcium hydroxide pretreatment on grass (7.5%) and corn straw (8%) resulted in 37% [74] and 105% increased methane yield, while 10% ammonia pretreatment gave an enhancement of 67% instead for the latter [78]. The same concentration of ammonia pretreatment resulted in sugarcane bagasse yielding 1.4 times more methane [106], and 15% on maize bran enhanced yield by more than three times [107]. 2.5% potassium hydroxide was used by Li et al. [108] to pretreat corn stover, and they recovered almost double the methane amount, while 1% urea was used to pretreat wheat straw by Yao et al. [109] and resulted in 45% increased methane yield.

2.2.2.3 Ionic Liquids Pretreatment

Ionic liquids are molten salts that are liquid at near ambient temperatures because of a low melting point. These salts can penetrate the lignocellulosic material and disrupt the intermolecular hydrogen bonds between cellulose fibrils, leading to lignocellulose dissolution [110–112]. Besides that, cellulose crystallinity is also reduced, although the degree of polymerization is unaffected [113] and the cellulose can be recovered by the use of anti-solvents such as alcohol, acetone, or water. The low volatility allows ionic liquids to be recovered [114], although the initial investment cost is extremely high. Commonly used ionic liquids include *N*-methyl morpholine-*N*-oxide, 1-ethyl-3-methylimidazolium acetate, and 1-*N*-butyl-imidazolium chloride [115].

With good recovery [116], most of the cons of ionic liquid pretreatment can be avoided, e.g. the high cost and impact to the environment. However, there are relatively few studies on its use for the purpose of enhancing biogas recovery from lignocellulosic materials, with consequently little information on toxicity or inhibitory products generation.

Among the studies that reported on methane generation, Perez-Pimenta et al. [117] reported that methane recovery was improved five times when Agave bagasse was treated with ionic liquids, while Padrino et al. [118] utilized 1-ethyl-3-methylimidazolium acetate to pretreat barley straw and obtained a 28% and 80% increase in methane under different conditions of anaerobic digestion. Kabir et al. [119] reported using *N*-methyl morpholine-*N*-oxide to pretreat the same substrate as well as pine and spruce residues with 83% and 88% increased biomethane yield, respectively. The same ionic liquid was utilized by Mancini et al. [120] to pretreat rice straw and cocoa shell with an enhancement of 82% and 14%, respectively. Allison et al. [112] pretreated tomato pomace with 1-ethyl-3-methylimidazolium acetate and achieved a 12% gain in methane yield. Gao et al. [121] pretreated mango leaves, rice straw, spruce, and water hyacinth with 1-*N*-butyl-3-methylimidazolium chloride and reported that the biogas yield was 23%, 41%, 57%, and 98%, respectively.

2.2.2.4 Deep Eutectic Solvents Pretreatment

Deep eutectic solvents are eutectic mixtures of Brønsted–Lowry or Lewis acids and bases. This implies that the acid and base mixture is involved in complex hydrogen bonding that results in a much lower melting point than the pure acid or base [122]. There are generally four different types of deep eutectic solvents, Type I–IV, of which the first three consists of a quarternary ammonium salt paired with a metal chloride, a hydrated metal chloride, or a hydrogen bond donor. The last type is a mixture of the latter two compounds [123]. Similar in principle to ionic liquids, deep eutectic solvents are able to penetrate lignocellulosic materials and aid in cellulose dissolution, overcoming the recalcitrance of the superstructure.

Deep eutectic solvents are similar to ionic liquids in their pros and cons, including a lack of volatility, flammability, and toxicity as compared to organic solvents [124]. On the other hand, they are easier to synthesize, cheaper, and biodegradable and have less of an impact on the environment [125, 126].

The bulk of literature available on deep eutectic solvents utilized for biogas recovery is on the upgrading of biogas to a higher purity for injection into the natural gas grid [127]. On its use as a pretreatment method, the thrust has mainly been in the extraction of lignin or cellulose as part of a biorefinery [128–130]. Concerning pretreatment for biogas production, Olugbemide et al. [131] utilized ammonium thiocyanate and urea on corn stover and reported an increase of 48% biomethane yield. With a prehydrolyzed OF-MSW that might contain the more recalcitrant lignocellulosic components like fruit and vegetable waste, Lima et al. [132] applied calcium chloride and oxalic acid before anaerobic digestion, presenting a biogas accumulation three times more than the control. Yu et al. [88] claimed that their modified liquid hot water pretreatment with choline chloride amounted to an in situ deep eutectic solvent pretreatment and reported that the leaf sheaths of *Roystonea regia* yielded 309% more biomethane after the process.

2.2.2.5 Organosolvents Pretreatment

Rounding up the chemical pretreatments, organosolvs are organic solvents which have traditionally been used extensively in delignification of wood for paper production. The organic solvents include methanol, ethanol, aqueous *n*-butanol, tetrahydrofurfuryl alcohol, aqueous phenol, glycerol, ethylene glycol, triethylene glycol, and acetone [133] and are used to dissolve lignin as well as hemicellulose [134]. The solutes can then be precipitated when the organic solvents are distilled for recycling.

Compared to ionic liquids and deep eutectic solvents which function by a similar principle, organosolvs are relatively cheaper although still costly. They still require recovery of the solvent, adding to operational complexity, although it is a relatively simpler distillation step. Organosolvs are generally toxic to the microbes in anaerobic digestion, as well as to human health and the environment. Additionally, they are usually volatile and highly flammable [54]. Similar to Sections 2.2.2.3 and 2.2.2.4, the majority of literature on organosolvs has been on the extraction of chemical components [135]. Among the studies for organosolv pretreatment of lignocellulosic materials for biogas recovery, the solvent most commonly used is ethanol. It was applied to pretreat wheat straw by Mancini et al. [92] at 50% concentration with a biomethane yield increase of 15%, rice straw at the same concentration for an enhancement of 42% [104], forest residues at the same concentration for five times more methane [136], and rubber wood waste at 75% for 175% increase [137].

2.2.3 Biological Pretreatment

Biological agents mediate anaerobic digestion, and hence biological pretreatments use these products to concentrate on mitigating rate-limiting steps before the actual production of biogas. Figure 2.6 presents the schematic diagram displaying how biological pretreatment technologies affect the lignocellulosic structure of substrates. They are characterized by very low environmental impact and production of inhibitory factors but generally require long periods of time as well as specialized knowledge for successful operation.

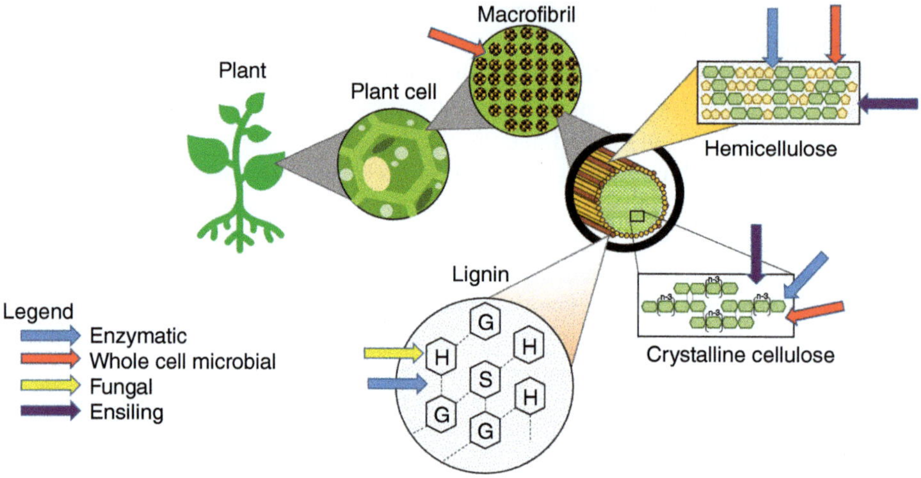

Figure 2.6 Mechanism of action of biological pretreatments (G, H, and S refer to guaiacyl-, hydroxyphenyl-, and syringyl-phenylpropanoid precursors, respectively)

2.2.3.1 Enzymatic Pretreatment

Enzymes are the tools by which the microbial community effects the anaerobic digestion process, and hence using them to pretreat lignocellulosic material affords good results without producing refractory compounds. These enzymes are isolated from lignocellulolytic cultures of microbes and added exogenously for the pretreatment. The enzymes involved in degradation of cellulose are broadly termed cellulases: β-1–4-endoglucanases randomly cleave β-1–4-glycosidic bonds to release smaller fragments of cellulose, while cellobiohydrolase works systematically from the ends of the cellulose chains to hydrolyze and release cellobiose. β-glucosidases then catalyze the hydrolysis of the cellobiose residues to glucose [138]. There are a myriad of enzymes that catalyze the degradation of hemicellulose because of the heterogeneity and complexity of the substrate, for example, the deconstruction of glucuronoarabinoxylan found in grasses requires at least six different types of enzymes: endoxylanase, β-xylosidase, α-arabinofuranosidase, α-glucuronidase, acetylxylan esterase, and ferulic acid esterase [14]. The known cellulases and hemicellulases can be found in the carbohydrate-active enzyme database known as CAZy [139]. Finally, lignin-degrading enzymes include laccase, manganese-peroxidase, and lignin-peroxidase, although oxidative conditions which are not found in anaerobic systems are required for lignin metabolism, leading some to introduce micro-aeration to aid in enzymatic as well as whole-cell microbial pretreatments.

Enzymatic pretreatment is extremely safe and, like all biological pretreatments, has a very low impact on the environment. The rate-limiting step can also be targeted with pinpoint accuracy [140]. However, incubation time is longer and has to be optimized, and the cost of isolated enzymes are extremely high.

Michalska et al. [141] pretreated energy crops with a mixture of cellulases and recovered 30% more methane. Corn stover pretreated with laccase and peroxidases increased methane yield by 12% and 17% [142] and wheat straw pretreated for six hours presented with 12% increase over the control [143]. When two different peroxidases were used on switchgrass, 29–42% more biomethane was harvested [144]. Corncob pretreated with a mixture of cellulase, xylanases, and endoglucanases produced 16% more methane [145]. Kucuker et al. [146] reported that chicken manure pretreated with a commercial enzyme resulted in 54% increased biomethane yield. Maize straw pretreated with enzymes obtained from *Trichoderma harzianum* by Zhao, X. et al. [147] was able to produce 9% more methane during anaerobic digestion.

2.2.3.2 Whole-cell Microbial Pretreatment

Instead of the prohibitive cost of isolated enzymes and the restrictive optimal conditions they operate in, whole-cell microbial pretreatment uses the microbes directly, allowing the cells to proliferate and produce the requisite intra- and extracellular enzymes. The idea of studying microbial degradation of lignocellulosic material with defined cultures is not new, with many studies focusing on pure cultures [148]. A few of these microorganisms possessed strong lignocellulolytic activities and could degrade substances with relatively simple composition and structure but were unable to fully digest natural lignocellulosic compounds that are more complex [149, 150]. Thus, attention turned toward developing symbiotic consortia able to degrade cellulosic material well and produce biofuels efficiently. Where monocultures would face product inhibition or be unable to degrade certain components, a symbiotic consortia would be able to engage in a "division of labour" to overcome these difficulties [151] or specifically to have a combined potential to produce catabolic enzyme activities [152]. These consortia can be based off engineered consortia of pure cultures, enriched cultures with lignocellulolytic activities, as well as simply an augmentation of bioactive inoculum. In some cases, researchers have claimed that separating the anaerobic digestion into multiple stages allows the first stage to be a pretreatment; however, in this chapter, those studies are not included.

This method shares traits with Section 2.2.3.1 enzymatic pretreatment in terms of safety and environmental impact but cuts down greatly on cost and generally on operational complexity with the trade-off in terms of time requirements. The formidable issue is to select the optimal source of whole-cell microbes as well as the pretreatment length of time, as periods that are too extensive cause substrates to be consumed by the microbes themselves, thereby reducing the amount of nutrients available for conversion in anaerobic digestion.

Pretreating water hyacinth with a single bacterial strain isolated from silver fish for four days increased biogas production by 23% [153]. With a thermophilic screened microbial consortium, Hua et al. [154] managed to increase biogas recovery of pretreated maize straw by 75%. Tantayotai et al. [155] recovered 3.2 and 5.2 times more methane when treating rice straw with microbes obtained from horse manure or decomposed wood, respectively, while Kong et al. [156] reported an increase of 37% biomethane from pretreated wheat straw. Rumen fluid from sheep was used to pretreat the same substrate, with a 27% enhancement in methane production [157]. A barley straw-adapted microbial consortium was applied to pretreat barley straw and hay for co-digestion and produced 40 times more methane yield as reported by Raut et al. [158].

2.2.3.3 Fungal Pretreatment

Fungal pretreatment stands in contrast to the previous sections due to its pertinence with regard to lignocellulosic materials. In the terrestrial ecosystem, lignin encrustation can be circumvented by two main methods – white- and brown-rot strategies utilized by different members of the Basidiomycetes fungal family. White-rot fungi cleave lignin by the fairly energetically expensive method of producing lignin-degrading enzymes introduced in Section 2.2.3.1, while brown-rot fungi utilize an oxidative radical-based step to depolymerize and repolymerize lignin. The oxygen-derived radicals are produced via extracellular Fenton chemistry and disrupt the lignocellulose matrix by "digging holes" instead of mineralizing the lignin in order to open up voids large enough for carbohydrate hydrolases to enter [159]. This focus on carbohydrate metabolism causes brown-rot fungi to be shunned as the subsequent methane yield from anaerobic digestion of pretreated lignocellulosic materials is thus reduced.

This method is similar to the pros and cons of Section 2.2.3.2. Whole-cell microbial pretreatment except that the substrate should be sterilized, the time required for treatment is much longer, and in the case of white-rot fungi, minor amounts of oxygen has to be supplemented.

Ghosh and Bhattacharyya [160] reported methane recovery of 32% and 46% more when brown-rot and white-rot fungi were used, respectively. Sisal leaves pretreated with the white-rot fungus *Trichoderma reesei* increased biogas production by 30–40% [161]. Using the same fungal pretreatment, Mustafa et al. [162] reported 78% increased methane production from rice straw, and 120% when using a different white-rot fungus *Pleurotus ostreatus*. The white-rot fungus *Flammulina velutipes* was able to increase biomethane yield by 65% [163], while a different white-rot fungus named *Polyporus brumalis* was reported to increase biogas yield of wheat straw by 52% [164]. Cattle manure was pretreated by *T. versicolor* and reportedly increased biomethane yield by 41% [165], while corn stover treated with *P. chrysosporium* and *Pleurotus eyngii* produced 23% [166] and 20% [167] increased methane, respectively.

2.2.3.4 Ensiling

Ensiling is one of the oldest pretreatment methods, as it has been used for storage of lignocellulosic materials for over a hundred years [168]. The storage leads to a growth of lactic acid bacteria that produce organic acids, leading to lowered pH [169, 170]. This decreases the crystallinity and degree of polymerization of the holocellulose, with no significant effect on lignin [171]. Factors affecting the process of ensiling have to be taken into consideration, including the availability of microorganisms, water content, and particle size [172].

Ensiling is the easiest and costs the least among all biological pretreatment methods, as it simply takes up space for storage. Depending on the moisture content and subsequent leachate production, there can be an impact to the environment if it is not properly managed. These pros are balanced by the significantly extended period required for it [173, 174].

Zhao, Y. et al. [48] reported that a 30-day ensiling period for switchgrass improved methane yield by 33%. Maize stover ensiled for double the period of time produced 11%

[175] and 17% [176] more methane; however, a 30-day pretreatment resulted in negative yield instead [177].

2.2.3.5 Summary of Individual Pretreatment Efficiencies

Figure 2.7 presents the biogas or methane yield enhancement for the three most common and abundant lignocellulosic materials of rice straw, wheat straw, and corn stover/straw, as well as two groups of pertinent substrates that are the grasses and various animal manures. The average reported value of the gas production from multiple studies is displayed, without taking into consideration factors that will influence the yield such as the amount and time of pretreatment, inoculum and substrate sources, and biogas measurement methods. This figure is meant as an estimation for comparison, and more details can be found in the references cited. It should also be noted that significantly large numbers, e.g. rice straw results

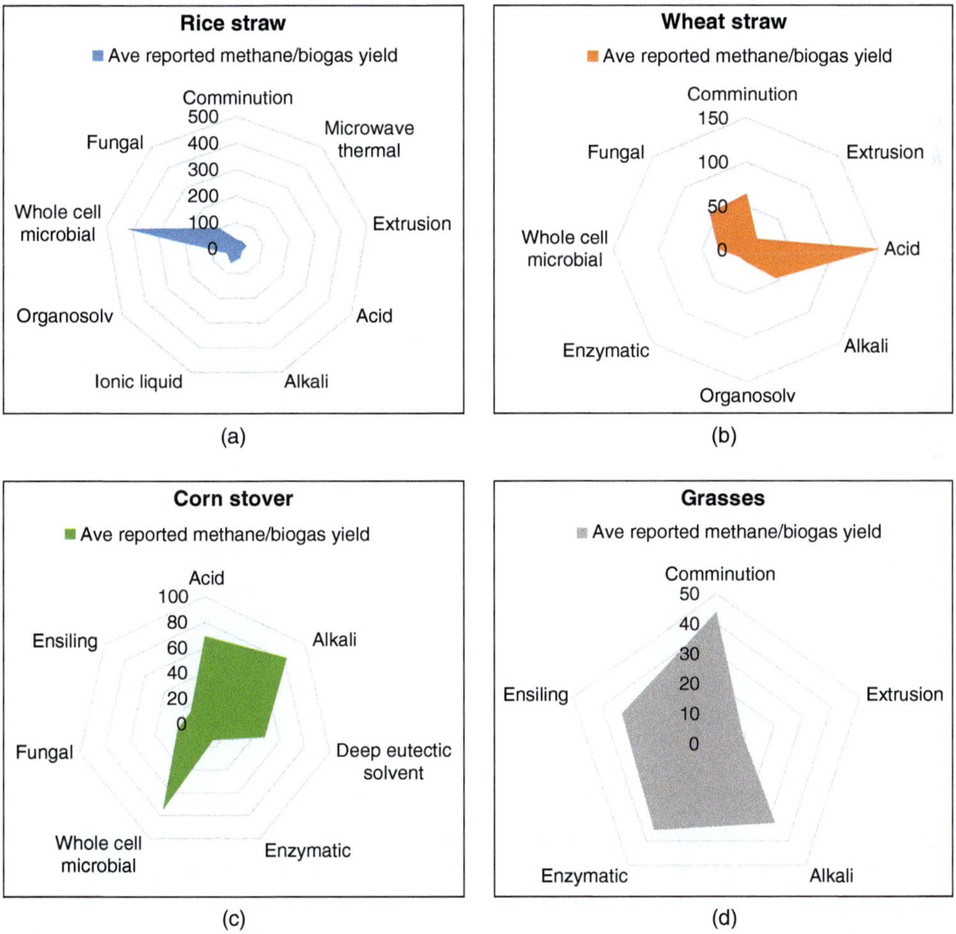

Figure 2.7 Summary of biogas production enhancement in percentages for individual pretreatment methods on (a) rice straw, (b) wheat straw, (c) corn stover, (d) grasses, and (e) animal manures.

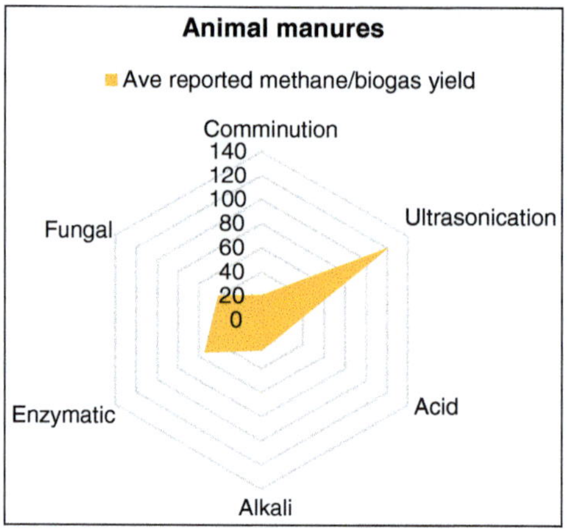

(e)

Figure 2.7 (Continued)

display an overwhelming production by whole-cell microbial pretreatment of an average of 420% increased methane production, skew some of the graphs.

2.2.4 Physiochemical Pretreatment of Lignocellulosic Biomass in the Production of Biogas

It is difficult to enhance the recovery processes pertaining to reducing/fermentable sugars vis-a-vis mitigate the extent of biocatalytic inhibitors during pretreatment in the biogas generation process [45]. The pretreatment method converts the complex polysaccharide hemicellulose and cellulose moieties into simpler monosaccharides [42]. The solid lignin is duly removed using a membrane filtering technique, and the monosaccharides can be treated subsequently via anaerobic fermentation and enzyme pyrolysis to produce biogas. The main drawbacks of anaerobic bioreactors include bigger digester capacity and longer retention times in the bioreactor compared to the typical anaerobic digestion system [178]. Lignocellulosic preparation is necessary due to the absence and slow digestion of organics, coupled with the biodegradable organic counterparts. With advancements in a variety of pretreatment machineries pertaining to bio-thermo-physicochemical and mechanical techniques, extensive research has been conducted worldwide to determine a sound biomass pretreatment knowhow, as summarized in Table 2.1.

2.2.4.1 Hybrid State of Art Lignocellulosic Pretreatments

2.2.4.1.1 Advanced Wet Oxidation and Steam Explosion Pretreatment (AWOEx)

AWOEX in its core is a thermochemical process that combines wet oxidation in tandem with steam explosion [130, 187–189]. This process exposes biomass to oxidizing chemicals such as hydrogen peroxide, air, pure oxygen, and others for 15–45 minutes at temperature extremes (above 140 °C) and 10 bars atmospheric pressure [128, 179, 187]. AWOEx preparation of lignocellulosic materials considerably improves lignin solubilization.

Table 2.1 Integrated Pretreatment Approaches.

Hybrid strategies	Guiding principle	Advantages	Disadvantages	Recovery	References
AWOEx	Thermochemical method that combines wet oxidation with steam explosion	Environmentally friendly, increased carbon conversion efficiency, and time-efficient	—	56% more methane yield	[130, 179]
Liquid hot water pretreatment with chemicals	Two-phased alkaline oxygen-aided technique aids in the isolation of biomass content	Safe process with no hazardous chemicals	Only limited to water-soluble components	31% increased sugar recovery	[180–183]
Ultrasound-aided salt an ammonia pretreatment	Ammonia promotes swelling of the biomass in the presence of ultrasound, disrupting the internal connections in a variety of ways	Aqueous media and low temperature	Unchecked high ammonia release detrimental for the environment	Recovery for cellulose ~90% and lignin~50%	[184, 185]
Ultrasound-aided dilute acid treatment	Diluted acidic solutions in the presence of ultrasound are utilized as a biomass extractors	Reagents readily available	Lowered yield with stringent pH maintenance	~60% ethanol yield	[181–183]
Ultrasound-aided organic solvent	Biomass treatments used in the presence of ultrasonic energy mediated through various organic solvents	There are numerous options to consider for various implications	(1) Difficulties linked with maintaining optimal condition (2) Specified solvents required	70% hemicellulose recovery	[181–183]
Irradiation using an electron beam in conjunction with physiochemical treatments	When an electron beam is combined with several physiochemical pretreatment processes, high-level energized electrons are released, which form highly sensitive biomass components	(1) Eco-friendly (2) By making simple adjustments to the electron beam components, yield may be dramatically boosted	(1) In a closed system, regulations must be scrupulously followed (2) Carcinogenic compounds may develop	~30% yield in biogas	[180, 182, 186]
Integrating gamma irradiation with physiochemical treatments	Several physiochemical pretreatment processes in tandem with gamma irradiation aids in increasing yield during pretreatment	Heightened enzymatic hydrolysis with increased energy efficiency	(1) Excessive irradiation reduces yield (2) The closed system must be rigorously adhered to	~72% yield	[180, 182, 184]

Furthermore, crystalline cellulose is retained in the solid fraction with the majority of hemicellulose components being solubilized throughout the operating period [179]. The biomass surface area harboring the carbohydrates is considerably enhanced as a consequence of AWOEx pretreatment, which provides unhindered access to biocatalytic and microbial agents [190]. Furthermore, acetic acid which is produced following acetyl group de-esterification impacts the hemicellulosic dissolution into the aqueous phase [128, 179]. The benefits of AWOEx pretreatment include, in addition to its efficiency, the absence of chemicals that must be recovered and the minimal development of inhibitory intermediates. AWOEx pretreatment is preferably conducted at high dry matter concentrations in upward of 40%, lowering the total pretreatment costs [179].

2.2.4.1.2 Chemical Pretreatment Mediated Through Microwave

A moderate-frequency electromagnetic radiation without ionizing characteristics is referred to as a microwave. Under some circumstances, the microwave radiation's short wavelength allows for material modification at the electron level. For lignocellulosic pretreatment, the microwave-assisted chemical treatment process may be used to provide a significant reducing sugar yield [180, 181]. The lignocellulosic biomass' dielectric factors enable the heating process of the microwave. The biomass dipolar components eventually heats up the biomass when exposed to microwave radiation. The quantity of hydroxyl linkages and lignocellulosic fibers in the untreated biomass are often responsible for controlling this dipole formation [182, 191]. The intensity of the biomass depolarization is further influenced by the presence of alkaline or acidic functional groups found in lignocellulosic biomass and their hydrophobic/hydrophilic core. During microwave radiation, these groups collectively produce tremendous heat and molecular vibration that destabilize and dissolve several bonds. Since the lignocellulosic biomass will subsequently merge with other liquid, these chemicals will be released. The high temperature within the lignocellulosic biomass also causes steam to be created [192]. Pyrolysis can be used to separate valuable items in addition to such chemical treatment via microwave-assisted processes, albeit its commercial applicability is still in the early stages [181, 193].

2.2.4.1.3 Liquid Hot Water Pretreatment with Chemicals

Among the most ecofriendly biomass pretreatment methods is the implementation of hot water, and when used in conjunction with chemically aided procedures, it helps to reduce any potential environmental harm that the hazardous chemicals may cause [186]. The procedure is comparable to the steam pretreatment approach on its own [183]. However, this technique raises the sustainability index of several processes when combined with a chemical process. In this procedure, water and other pH-neutral liquid substances ensure increased lignocellulosic biomass solubilization under specified conditions. Most of the time, ethanol extraction from biomass makes use of such hot water-based treatment methods. Being when the process is integrated with other chemically driven processes, its efficiency also rises.

2.2.4.1.4 Ultrasound-aided Ammonia Pretreatment

Biomass pretreatment may be accomplished using a variety of ultrasound-assisted techniques. The biomass molecules are energized by ultrasound during the salt and ammonia pretreatment process, which is then combined with an ammonia-mediated treatment procedure to break down the bonds. Deacetylation of hemicelluloses vis-a-vis decreased crystallinity and lignin–carbohydrate bond breakage, and other mechanisms contribute to this breakdown [181]. The majority of the carbonyl group in the biomass are broken down by the ammonia components in the aqueous solution, and the addition of salt and ultrasonic components speeds up this process [183]. Both the ammonia and ultrasound-assisted procedures may separate valuable components from lignocellulosic biomass, but when they are combined, they increase sugar recovery from biomass. Though this method has a lot of capability for commercial purposes, its process efficacy depends on the biocatalyst type. For instance, potassium permanganate increases the pretreatment efficiency of lignocellulosic biomass that has co-treated with coffee by 1.7 times. However, this particular reagent does not exhibit the same level of effectiveness in other situations, necessitating the selection of an appropriate reagent in order to obtain the desired yield [181]. Despite the fact that this hybrid pretreatment technique allows for the recovery of several types of carbohydrates, it is more suited for the recovery of cellulose [180].

2.2.4.1.5 Ultrasound-assisted Acidic and Alkaline Pretreatments

Another important kind of lignocellulosic biomass treatment technique is acid-mediated saccharification and biofuel production. This synergistic technology is used to enhance biomass solubilization, which disturbs the substance's original shape [184]. Ultrasound energy enhances process effectiveness when used with this procedure. The primary end product of this process is hydrogen [182]. The procedure is dependent on many widely accessible products, which qualifies it for broad commercial applications.

The procedure of ultrasound-assisted alkaline pretreatment is identical to that of ultrasound mediated ammonia and salt pretreatment. Alkaline-mediated treatment increases biomass in a variety of ways, including boosted solubilization, lowered crystallinity and enhanced cellulose digestibility [181]. Several investigations have indicated that the two principal components recovered from this technique are lignin and hemicellulose. Furthermore, bioethanol is a popular biofuel retrieved from this procedure [183]. Mostly, the alkaline reagents that stimulate their hydrolysate group for sugar recovery are NaOH and $Ca(OH)_2$. Ultrasound boosts hydroxyl radical interaction, decreasing reaction energy [182]. This increases the energy efficiency of the alkaline group-mediated practices. This mechanism's primary target the hydrogen bonds found in lignin-rich biomass. Biofuels can be generated by breaking these connections [180].

2.2.4.1.6 Ultrasound-aided Ionic Liquid Pretreatment

As previously mentioned, ultrasound-aided activities usually reduce cellulose crystallinity, resulting in the breakdown of biomass bonds. Another approach facilitating the mining of usable biomass components require liquefied polar stimulant medium implementing the ultrasound-mediated ionic liquid pretreatment technique. In the presence of ultrasonic energy, 1-butyl-3-methylimidazolium acetate and 1-butyl-3-methylimidazolium chloride constitute the common ionic liquid medium exhibiting substantial ability for monosaccharide recovery from biomass [181]. When evaluated against chloride anions, the acetate anions in comparison promote more efficient sugar extraction. The anions' combined actions eventually break biomass connections, allowing more room for catalytic couplings between monosaccharides and enzymes.

2.2.4.1.7 *Ultrasound-aided Organic Solvents*

Organic solvents behave similarly to ionic liquids. Ultrasound disrupts the crystallinity of biomass, allowing organic solvents to access biomass associations with reduced energy. Organic solvents having sufficient diacylglycerol-beta-aryl and alpha-aryl groups eventually disrupt various biolinkages [181]. The time frame required for lignocellulosic biomass and ultrasonic radiation interaction determines process efficiency. Traditional organosolv pretreatment techniques are not at par with ultrasound-mediated organosolv procedures in terms of cost-effectiveness or energy efficiency when it is being used on a big scale.

2.2.4.1.8 *Chemical and Physical Pretreatment Mediated Through Gamma Irradiation*

Gamma irradiation may be used for biomass pretreatment which targets the cellulosic carboxyl and carbonyl functional groups [180, 182]. As gamma irradiation damages certain atoms, the recovered product has a varied orientation in terms of its carbon, oxygen, and hydrogen ratio compared to its previous structure. Relatively, low dosage of gamma irradiation combined with other physicochemical treatments might result in a large yield of carbohydrates. To increase the yield of the targeted sugar moiety, gamma irradiation can be combined with physical and ionic liquid treatments [184].

2.3 Pertinent Perspectives

2.3.1 Integrated Biorefinery While Treating Various Wastes

Lignocellulosic biomass is made up of dry plant materials and so encompasses a wide range of substrates, including various grasses, plant stems, trees, and wastes from contemporary paper mills and sawmills. It is generically characterized as waste and virgin biomass, and energy crops. Every year, approximately 200 billion tons of lignocellulosic biomass waste are generated worldwide [194], the majority of which are low-value byproducts of human activities, or the natural environment, various industrial sectors, such as agro waste comprising of crop stalk and straw, forestry and municipal waste. The US Department of Energy has identified these lignocellulosic biomass wastes as important reserves.

2.3.1.1 *Municipal Solid Waste (MSW)*

MSW consists mostly of residential, commercial, and yard wastes produced in municipalities in either solid or semisolid form, omitting agro industrial hazardous wastes but containing pretreated biomedical wastes. Worldwide MSW generation is expected to reach around 2.0 billion tonnes by 2024 [195]. The majority of MSW is produced in middle- and low-income regions, accounting up to 91% of the total production [195]. America and China generate the most MSW in the world, with ~290 Mt (2.5 kg per capita/day) and 230 Mt (0.45 kg per capita/day), respectively. Faster lignin breakdown with concomitant methanogenesis in lignocellulosic biomass waste is necessary in constraining methane generation to a lesser duration of higher concentration release, permitting additional landfill biogas to be obtained as energy and limiting long-term lowered GHG emissions. Biotechnological approaches can solve this dually: firstly by the use of extracellular catalysts [143, 196, 197] or enzyme-producing microorganisms [198–200]. However, because the previously described investigations were conducted in a laboratory setting with carefully controlled circumstances or standard materials, implementation of these practices in landfills will be met with great difficulty.

2.3.1.2 *Forestry Waste*

Forests constitute the greatest terrestrial carbon sink and are instrumental in maintenance of carbon cycle by absorbing energy [201, 202]. The present carbon store in forests is projected to be 860 billion tonnes, with soil (40%) and living biomass (44%) accounting for the great majority [203]. Hardwood and softwood are the two primary varieties of wood. Hardwoods are produced by deciduous trees and softwoods are produced by coniferous trees.

Large-scale burning of wood waste to create electricity or energy is no more seen to be an effective or ecologically friendly solution. As a result, emphasis must be directed to the development of alternate solutions for renewable biofuel [204]. The lignin concentration of wood is relatively high, 39% in softwood and 25% in hardwood [205], which does not allow microbes or enzymes to degrade. Researchers have screened and isolated novel high-efficiency lignin-degrading microbes [206, 207] to manifest a favorable situation for fermentation to make better use of wood waste.

2.3.1.3 *Crop Straw*

The penchant of shifting to circular economy entailing limiting resource inputs and structured waste outputs within a closed environment is a key sustainable development aim [208, 209]. The application of agricultural straw as a feedstock in generating biogas via anaerobic digestion is consistent with the corresponding goal of developing a circular economy. Biogas, as a clean renewable energy, has the potential to ease energy shortages while also reducing the danger of air pollution caused by inappropriate crop straw management. Currently, biogas generation from agricultural waste is limited, despite an abundant resource of crop straw. Only 2.05 billion metric tons of biogas are now generated in India per year, despite the potential for 48 billion metric tons per year based on straw volume [210]. Due to a large volume of agricultural straw, China's biogas sector is seen to have considerable potential. Despite this, the eventual biogas output to the calculated biogas potential ratio is around 6.17% [211]. Crop straw's biogas potential remains untapped owing to an inefficient resource chain and sustainable economic models, dearth of easy pretreatment techniques, inadequate limited period returns, and a scarcity of sophisticated equipment [212]. A number of small-scale biogas facilities have been functioning seamlessly for decades, whereas large-scale technologically sophisticated plants are rare and relatively new [213].

2.3.2 Biogas Production from Lignocellulosic Waste and Its Economic Viability

Many factors influence waste management costs throughout the economic chain of garbage manipulation, including collection and transportation to processing plants. Although lignocellulosic biomass wastes are uneven in orientation, proper processing might mitigate waste amount and enhance handling, resulting in cheaper transportation costs [214]. Because of its varied and complicated composition, MSW collection and transportation are challenging to price. Forestry trash does not yet have a market price, since it is rarely collected. As a result, only an estimated price ranging from 50 UK £ per oven dry tonne [215] could be offered.

Biogas is a gaseous combination that mostly consists of methane (50–70%), carbon dioxide (20–30%), and trace gases such as hydrogen sulfide (H_2S) [216]. Biogas produced by anaerobic digestion could be employed directly for household cooking or utilized to generate electricity, emitting less GHG than fossil fuels [217]. The addition of CO_2 to biogas decreases its calorific value constraining its usefulness. Minimal levels of H_2S may be detrimental for sophisticated machinery and diesel engines. As a result, biogas quality must be improved before it can be utilized as a vehicle fuel [218]. The upgrading process generates a concentrated CO_2 stream, resulting in CO_2 collection prices as low as $20/tonne [219]. Energy prices increase the economic justification for methane usage, promoting anaerobic digestion of lignocellulosic biomass wastes and offering an extra source of revenue for remote regions. By the end of 2015, the International Energy Agency (IEA) has registered 430 biogas facilities globally. EPA data suggest that biogas utilization will reach 14 EJ in 2050, which will play an essential part in achieving net-zero emissions in the worldwide energy industry by 2050. Furthermore, family and community digesters in rural regions will supply renewable energy and clean cooking to almost 500 million families by 2030 [220]. The limited amount of fossil fuel reserves accentuates the need for biogas production as an alternative energy source with investment in biogas and biomethane estimated to reach $15 billion by 2040. Approximately 30 Mtoe of biomethane may now be produced at a cheaper cost than natural gas. Methane contributes to the greenhouse effect substantially, and if policymakers acknowledge the benefit of preventing methane release from feedstock breakdown, large-scale biogas production will be economically viable.

2.4 Conclusions

As previously stated, determining optimal pretreatment procedures for lignocellulosic biomass is challenging, since their efficiency is dependent on the composition of the feedstock. Nonetheless, differential pretreatment procedures provide sustained benefits confronting certain substantial constraints that may be used to guide technology selection. As stated in depth in the preceding sections, energy prices and logistics are significant barriers for a number of pretreatment procedures. Furthermore, some processes may be ecologically untenable due to hazardous reagent consumption and waste creation. Eco-friendly procedures, namely ultrasound-assisted methods and/or microwave or [180, 182], on the other hand, are not always feasible on a wide scale [221]. Hybrid approaches combine the benefits of several pretreatment techniques and show better assurance than traditional preprocessing methodologies in terms of guaranteeing a good yield in following bioprocesses [222], while reliably decreasing energy, cost, and time.

Despite their high cost, hybrid pretreatment approaches have showed significant promise in increasing yield. In terms of pretreatment of lignocellulosic biomass, chemically assisted liquid hot water or steam pretreatment is highly successful. Alkaline-coupled physical pretreatment procedures like microwave and ultrasonication have also yielded satisfactory results. In spite of the pretreatment adapted for specific biomass types and the preferred uses, efforts must be undertaken to integrate processes and standardize and optimize process parameters in order to make the eventual process techno-economically viable. Furthermore, thorough fundamental research of each pretreatment method is essential. This methodology can help technical professionals improve and create pretreatment procedures for various types of lignocellulosic biomass with a wide range of applications.

Acknowledgments

This project was funded by the National Research Foundation, Prime Minister's Office, Singapore, under its Campus for Research Excellence and Technological Enterprise (CREATE) program.

References

1. Gerber, P.J., Steinfeld, H., Henderson, B. et al. (2013). *Tackling Climate Change Through Livestock: A Global Assessment of Emissions and Mitigation Opportunities*. Food and Agriculture Organization of the United Nations (FAO).
2. De Schamphelaire, L. and Verstraete, W. (2009). Revival of the biological sunlight-to-biogas energy conversion system. *Biotechnology and Bioengineering* 103 (2): 296–304.
3. Börjesson, P. and Mattiasson, B. (2008). Biogas as a resource-efficient vehicle fuel. *Trends in Biotechnology* 26 (1): 7–13.
4. Dashtban, M., Schraft, H., Syed, T.A., and Qin, W. (2010). Fungal biodegradation and enzymatic modification of lignin. *International Journal of Biochemistry and Molecular Biology* 1 (1): 36–50.
5. Oh, J.-I., Lee, J., Lin, K.-Y.A. et al. (2018). Biogas production from food waste via anaerobic digestion with wood chips. *Energy and Environment* 29 (8): 1365–1372.
6. Hashemi, B., Sarker, S., Lamb, J.J., and Lien, K.M. (2021). Yield improvements in anaerobic digestion of lignocellulosic feedstocks. *Journal of Cleaner Production* 288: 125447.
7. Paudel, S.R., Banjara, S.P., Choi, O.K. et al. (2017). Pretreatment of agricultural biomass for anaerobic digestion: current state and challenges. *Bioresource Technology* 245: 1194–1205.
8. Kang, X., Xu, C., Lin, R. et al. (2022). Feedstock pretreatment for enhanced anaerobic digestion of lignocellulosic residues for bioenergy production. In: *Biomass, Biofuels, Biochemicals*, 253–282. Elsevier.
9. Bandgar, P., Jain, S., and Panwar, N. (2022). A comprehensive review on optimization of anaerobic digestion technologies for lignocellulosic biomass available in India. *Biomass and Bioenergy* 161: 106479.
10. Cai, Y., Zheng, Z., Schäfer, F. et al. (2021). A review about pretreatment of lignocellulosic biomass in anaerobic digestion: Achievement and challenge in Germany and China. *Journal of Cleaner Production* 299: 126885.
11. Himmel, M.E., Ding, S.-Y., Johnson, D.K. et al. (2007). Biomass recalcitrance: engineering plants and enzymes for biofuels production. *Science* 315 (5813): 804–807.
12. Vogel, J. (2008). Unique aspects of the grass cell wall. *Current Opinion in Plant Biology* 11 (3): 301–307.
13. Chundawat, S.P., Beckham, G.T., Himmel, M.E., and Dale, B.E. (2011). Deconstruction of lignocellulosic biomass to fuels and chemicals. *Annual Review of Chemical and Biomolecular Engineering* 2: 121–145.
14. Jordan, D.B., Bowman, M.J., Braker, J.D. et al. (2012). Plant cell walls to ethanol. *Biochemical Journal* 442 (2): 241–252.
15. Mansfield, S.D., Mooney, C., and Saddler, J.N. (1999). Substrate and enzyme characteristics that limit cellulose hydrolysis. *Biotechnology Progress* 15 (5): 804–816.

16. Anwar, Z., Gulfraz, M., and Irshad, M. (2014). Agro-industrial lignocellulosic biomass a key to unlock the future bio-energy: a brief review. *Journal of Radiation Research and Applied Sciences* 7 (2): 163–173.
17. Park, S., Baker, J.O., Himmel, M.E. et al. (2010). Cellulose crystallinity index: measurement techniques and their impact on interpreting cellulase performance. *Biotechnology for Biofuels* 3: 1–10.
18. Wood, T.M., McCRAE, S.I., and Bhat, K.M. (1989). The mechanism of fungal cellulase action. Synergism between enzyme components of *Penicillium pinophilum* cellulase in solubilizing hydrogen bond-ordered cellulose. *Biochemical Journal* 260 (1): 37–43.
19. Laureano-Perez, L., Teymouri, F., Alizadeh, H., and Dale, B.E. (2005). Understanding factors that limit enzymatic hydrolysis of biomass: characterization of pretreated corn stover. *Applied Biochemistry and Biotechnology* 124: 1081–1099.
20. Scheller, H.V. and Ulvskov, P. (2010). Hemicelluloses. *Annual review of Plant Biology* 61: 263–289.
21. Yang, L., Xu, F., Ge, X., and Li, Y. (2015). Challenges and strategies for solid-state anaerobic digestion of lignocellulosic biomass. *Renewable and Sustainable Energy Reviews* 44: 824–834.
22. Macias-Corral, M., Samani, Z., Hanson, A. et al. (2008). Anaerobic digestion of municipal solid waste and agricultural waste and the effect of co-digestion with dairy cow manure. *Bioresource Technology* 99 (17): 8288–8293.
23. Bruni, E., Jensen, A.P., and Angelidaki, I. (2010). Comparative study of mechanical, hydrothermal, chemical and enzymatic treatments of digested biofibers to improve biogas production. *Bioresource Technology* 101 (22): 8713–8717.
24. Zhong, W., Zhang, Z., Luo, Y. et al. (2012). Biogas productivity by co-digesting Taihu blue algae with corn straw as an external carbon source. *Bioresource Technology* 114: 281–286.
25. Batstone, D.J., Keller, J., Angelidaki, I. et al. (2002). The IWA anaerobic digestion model no 1 (ADM1). *Water Science and Technology* 45 (10): 65–73.
26. Ziemiński, K. and Kowalska-Wentel, M. (2017). Effect of different sugar beet pulp pretreatments on biogas production efficiency. *Applied Biochemistry and Biotechnology* 181: 1211–1227.
27. Kumari, D. and Singh, R. (2018). Pretreatment of lignocellulosic wastes for biofuel production: a critical review. *Renewable and Sustainable Energy Reviews* 90: 877–891.
28. Elliott, A. and Mahmood, T. (2012). Comparison of mechanical pretreatment methods for the enhancement of anaerobic digestion of pulp and paper waste activated sludge. *Waste Environment Research* 84 (6): 497–505.
29. De la Rubia, M., Fernández-Cegrí, V., Raposo, F., and Borja, R. (2011). Influence of particle size and chemical composition on the performance and kinetics of anaerobic digestion process of sunflower oil cake in batch mode. *Biochemical Engineering Journal* 58: 162–167.
30. Ferreira, L., Nilsen, P., Fdz-Polanco, F., and Pérez-Elvira, S.I. (2014). Biomethane potential of wheat straw: influence of particle size, water impregnation and thermal hydrolysis. *Chemical Engineering Journal* 242: 254–259.
31. Kang, X., Zhang, Y., Song, B. et al. (2019). The effect of mechanical pretreatment on the anaerobic digestion of *Hybrid Pennisetum*. *Fuel* 252: 469–474.
32. Stanley, J.T., Thanarasu, A., Kumar, P.S. et al. (2022). Potential pre-treatment of lignocellulosic biomass for the enhancement of biomethane production through anaerobic digestion-a review. *Fuel* 318: 123593.
33. Sharma, S.K., Mishra, I., Sharma, M., and Saini, J.S. (1988). Effect of particle size on biogas generation from biomass residues. *Biomass* 17 (4): 251–263.
34. Chandra, R., Takeuchi, H., Hasegawa, T., and Vijay, V.K. (2015). Experimental evaluation of substrate's particle size of wheat and rice straw biomass on methane production yield. *Agricultural Engineering International: CIGR Journal* 17 (2): 93–104.
35. Dell'Omo, P.P. and Spena, V.A. (2020). Mechanical pretreatment of lignocellulosic biomass to improve biogas production: Comparison of results for giant reed and wheat straw. *Energy* 203: 117798.

36. Tsapekos, P., Kougias, P.G., and Angelidaki, I. (2015). Anaerobic mono-and co-digestion of mechanically pretreated meadow grass for biogas production. *Energy and Fuels* 29 (7): 4005–4010.
37. Mshandete, A., Björnsson, L., Kivaisi, A.K. et al. (2006). Effect of particle size on biogas yield from sisal fibre waste. *Renewable Energy* 31 (14): 2385–2392.
38. Angelidaki, I. and Ahring, B.K. (2000). Methods for increasing the biogas potential from the recalcitrant organic matter contained in manure. *Water Science and Technology* 41 (3): 189–194.
39. Hidalgo, D., Martín-Marroquín, J., Castro, J. et al. (2022). Influence of cavitation, pelleting, extrusion and torrefaction petreatments on anaerobic biodegradability of barley straw and vine shoots. *Chemosphere* 289: 133165.
40. Mönch-Tegeder, M., Lemmer, A., and Oechsner, H. (2014). Enhancement of methane production with horse manure supplement and pretreatment in a full-scale biogas process. *Energy* 73: 523–530.
41. Sapci, Z. (2013). The effect of microwave pretreatment on biogas production from agricultural straws. *Bioresource Technology* 128: 487–494.
42. Abraham, A., Mathew, A.K., Park, H. et al. (2020). Pretreatment strategies for enhanced biogas production from lignocellulosic biomass. *Bioresource Technology* 301: 122725.
43. Singh, R., Krishna, B.B., Kumar, J., and Bhaskar, T. (2016). Opportunities for utilization of non-conventional energy sources for biomass pretreatment. *Bioresource Technolgy* 199: 398–407.
44. Kainthola, J., Shariq, M., Kalamdhad, A.S., and Goud, V.V. (2019). Enhanced methane potential of rice straw with microwave assisted pretreatment and its kinetic analysis. 232: 188–196.
45. Amin, F.R., Khalid, H., Zhang, H. et al. (2017). Pretreatment methods of lignocellulosic biomass for anaerobic digestion. *AMB Express* 7 (1): 72.
46. Paul, S. and Dutta, A. (2018). Challenges and opportunities of lignocellulosic biomass for anaerobic digestion. *Resources, Conservation and Recycling* 130: 164–174.
47. Jackowiak, D., Bassard, D., Pauss, A., and Ribeiro, T. (2011). Optimisation of a microwave pretreatment of wheat straw for methane production. *Bioresource Technology* 102 (12): 6750–6756.
48. Zhao, Y., Wu, J., Yuan, X. et al. (2017). The effect of mixing intensity on the performance and microbial dynamics of a single vertical reactor integrating acidogenic and methanogenic phases in lignocellulosic biomass digestion. *Bioresource Technology* 238: 542–551.
49. Feng, R., Zaidi, A.A., Zhang, K., and Shi, Y. (2019). Optimisation of microwave pretreatment for biogas enhancement through anaerobic digestion of microalgal biomass. *Periodica Polytechnica Chemical Engineering* 63 (1): 65–72.
50. Passos, F. and Ferrer, I. (2014). Microalgae conversion to biogas: thermal pretreatment contribution on net energy production. *Environmental Science and Technology* 48 (12): 7171–7178.
51. Panepinto, D. and Genon, G. (2016). Analysis of the extrusion as a pretreatment for the anaerobic digestion process. *Industrial Crops and Products* 83: 206–212.
52. Ravindran, R. and Jaiswal, A.K. (2016). A comprehensive review on pre-treatment strategy for lignocellulosic food industry waste: challenges and opportunities. *Bioresource Technology* 199: 92–102.
53. Hjorth, M., Gränitz, K., Adamsen, A.P., and Møller, H.B. (2011). Extrusion as a pretreatment to increase biogas production. *Bioresource Technology* 102 (8): 4989–4994.
54. Zheng, Y., Zhao, J., Xu, F., and Li, Y. (2014). Pretreatment of lignocellulosic biomass for enhanced biogas production. *Progress in Energy and Combustion Science* 42: 35–53.
55. Duque, A., Manzanares, P., and Ballesteros, M. (2017). Extrusion as a pretreatment for lignocellulosic biomass: Fundamentals and applications. *Renewable Energy* 114: 1427–1441.
56. Lamsal, B., Yoo, J., Brijwani, K., and Alavi, S. (2010). Extrusion as a thermo-mechanical pre-treatment for lignocellulosic ethanol. *Biomass and Bioenergy* 34 (12): 1703–1710.
57. Chen, X., Gu, Y., Zhou, X., and Zhang, Y. (2014). Asparagus stem as a new lignocellulosic biomass feedstock for anaerobic digestion: Increasing hydrolysis rate, methane production and biodegradability by alkaline pretreatment. *Bioresource Technology* 164: 78–85.

58. Miller, S. and Hester, R. (2007). Concentrated acid conversion of pine softwood to sugars. Part I: use of a twin-screw reactor for hydrolysis pretreatment. *Chemical Engineering Communications* 194 (1): 85–102.
59. Kumar, M., Oyedun, A.O., and Kumar, A. (2018). A review on the current status of various hydrothermal technologies on biomass feedstock. *Renewable and Sustainable Energy Reviews* 81: 1742–1770.
60. Menardo, S., Cacciatore, V., and Balsari, P. (2015). Batch and continuous biogas production arising from feed varying in rice straw volumes following pre-treatment with extrusion. *Bioresource Technology* 180: 154–161.
61. Wahid, R., Hjorth, M., Kristensen, S., and Møller, H.B. (2015). Extrusion as pretreatment for boosting methane production: effect of screw configurations. *Energy & Fuels* 29 (7): 4030–4037.
62. Khor, W.C., Rabaey, K., and Vervaeren, H. (2015). Low temperature calcium hydroxide treatment enhances anaerobic methane production from (extruded) biomass. *Bioresource Technology* 176: 181–188.
63. Bussemaker, M.J. and Zhang, D. (2013). Effect of ultrasound on lignocellulosic biomass as a pretreatment for biorefinery and biofuel applications. *Industrial & Engineering Chemistry Research* 52 (10): 3563–3580.
64. Patil, P.N., Gogate, P.R., Csoka, L. et al. (2016). Intensification of biogas production using pretreatment based on hydrodynamic cavitation. *Ultrasonics Sonochemistry* 30: 79–86.
65. Lamb, J.J., Islam, M.H., Hjelme, D.R. et al. (2019). Effect of power ultrasound and Fenton reagents on the biomethane potential from steam-exploded birchwood. *Ultrasonics Sonochemistry* 58: 104675.
66. Luo, J., Fang, Z., and Smith, R.L. Jr. (2014). Ultrasound-enhanced conversion of biomass to biofuels. *Progress in Energy and Combustion Science* 41: 56–93.
67. Terán Hilares, R., Ramos, L., da Silva, S.S. et al. (2018). Hydrodynamic cavitation as a strategy to enhance the efficiency of lignocellulosic biomass pretreatment. *Critical Reviews in Biotechnology* 38 (4): 483–493.
68. Castrillón, L., Fernández-Nava, Y., Ormaechea, P., and Marañón, E. (2011). Optimization of biogas production from cattle manure by pre-treatment with ultrasound and co-digestion with crude glycerin. *Bioresource Technology* 102 (17): 7845–7849.
69. Zieliński, M., Dębowski, M., Kisielewska, M. et al. (2019). Cavitation-based pretreatment strategies to enhance biogas production in a small-scale agricultural biogas plant. *Energy for Sustainable Development* 49: 21–26.
70. Zou, S., Wang, X., Chen, Y. et al. (2016). Enhancement of biogas production in anaerobic co-digestion by ultrasonic pretreatment. *Energy Conversion and Management* 112: 226–235.
71. Yao, Y. and Chen, S. (2016). A novel and simple approach to the good process performance of methane recovery from lignocellulosic biomass alone. *Biotechnology for Biofuels* 9 (1): 1–9.
72. Tu, W.-C. and Hallett, J.P. (2019). Recent advances in the pretreatment of lignocellulosic biomass. *Current Opinion in Green and Sustainable Chemistry* 20: 11–17.
73. Keskin, T., Abubackar, H.N., Arslan, K., and Azbar, N. (2019). Biohydrogen production from solid wastes. In: *Biohydrogen*, 2e, 321–346. Elsevier.
74. Jiang, D., Ge, X., Zhang, Q. et al. (2017). Comparison of sodium hydroxide and calcium hydroxide pretreatments of giant reed for enhanced enzymatic digestibility and methane production. *Bioresource Technology* 244: 1150–1157.
75. Fu, S.-F., Chen, K.-Q., Zhu, R. et al. (2018). Improved anaerobic digestion performance of Miscanthus floridulus by different pretreatment methods and preliminary economic analysis. *Energy Conversion and Management* 159: 121–128.
76. Monlau, F., Barakat, A., Steyer, J.-P., and Carrère, H. (2012). Comparison of seven types of thermo-chemical pretreatments on the structural features and anaerobic digestion of sunflower stalks. *Bioresource Technology* 120: 241–247.

77. Passos, F., Ortega, V., and Donoso-Bravo, A. (2017). Thermochemical pretreatment and anaerobic digestion of dairy cow manure: experimental and economic evaluation. *Bioresource Technology* 227: 239–246.
78. Song, Z., Liu, X., Yan, Z. et al. (2014). Comparison of seven chemical pretreatments of corn straw for improving methane yield by anaerobic digestion. *PloS One* 9 (4): e93801.
79. Taherdanak, M., Zilouei, H., and Karimi, K. (2016). The influence of dilute sulfuric acid pretreatment on biogas production from wheat plant. *International Journal of Green Energy* 13 (11): 1129–1134.
80. Syaichurrozi, I., Villta, P.K., Nabilah, N., and Rusdi, R. (2019). Effect of sulfuric acid pretreatment on biogas production from *Salvinia molesta*. *Journal of Environmental Chemical Engineering* 7 (1): 102857.
81. Sarto, S., Hildayati, R., and Syaichurrozi, I. (2019). Effect of chemical pretreatment using sulfuric acid on biogas production from water hyacinth and kinetics. *Renewable Energy* 132: 335–350.
82. Nair, R.B., Kabir, M.M., Lennartsson, P.R. et al. (2018). Integrated process for ethanol, biogas, and edible filamentous fungi-based animal feed production from dilute phosphoric acid-pretreated wheat straw. *Applied Biochemistry and Biotechnology* 184: 48–62.
83. Saha, S., Jeon, B.-H., Kurade, M.B. et al. (2018). Optimization of dilute acetic acid pretreatment of mixed fruit waste for increased methane production. *Journal of Cleaner Production* 190: 411–421.
84. Peng, J., Abomohra, A.E.-F., Elsayed, M. et al. (2019). Compositional changes of rice straw fibers after pretreatment with diluted acetic acid: towards enhanced biomethane production. *Journal of Cleaner Production* 230: 775–782.
85. Pellera, F.-M. and Gidarakos, E. (2018). Chemical pretreatment of lignocellulosic agroindustrial waste for methane production. *Waste Management* 71: 689–703.
86. Carlsson, M., Lagerkvist, A., and Morgan-Sagastume, F. (2012). The effects of substrate pre-treatment on anaerobic digestion systems: a review. *Waste Management* 32 (9): 1634–1650.
87. Modenbach, A.A. and Nokes, S.E. (2012). The use of high-solids loadings in biomass pretreatment—a review. *Biotechnology and Bioengineering* 109 (6): 1430–1442.
88. Yu, Q., Qin, L., Liu, Y. et al. (2019). In situ deep eutectic solvent pretreatment to improve lignin removal from garden wastes and enhance production of bio-methane and microbial lipids. *Bioresource Technology* 271: 210–217.
89. Zhang, C., Su, H., Baeyens, J., and Tan, T. (2014). Reviewing the anaerobic digestion of food waste for biogas production. *Renewable and Sustainable Energy Reviews* 38: 383–392.
90. Koyama, M., Watanabe, K., Kurosawa, N. et al. (2017). Effect of alkaline pretreatment on mesophilic and thermophilic anaerobic digestion of a submerged macrophyte: inhibition and recovery against dissolved lignin during semi-continuous operation. *Bioresource Technology* 238: 666–674.
91. Sabeeh, M., Liaquat, R., and Maryam, A. (2020). Effect of alkaline and alkaline-photocatalytic pre-treatment on characteristics and biogas production of rice straw. *Bioresource Technology* 309: 123449.
92. Mancini, G., Papirio, S., Lens, P.N., and Esposito, G. (2018). Increased biogas production from wheat straw by chemical pretreatments. *Renewable Energy* 119: 608–614.
93. Shetty, D.J., Kshirsagar, P., Tapadia-Maheshwari, S. et al. (2017). Alkali pretreatment at ambient temperature: A promising method to enhance biomethanation of rice straw. *Bioresource Technology* 226: 80–88.
94. Tsapekos, P., Kougias, P.G., Frison, A. et al. (2016). Improving methane production from digested manure biofibers by mechanical and thermal alkaline pretreatment. *Bioresource Technology* 216: 545–552.
95. Pang, Y., Liu, Y., Li, X. et al. (2008). Improving biodegradability and biogas production of corn stover through sodium hydroxide solid state pretreatment. *Energy & Fuels* 22 (4): 2761–2766.
96. Xu, H., Li, Y., Hua, D. et al. (2020). Enhancing the anaerobic digestion of corn stover by chemical pretreatment with the black liquor from the paper industry. *Bioresource Technology* 306: 123090.
97. Zheng, M., Li, X., Li, L. et al. (2009). Enhancing anaerobic biogasification of corn stover through wet state NaOH pretreatment. *Bioresource Technology* 100 (21): 5140–5145.

98. Dong, C., Chen, J., Guan, R. et al. (2018). Dual-frequency ultrasound combined with alkali pretreatment of corn stalk for enhanced biogas production. *Renewable Energy* 127: 444–451.
99. Sambusiti, C., Monlau, F., Ficara, E. et al. (2013). A comparison of different pre-treatments to increase methane production from two agricultural substrates. *Applied Energy* 104: 62–70.
100. Xie, S., Frost, J., Lawlor, P.G. et al. (2011). Effects of thermo-chemical pre-treatment of grass silage on methane production by anaerobic digestion. *Bioresource Technology* 102 (19): 8748–8755.
101. Chufo, A., Yuan, H., Zou, D. et al. (2015). Biomethane production and physicochemical characterization of anaerobically digested teff (Eragrostis tef) straw pretreated by sodium hydroxide. *Bioresource Technology* 181: 214–219.
102. Yao, Y., He, M., Ren, Y. et al. (2013). Anaerobic digestion of poplar processing residues for methane production after alkaline treatment. *Bioresource Technology* 134: 347–352.
103. Khatri, S., Wu, S., Kizito, S. et al. (2015). Synergistic effect of alkaline pretreatment and Fe dosing on batch anaerobic digestion of maize straw. *Applied Energy* 158: 55–64.
104. Mancini, G., Papirio, S., Riccardelli, G. et al. (2018). Trace elements dosing and alkaline pretreatment in the anaerobic digestion of rice straw. *Bioresource Technology* 247: 897–903.
105. Chandra, R., Takeuchi, H., Hasegawa, T., and Kumar, R. (2012). Improving biodegradability and biogas production of wheat straw substrates using sodium hydroxide and hydrothermal pretreatments. *Energy* 43 (1): 273–282.
106. Hashemi, S.S., Karimi, K., and Karimi, A.M. (2019). Ethanolic ammonia pretreatment for efficient biogas production from sugarcane bagasse. *Fuel* 248: 196–204.
107. Cayetano, R.D.A., Oliwit, A.T., Kumar, G. et al. (2019). Optimization of soaking in aqueous ammonia pretreatment for anaerobic digestion of African maize bran. *Fuel* 253: 552–560.
108. Li, L., Chen, C., Zhang, R. et al. (2015). Pretreatment of corn stover for methane production with the combination of potassium hydroxide and calcium hydroxide. *Energy & Fuels* 29 (9): 5841–5846.
109. Yao, Y., Bergeron, A.D., and Davaritouchaee, M. (2018). Methane recovery from anaerobic digestion of urea-pretreated wheat straw. *Renewable Energy* 115: 139–148.
110. Moulthrop, J.S., Swatloski, R.P., Moyna, G., and Rogers, R.D. (2005). High-resolution 13C NMR studies of cellulose and cellulose oligomers in ionic liquid solutions. *Chemical Communications* 12: 1557–1559.
111. Feng, L. and Chen, Z.-L. (2008). Research progress on dissolution and functional modification of cellulose in ionic liquids. *Journal of Molecular Liquids* 142 (1-3): 1–5.
112. Allison, B.J., Cádiz, J.C., Karuna, N. et al. (2016). The effect of ionic liquid pretreatment on the bioconversion of tomato processing waste to fermentable sugars and biogas. *Applied Biochemistry and Biotechnology* 179: 1227–1247.
113. Zhu, S. (2008). Use of ionic liquids for the efficient utilization of lignocellulosic materials. *Biotechnology: International Research in Process, Environmental & Clean Technology* 83 (6): 777–779.
114. Heinze, T., Schwikal, K., and Barthel, S. (2005). Ionic liquids as reaction medium in cellulose functionalization. *Macromolecular Bioscience* 5 (6): 520–525.
115. Mora-Pale, M., Meli, L., Doherty, T.V. et al. (2011). Room temperature ionic liquids as emerging solvents for the pretreatment of lignocellulosic biomass. *Biotechnology and Bioengineering* 108 (6): 1229–1245.
116. Cheng, J., Zhang, J., Lin, R. et al. (2017). Ionic-liquid pretreatment of cassava residues for the cogeneration of fermentative hydrogen and methane. *Bioresource Technology* 228: 348–354.
117. Pérez-Pimienta, J.A., Icaza-Herrera, J.P., Méndez-Acosta, H.O. et al. (2020). Bioderived ionic liquid-based pretreatment enhances methane production from *Agave tequilana* bagasse. *RSC Advances* 10 (24): 14025–14032.
118. Padrino, B., Lara-Serrano, M., Morales-delaRosa, S. et al. (2018). Resource recovery potential from lignocellulosic feedstock upon lysis with ionic liquids. *Frontiers in Bioengineering and Biotechnology* 6: 119.

119. Kabir, M.M., Niklasson, C., Taherzadeh, M.J., and Horváth, I.S. (2014). Biogas production from lignocelluloses by *N*-methylmorpholine-*N*-oxide (NMMO) pretreatment: effects of recovery and reuse of NMMO. *Bioresource Technology* 161: 446–450.
120. Mancini, G., Papirio, S., Lens, P.N., and Esposito, G. (2016). Effect of *N*-methylmorpholine-*N*-oxide pretreatment on biogas production from rice straw, cocoa shell, and hazelnut skin. *Environmental Engineering Science* 33 (11): 843–850.
121. Gao, J., Chen, L., Yuan, K. et al. (2013). Ionic liquid pretreatment to enhance the anaerobic digestion of lignocellulosic biomass. *Bioresource Technology* 150: 352–358.
122. Smith, E.L., Abbott, A.P., and Ryder, K.S. (2014). Deep eutectic solvents (DESs) and their applications. *Chemical Reviews* 114 (21): 11060–11082.
123. Abbott, A.P., Barron, J.C., Ryder, K.S., and Wilson, D.J.C.A.E.J. (2007). *Eutectic-based ionic liquids with metal-containing anions and cations.* 13 (22): 6495–6501.
124. Garcia, G., Aparicio, S., Ullah, R., and Atilhan, M. (2015). Deep eutectic solvents: physicochemical properties and gas separation applications. *Energy & Fuels* 29 (4): 2616–2644.
125. Xu, G.-C., Ding, J.-C., Han, R.-Z. et al. (2016). Enhancing cellulose accessibility of corn stover by deep eutectic solvent pretreatment for butanol fermentation. *Bioresource Technology* 203: 364–369.
126. Mbous, Y.P., Hayyan, M., Hayyan, A. et al. (2017). Applications of deep eutectic solvents in biotechnology and bioengineering—Promises and challenges. *Biotechnology Advances* 35 (2): 105–134.
127. Khan, M.U., Lee, J.T.E., Bashir, M.A. et al. (2021). Current status of biogas upgrading for direct biomethane use: A review. *Renewable and Sustainable Energy Reviews* 149: 111343.
128. Dutta, N., Giduthuri, A.T., Khan, M.U. et al. (2022). Improved valorization of sewage sludge in the circular economy by anaerobic digestion: Impact of an innovative pretreatment technology. *Waste Management* 154: 105–112.
129. Dutta, N., Usman, M., Ashraf, M.A. et al. (2022). Methods to convert lignocellulosic waste into biohydrogen, biogas, bioethanol, biodiesel and value-added chemicals: a review. *Environmental Chemistry Letters* 21: 803–820.
130. Dutta, N., Garrison, R., Usman, M., and Ahring, B.K. (2022). Enhancing methane production of anaerobic digested sewage sludge by advanced wet oxidation & steam explosion pretreatment. *Environmental Technology & Innovation* 28: 102923.
131. Olugbemide, A.D., Oberlintner, A., Novak, U., and Likozar, B.J.S. (2021). Lignocellulosic corn stover biomass pre-treatment by deep eutectic solvents (Des) for biomethane production process by bioresource anaerobic digestion. *Sustainability* 13 (19): 10504.
132. Lima, F., Branco, L.C., Lapa, N., and Marrucho, I.M. (2021). Beneficial and detrimental effects of choline chloride–oxalic acid deep eutectic solvent on biogas production. *Waste Management* 131: 368–375.
133. Taherzadeh, M.J. and Karimi, K. (2008). Pretreatment of lignocellulosic wastes to improve ethanol and biogas production: a review. *International Journal of Molecular Sciences* 9 (9): 1621–1651.
134. Zhao, X., Cheng, K., and Liu, D. (2009). Organosolv pretreatment of lignocellulosic biomass for enzymatic hydrolysis. *Applied Microbiology and Biotechnology* 82: 815–827.
135. Zhang, K., Pei, Z., and Wang, D. (2016). Organic solvent pretreatment of lignocellulosic biomass for biofuels and biochemicals: a review. *Bioresource Technology* 199: 21–33.
136. Kabir, M.M., Rajendran, K., Taherzadeh, M.J., and Horváth, I.S. (2015). Experimental and economical evaluation of bioconversion of forest residues to biogas using organosolv pretreatment. *Bioresource Technology* 178: 201–208.
137. Tongbuekeaw, T., Sawangkeaw, R., Chaiprapat, S., and Charnnok, B. (2021). Conversion of rubber wood waste to methane by ethanol organosolv pretreatment. *Biomass Conversion and Biorefinery* 11: 999–1011.
138. Wood, T. and McCrae, S.I. (1979). Synergism between enzymes involved in the solubilization of native cellulose. In: *Advances in Chemistry*, vol. 181, 181–209. ACS Publications.

139. Cantarel, B.L., Coutinho, P.M., Rancurel, C. et al. (2009). The Carbohydrate-Active EnZymes database (CAZy): an expert resource for glycogenomics. *Nucleic Acid Research* 37 (suppl_1): D233–D238.
140. Ngumah, C.C., Ogbulie, J.N., Orji, J.C., and Amadi, E.S. (2013). Biogas potential of organic waste in Nigeria. *Journal of Urban and Environmental Engineering* 7 (1): 110–116.
141. Michalska, K., Bizukojć, M., and Ledakowicz, S. (2015). Pretreatment of energy crops with sodium hydroxide and cellulolytic enzymes to increase biogas production. *Biomass and Bioenergy* 80: 213–221.
142. Schroyen, M., Vervaeren, H., Van Hulle, S.W., and Raes, K. (2014). Impact of enzymatic pretreatment on corn stover degradation and biogas production. *Bioresource Technology* 173: 59–66.
143. Schroyen, M., Vervaeren, H., Vandepitte, H. et al. (2015). Effect of enzymatic pretreatment of various lignocellulosic substrates on production of phenolic compounds and biomethane potential. *Bioresource Technology* 192: 696–702.
144. Frigon, J.-C., Mehta, P., and Guiot, S.R. (2012). Impact of mechanical, chemical and enzymatic pre-treatments on the methane yield from the anaerobic digestion of switchgrass. *Biomass and Bioenergy* 36: 1–11.
145. Pérez-Rodríguez, N., Garcia-Bernet, D., and Dominguez, J.M. (2017). Extrusion and enzymatic hydrolysis as pretreatments on corn cob for biogas production. *Renewable Energy* 107: 597–603.
146. Kucuker, M., Demirel, B., and Onay, T. (2020). Enhanced biogas production from chicken manure via enzymatic pretreatment. *Journal of Material Cycles and Waste Management* 22: 1521–1528.
147. Zhao, J., Dong, Z., Li, J. et al. (2018). Ensiling as pretreatment of rice straw: The effect of hemicellulase and Lactobacillus plantarum on hemicellulose degradation and cellulose conversion. *Bioresource Technology* 266: 158–165.
148. Lynd, L.R., Weimer, P.J., Van Zyl, W.H., and Pretorius, I.S. (2002). Microbial cellulose utilization: fundamentals and biotechnology. *Microbiology and Molecular Biology Reviews* 66 (3): 506–577.
149. Koullas, D., Christakopoulos, P., Kekos, D. et al. (1992). Correlating the effect of pretreatment on the enzymatic hydrolysis of straw. *Biotechnology and Bioengineering* 39 (1): 113–116.
150. Kim, T.H., Lee, Y.Y., Sunwoo, C., and Kim, J.S. (2006). Pretreatment of corn stover by low-liquid ammonia recycle percolation process. *Applied Biochemistry and Biotechnology* 133: 41–57.
151. Brenner, K., You, L., and Arnold, F.H. (2008). Engineering microbial consortia: a new frontier in synthetic biology. *Trends in Biotechnology* 26 (9): 483–489.
152. Mikesková, H., Novotný, Č., and Svobodová, A.K. (2012). Interspecific interactions in mixed microbial cultures in a biodegradation perspective. *Applied Biochemistry and Biotechnology* 95: 861–870.
153. Barua, V.B., Goud, V.V., and Kalamdhad, A.S. (2018). Microbial pretreatment of water hyacinth for enhanced hydrolysis followed by biogas production. *Renewable Energy* 126: 21–29.
154. Hua, B., Dai, J., Liu, B. et al. (2016). Pretreatment of non-sterile, rotted silage maize straw by the microbial community MC1 increases biogas production. *Bioresource Technology* 216: 699–705.
155. Tantayotai, P., Pornwongthong, P., Muenmuang, C. et al. (2017). Effect of cellulase-producing microbial consortium on biogas production from lignocellulosic biomass. *Energy Procedia* 141: 180–183.
156. Kong, X., Du, J., Ye, X. et al. (2018). Enhanced methane production from wheat straw with the assistance of lignocellulolytic microbial consortium TC-5. *Bioresource Technology* 263: 33–39.
157. Ozbayram, E.G., Kleinsteuber, S., Nikolausz, M. et al. (2017). Effect of bioaugmentation by cellulolytic bacteria enriched from sheep rumen on methane production from wheat straw. *Anaerobe* 46: 122–130.
158. Raut, M.P., Pandhal, J., and Wright, P.C. (2021). Effective pretreatment of lignocellulosic co-substrates using barley straw-adapted microbial consortia to enhanced biomethanation by anaerobic digestion. *Bioresource Technology* 321: 124437.

159. Arantes, V., Jellison, J., and Goodell, B. (2012). Peculiarities of brown-rot fungi and biochemical Fenton reaction with regard to their potential as a model for bioprocessing biomass. *Applied Microbiology and Biotechnology* 94 (2): 323–338.
160. Ghosh, A. and Bhattacharyya, B. (1999). Biomethanation of white rotted and brown rotted rice straw. *Bioprocess Engineering* 20: 297–302.
161. Muthangya, M., Mshandete, A.M., and Kivaisi, A.K. (2009). Two-stage fungal pre-treatment for improved biogas production from sisal leaf decortication residues. *International Journal of Molecular Sciences* 10 (11): 4805–4815.
162. Mustafa, A.M., Poulsen, T.G., and Sheng, K. (2016). Fungal pretreatment of rice straw with Pleurotus ostreatus and Trichoderma reesei to enhance methane production under solid-state anaerobic digestion. *Applied Energy* 180: 661–671.
163. Lalak, J., Kasprzycka, A., Martyniak, D., and Tys, J. (2016). Effect of biological pretreatment of Agropyron elongatum 'BAMAR'on biogas production by anaerobic digestion. *Bioresource Technology* 200: 194–200.
164. Rouches, E., Zhou, S., Sergent, M. et al. (2018). Influence of white-rot fungus Polyporus brumalis BRFM 985 culture conditions on the pretreatment efficiency for anaerobic digestion of wheat straw. *Biomass and Bioenergy* 110: 75–79.
165. Tišma, M., Planinić, M., Bucić-Kojić, A. et al. (2018). Corn silage fungal-based solid-state pretreatment for enhanced biogas production in anaerobic co-digestion with cow manure. *Bioresource Technology* 253: 220–226.
166. Liu, S., Li, X., Wu, S. et al. (2014). Fungal pretreatment by Phanerochaete chrysosporium for enhancement of biogas production from corn stover silage. *Applied Biochemistry and Biotechnology* 174: 1907–1918.
167. Wyman, V., Henríquez, J., Palma, C., and Carvajal, A. (2018). Lignocellulosic waste valorisation strategy through enzyme and biogas production. *Bioresource Technology* 247: 402–411.
168. Ambye-Jensen, M., Thomsen, S.T., Kádár, Z., and Meyer, A.S. (2013). Ensiling of wheat straw decreases the required temperature in hydrothermal pretreatment. *Biotechnology for Biofuels* 6: 1–9.
169. Feng, L., Kristensen, E.F., Moset, V. et al. (2018). Ensiling of tall fescue for biogas production: Effect of storage time, additives and mechanical pretreatment. *Energy for Sustainable Development* 47: 143–148.
170. Zhang, H., Wu, J., Gao, L. et al. (2018). Aerobic deterioration of corn stalk silage and its effect on methane production and microbial community dynamics in anaerobic digestion. *Bioresource Technology* 250: 828–837.
171. Sun, H., Cui, X., Stinner, W. et al. (2020). Ensiling excessively wilted maize stover with biogas slurry: Effects on storage performance and subsequent biogas potential. *Bioresource Technology* 305: 123042.
172. Franco, R.T., Buffière, P., and Bayard, R. (2016). Ensiling for biogas production: critical parameters. A review. *Biomass and Bioenergy* 94: 94–104.
173. Zhao, X., Luo, K., Zhang, Y. et al. (2018). Improving the methane yield of maize straw: focus on the effects of pretreatment with fungi and their secreted enzymes combined with sodium hydroxide. *Bioresource Technology* 250: 204–213.
174. Li, J.H., Yang, L.X., Li, J.Q. et al. (2019). Anchoring nZVI on metal-organic framework for removal of uranium (VI) from aqueous solution. *Journal of Solid State Chemistry* 269: 16–23.
175. Guo, J., Cui, X., Sun, H. et al. (2018). Effect of glucose and cellulase addition on wet-storage of excessively wilted maize stover and biogas production. *Bioresource Technology* 259: 198–206.
176. Sun, H., Cui, X., Stinner, W. et al. (2019). Synergetic effect of combined ensiling of freshly harvested and excessively wilted maize stover for efficient biogas production. *Bioresource Technology* 285: 121338.
177. Zhao, Y., Yu, J., Liu, J. et al. (2016). Material and microbial changes during corn stalk silage and their effects on methane fermentation. *Bioresource Technology* 222: 89–99.

178. Paritosh, K., Yadav, M., Kesharwani, N. et al. (2021). Strategies to improve solid state anaerobic bioconversion of lignocellulosic biomass: an overview. *Bioresource Technology* 331: 125036.
179. Biswas, R., Teller, P.J., Khan, M.U., and Ahring, B.K.J.M. (2020). Sugar production from hybrid poplar sawdust: optimization of enzymatic hydrolysis and wet explosion pretreatment. *Molecules* 25 (15): 3396.
180. Hassan, S.S., Williams, G.A., and Jaiswal, A.K.J.B.T. (2018). Emerging technologies for the pretreatment of lignocellulosic biomass. *Bioresource Technology* 262: 310–318.
181. Kumar, B., Bhardwaj, N., Agrawal, K. et al. (2020). Current perspective on pretreatment technologies using lignocellulosic biomass: an emerging biorefinery concept. *Fuel Processing Technology* 199: 106244.
182. Haldar, D. and Purkait, M.K.J.C. (2021). A review on the environment-friendly emerging techniques for pretreatment of lignocellulosic biomass: Mechanistic insight and advancements. *Chemosphere* 264: 128523.
183. Das, N., Jena, P.K., Padhi, D. et al. (2021). Biorefinery. A comprehensive review of characterization, pretreatment and its applications on different lignocellulosic biomass for bioethanol production. *Biomass Conversion and Biorefinery* 1–25.
184. Baruah, J., Nath, B.K., Sharma, R. et al. (2018). Recent trends in the pretreatment of lignocellulosic biomass for value-added products. *Frontiers in Energy Research* 6: 141.
185. Ethaib, S., Omar, R., Mazlina, M.S. et al. (2020). Toward sustainable processes of pretreatment technologies of lignocellulosic biomass for enzymatic production of biofuels and chemicals: a review. *BioResources* 15 (4): 10063.
186. Bhutto, A.W., Qureshi, K., Harijan, K. et al. (2017). Insight into progress in pre-treatment of lignocellulosic biomass. *Energy* 122: 724–745.
187. Ahring, B.K. and Munck, J. (2013). Method for treating biomass and organic waste with the purpose of generating desired biologically based products. Google Patents.
188. Miklos, D.B., Remy, C., Jekel, M. et al. (2018). Evaluation of advanced oxidation processes for water and wastewater treatment–a critical review. *Water Research* 139: 118–131.
189. Wu, B., Dai, X., and Chai, X.J.W. (2020). Critical review on dewatering of sewage sludge: Influential mechanism, conditioning technologies and implications to sludge re-utilizations. *Water Research* 180: 115912.
190. Dewil, R., Mantzavinos, D., Poulios, I., and Rodrigo, M.A. (2017). New perspectives for advanced oxidation processes. *Journal of Environmental Management* 195: 93–99.
191. Tayyab, M., Noman, A., Islam, W. et al. (2018). Bioethanol production from lignocellulosic biomass by environment-friendly pretreatment methods: a review. *Applied Ecology & Environmental Research* 16 (1): 225–249.
192. Jacquet, N., Maniet, G., Vanderghem, C. et al. (2015). Application of steam explosion as pretreatment on lignocellulosic material: a review. *Industrial & Engineering Chemistry Research* 54 (10): 2593–2598.
193. Dhyani, V. and Bhaskar, T.J. (2018). A comprehensive review on the pyrolysis of lignocellulosic biomass. *Renewable Energy* 129: 695–716.
194. Zhang, Y.P. (2008). Reviving the carbohydrate economy via multi-product lignocellulose biorefineries. *Journal of Industrial Microbiology and Biotechnology* 35 (5): 367–375.
195. Hoornweg D, Bhada-Tata P. (2012) What a Waste: A Global Review of Solid Waste Management. 2012. Urban development series;knowledge papers no. 15. © World Bank, Washington, DC. http://hdl.handle.net/10986/17388 License: CC BY 3.0 IGO.
196. Ufarté, L., Potocki-Veronese, G., Cecchini, D. et al. (2018). Highly promiscuous oxidases discovered in the bovine rumen microbiome. *Frontiers in Microbiology* 9: 861.
197. Schroyen, M., Van Hulle, S.W., Holemans, S. et al. (2017). Laccase enzyme detoxifies hydrolysates and improves biogas production from hemp straw and miscanthus. *Bioresource Technology* 244: 597–604.

198. Rahimi, H., Sattler, M.L., Hossain, M.S., and Rodrigues, J.L.J.W.M. (2020). Boosting landfill gas production from lignin-containing wastes via termite hindgut microorganism. *Waste management* 105: 299–308.
199. Dollhofer, V., Dandikas, V., Dorn-In, S. et al. (2018). Accelerated biogas production from lignocellulosic biomass after pre-treatment with *Neocallimastix frontalis*. *Bioresource technology* 264: 219–227.
200. Sanitha, M., Aliya Fathima, A., and Ramya, M.J.B. (2021). Microbial diversity analysis of wood degrading microbiome and screening of natural consortia for bioalcohol production. *Biofuels* 12 (6): 697–702.
201. Hardersen, S. and Zapponi, L.J.A.S.E. (2018). Wood degradation and the role of saproxylic insects for lignoforms. *Applied Soil Ecology* 123: 334–338.
202. Grassi, G., House, J., Dentener, F. et al. (2017). The key role of forests in meeting climate targets requires science for credible mitigation. *Nature* 7 (3): 220–226.
203. Pan X, Xie D, Gilkes N, Gregg DJ, Saddler JN. Strategies to enhance the enzymatic hydrolysis of pretreated softwood with high residual lignin content. *Applied Biochemistry and Biotechnology*; 2005, 121–124:1069-79. https://doi.org/10.1385/abab:124:1-3:1069.
204. Amirta, R., Herawati, E., Suwinarti, W. et al. (2016). Two-steps utilization of shorea wood waste biomass for the production of oyster mushroom and biogas–a zero waste approach. *Agriculture and Agricultural Science Procedia* 9: 202–208.
205. Millati, R., Cahyono, R.B., Ariyanto, T. et al. (2019). Agricultural, industrial, municipal, and forest wastes: an overview. *Sustainable Resource Recovery and Zero Waste Approaches* 1–22.
206. Ali, S.S., Abomohra, A.E.-F., and Sun, J.J. (2017). Effective bio-pretreatment of sawdust waste with a novel microbial consortium for enhanced biomethanation. *Bioresource Technology* 238: 425–432.
207. Akyol, Ç., Ince, O., Bozan, M. et al. (2019). Fungal bioaugmentation of anaerobic digesters fed with lignocellulosic biomass: what to expect from anaerobic fungus Orpinomyces sp. *Bioresource Technology* 277: 1–10.
208. Kirchherr, J., Reike, D., and Hekkert, M.J.R. (2017). Conceptualizing the circular economy: an analysis of 114 definitions. *Resources, Conservation and Recycling* 127: 221–232.
209. Ghisellini, P., Cialani, C., and Ulgiati, S. (2016). A review on circular economy: the expected transition to a balanced interplay of environmental and economic systems. *Journal of Cleaner Production* 114: 11–32.
210. Mittal, S., Ahlgren, E.O., and Shukla, P.J.E.P. (2018). Barriers to biogas dissemination in India: a review. *Energy Policy* 112: 361–370.
211. Chang, I.-S., Wu, J., Zhou, C. et al. (2014). A time-geographical approach to biogas potential analysis of China. *Renewable and Sustainable Energy Reviews* 37: 318–333.
212. Kapoor, R., Ghosh, P., Kumar, M. et al. (2020). Valorization of agricultural waste for biogas based circular economy in India: A research outlook. *Bioresource Technology* 304: 123036.
213. Igliński, B., Piechota, G., Iwański, P. et al. (2020). 15 Years of the Polish agricultural biogas plants: their history, current status, biogas potential and perspectives. *Clean Technologies and Environmental Policy* 22 (2): 281–307.
214. Gil, A. (2021). Current insights into lignocellulose related waste valorization. *Chemical Engineering Journal Advances* 8: 100186.
215. Giuntoli, J., Agostini, A., Edwards, R., and Marelli, L. (2015). Solid and gaseous bioenergy pathways: input values and GHG emissions. *Report EUR* 26696.
216. Pathak, B. and Chandel, A.K. (2017). Feedstock transportation, agricultural processing, logistic from farm to bio-refinery: recent developments, mechanization, and cost analysis. *Sustainable Biofuels Development in India* 207–221.
217. Koornneef, J., van Breevoort, P., Noothout, P. et al. (2013). Global potential for biomethane production with carbon capture, transport and storage up to 2050. *Energy Procedia* (37): 6043–6052.

218. Neshat, S.A., Mohammadi, M., Najafpour, G.D. et al. (2017). Anaerobic co-digestion of animal manures and lignocellulosic residues as a potent approach for sustainable biogas production. *Renewable and Sustainable Energy Reviews* 79: 308–322.
219. Parmar, K.R., Brown, A.E., Hammerton, J.M. et al. (2022). Co-processing lignocellulosic biomass and sewage digestate by hydrothermal carbonisation: influence of blending on product quality. *Energy* 15 (4): 1418.
220. IEA (ed.) (2020). *Outlook for Biogas and Biomethane: Prospects for Organic Growth*. France: IEA Paris.
221. Rooni, V., Raud, M., and Kikas, T.J.A.R. (2017). Technical solutions used in different pretreatments of lignocellulosic biomass: a review. *Agronomy Research* 15 (3): 848–858.
222. Theuretzbacher, F., Lizasoain, J., Lefever, C. et al. (2015). Steam explosion pretreatment of wheat straw to improve methane yields: Investigation of the degradation kinetics of structural compounds during anaerobic digestion. *Bioresource Technology* 179: 299–305.

3

Biogas Technology and the Application for Agricultural and Food Waste Treatment

Wei Qiao[1,*], Simon M. Wandera[2], Mengmeng Jiang[1], Yapeng Song[1], and Renjie Dong[1]

[1] *College of Engineering, China Agricultural University, Beijing, China*
[2] *Department of Civil, Construction & Environmental Engineering, Jomo Kenyatta University of Agriculture & Technology, Nairobi, Kenya*

3.1 Development of Biogas Plants

Anaerobic digestion technology has developed rapidly in China. At present, biogas plants fed on various substrates have been established. The scale and number of biogas plants in China have also significantly and rapidly increased. According to statistics from the Ministry of Agriculture and Rural Affairs, as of 2016, the number of household digesters was estimated at 43.8 million. The number of small-, medium-, and large-scale biogas plants utilizing agricultural waste as substrate was estimated at 95,183, 10,734, and 7,265, respectively. The division of biogas plant scales is shown in Table 3.1. The annual biogas production from household digesters was estimated at 11.7 billion m^3 [1].

*Corresponding author: qiaowei@cau.edu.cn; wayqiao@sina.cn

Biogas Plants: Waste Management, Energy Production and Carbon Footprint Reduction,
First Edition. Edited by Wojciech Czekała

Table 3.1 Scale classification of biogas plants (NY/T 667-2022).

Scale	Total digesters volume in a plant/(m^3)	Type and quantity of raw materials: Livestock and poultry in stock (pig equivalent)	Type and quantity of raw materials: Straw (*W*)/(t)
Large-scale bio-natural gas size	V≥10000	H≥50000	*W*≥45
Super large size	5000≤V<10000	H≥50000	*W*≥15
Large size	1000≤V<5000	5000≤V<50000	1.5≤*W*<15
Middle/Small size	V<1000	1500≤V<5000	0.5≤*W*<1.5

Note: V is the volume of the digester; H is equivalent number of pigs; *W* is dry straw weight.

3.1.1 Agricultural Waste

3.1.1.1 Livestock and Poultry Manure

In 2022, the number of pigs slaughtered was 671 million, cattle bred was 98 million, and slaughtered poultry reached 13 billion [2]. Accordingly, the annual production of livestock and poultry manure and urine was approximated to be 3.8 billion tons [1]. According to the FAO, in 2020, the nitrogen content from livestock and poultry manure in China reached 12.4 million tonnes [3]; this can be treated by the anaerobic digestion process and the resulting digestate used as fertilizers. Livestock and poultry manure contains high organic content, such as protein, cellulose, and fat with a high potential for biogas production (shown in Table 3.2). For instance, the methane yield potential of chicken manure was found to be 250 ~ 450 mL g-VS^{-1} [6–10]. According to previous studies, theoretical biogas production from livestock and poultry manure exceeds 147 billion m^3/annually in China, generating 160 billion kWh [11].

3.1.1.2 Crop Straw

In China, the crop straw mainly includes corn, wheat, rice, and a small proportion of legumes and tubers. Among the crops, corn, wheat, and rice yields account for 93% of the total crop yield [2]. As the worlds leading grain producing country, China is rich in crop straw resources. Among the crops, rice, corn, and wheat straw yields account for approximately 75% of the total crop straw. More than 85% of the straw was used as fertilizer, animal feed, industrial materials, biofuels, and base material. Anaerobic treatment and methane recovery by utilizing crop straw minimize some risks associated

Table 3.2 Properties of livestock manure and methane production potential.

Species	Livestock and poultry manure characteristics [4]: C (%TS)	N (%TS)	C/N	Anaerobic digestion [5]: Methane yields (mL g-VS^{-1})	VS removal efficiency/%
Chicken manure	25~40	2~5	3~10	250~450	45~55
Swine manure	35~45	2~6	10~20	200~400	45~55
Cattle dung	20~45	1~2	20~30	100~200	25~35

VS: volatile solids

Table 3.3 Characteristics of straw and methane production potential.

	Livestock and poultry manure characteristics [13]			Anaerobic digestion [14]
Species	C (%TS)	N (%TS)	C/N	Methane yields (mL g-VS^{-1})
Corn straw	40~45	0.5~1.5	45~60	210~310
Wheat straw	40~45	0.5~1	60~70	130~290
Rice straw	40~45	0.5~1.5	55~65	150~280

VS: volatile solids

with direct use, such as pest transmission, disease risks, and emissions of methane [12]. The characteristics of crop straw and methane production potential are shown in Table 3.3. The annual theoretical biomethane potential of straw is about 82 billion m^3, and this can cover 29.2% of China's annual natural gas requirement and 2.25% of energy consumption [12].

3.1.2 Municipal Solid Waste

Over the last few decades, China's level of urbanization has risen rapidly which has resulted in increase in amount of solid waste generated. In 2020, municipal solid waste generation in China was 235 million tons [2]. It is estimated that 60% of municipal waste is hotel food waste and kitchen waste.

3.1.2.1 Municipal Solid Waste

Municipal solid waste contains high levels of nutrients, particularly restaurant food waste and kitchen waste which contains organic components such as carbohydrates, lipids, and proteins. Due to the rapid urbanization, many cities separate food waste from municipal solid waste. These collected food wastes are highly biodegradable and suitable for anaerobic digestion. The methane yield of food waste is about 0.3~0.5 L g-VS^{-1} [15, 16]. The ratio of volatile solids to total solids in the food waste is approximately 90% which is an indication of suitability of the substrate as a fedstosck in anaerobic digestion. In addition, the biogas production potential of separately collected food waste can reach 104 m^3 per ton of wet materials [17]. Consequently, China's theoretical biogas production of food waste reaches 15 billion m^3 per year.

3.1.2.2 Sewage Sludge

In 2021, the number of wastewater treatment plants in China increased to 2,827 sites with a treatment capacity of 207.67 million m^3 d^{-1} [2]. According to data published by the Ministry of Housing and Urban-Rural Development of China, sludge production reached 14.2 million tons of dry matter, corresponding to wet sludge production (with a moisture content of 80%) of approximately 71 million tons [2]. The raw sludge from the wastewater treatment plant contains a large amount of water, volatile solids, pathogens, heavy metal ions, and certain organic pollutants which are highly biodegradable, smelly, and not conducive to be transported to the disposal sites. The organic matter content in dry matter sludge is about 45~72%. The biogas production potential of wet sludge (with 80% water) is 12~80 m^3 t^{-1}, with an average value of 34 m^3 t^{-1}-sludge [18]. The annual biogas production potential of sewage sludge is approximately 1.9 billion m^3 in China.

3.2 Anaerobic Digestion Process

During anaerobic digestion process, organic matter is converted to methane and carbon dioxide. Methane-rich biogas can be used to generate electricity and heat. Considering carbon emission and environmental protection, bioenergy is becoming more and more critical. Digestate that results from the degradation process can be used for soil improvement or as organic fertilizer. An anaerobic digestion process occurs through four main processes: hydrolysis, acidogenesis, acetogenesis, and methanogenesis. In hydrolysis and acidogenesis stage, complex organic matter like carbohydrates, proteins, and lipids are hydrolyzed into smaller fractions like monosaccharides, amino acids, and fatty acids, and further into volatile fatty acids (acetate, propionate, butyrate, lactic acid, pyruvic acid, etc.) and hydrogen [19]. The acetogenesis stage results into the generation of acetate, hydrogen, and other substances. The acetate are generated in two main ways by directly converting macromolecular acids and forming hydrogen and carbon dioxide [20]. The methanogenesis which is the final step involves the decomposition of acetate and other substances to produce methane. The acetoclastic bacteria converts acetic acid or methanol to methane. In contrast, hydgrogenetrophic bacteria utilizes hydrogen and carbondioxide to produce methane [20]. Methanogens, such as *Methanosaeta, Methanobacterium*, and *Methanosarcina*, and other archaea, play a significant role in methane formation. These microorganisms use acetate or hydrogen as a substrate to produce methane.

The main factors influencing anaerobic digestion are temperature, pH, C/N, pretreatment, and the possible existence of toxic substances. The AD process takes place in a wide range of temperatures which includes psychrophilic (15~20 °C), mesophilic (30~37 °C), and thermophilic (50~55 °C) [21]. For the optimization of the AD process, the suitable pH range should be maintained between 6.5 and 8.2, and the C/N should be maintained between 20 : 1 and 30 : 1 [14]. If the pH or C/N is beyond the appropriate scope, AD is adversely affected, especially the growth of microorganisms, and inhibits methane production. Some toxic substances, such as ammonia, nitrogen, and hydrogen sulfide, will also impede the process of AD.

The total solid of the feeding substrate in wet anaerobic digestion is generally less than 10%. While in the high solids anaerobic digestion, the feeding substrate can have a TS content between 10% and 20%. This value can increase to 20~40% when the dry anaerobic process is adopted [22]. In China, wet anaerobic digestion is the most widely used process, especially for treating livestock and poultry manure. If water goes into the pig or dairy manure, it sometimes makes the substrate much diluted. The diluted feedstock will negatively impact the performance of the biogas plant. In contrast, the cost of digestate transportation has increased, and the treatment of the high amount of effluent requires more investment. The high solids anaerobic digestion process has the advantages of low digestate production and high volumetric biogas production, low energy requirement, and then suitable for manure without dilution. While, the application of high solids anaerobic digestion in industry is still very lack.

3.3 Biogas Production from Livestock and Poultry Manure

3.3.1 Successful AD of Cattle and Swine Manure

3.3.1.1 Industrial-scale AD of Cattle Manure

A large-scale cattle manure biogas plant in China is located in Heilongjiang Provinces, and has been operational since 2015 [23]. Its design process adopts plug-flow reactors operating at mesophilic temperatures (37~40 °C). Its hydraulic retention time is 25 days with a TS concentration of 7~10% and the organic load rate of 1.1×10^5 kg·COD d^{-1}. The total working volume of the digester is 39,200 m^3 (16 × 2450 m^3). The dimensions of the reactor are length 99, width 6, and 4.8 m in height, and it treats approximately 548,960 tonnes of cattle dung a year. Biogas produced is approximated at 8.9 million m^3 y^{-1}. The biogas is mainly used for power generation and heating. The generated heat energy is mainly used for heating the digester, and the excess heat energy produced could be used to sterilize the solid fraction of digestate. The digestate is separated by a solid–liquid separator. The solid residue is used as cattle bedding material, and some are composted. The slurry fraction is returned to the croplands as an organic fertilizer.

3.3.1.2 Industrial-scale AD of Swine Manure

This plant is located in Jiangxi Province, China [24]. Its design process utilizes continuously stirred tank reactors technology, and the total working volume of the reactor is 20,010 m^3 (6 × 3335 m^3), operated under mesophilic temperature conditions. It can treat manure at total solids concentration (TS about 6%) that is generated by 600,000 live pigs and 100,000 sick pigs per year. According to statistics [24], 10 million m^3 of biogas with power generation of more than 20 million kWh, 30,000 tons of solid organic fertilizer, and 380,000 tons of organic liquid fertilizer are produced yearly.

The feeding substrate is pig manure that is collected and transported from the farm to a centralized AD plant where it is mixed with crushed agricultural and rural straw in a particular proportion for hydrolysis. After the AD treatment, the digestate is separated into solid and liquid. The solid residue is utilized as solid organic fertilizer, and the slurry is used as liquid organic fertilizer to supply the surrounding planting industry. The biogas is stored in the gas storage unit after dehydration and desulfurization. The biogas is then used for the gas supply, or power generation which is fed into the power grid.

3.3.2 Successful Anaerobic Digestion of Chicken Manure in a Large Plant

An optimal C/N ratio suitable for AD process should be between 20:1 and 30:1 [25]. However, chicken manure has a low C/N between 3:1 and 10:1. The nitrogen content of chicken manure is approximated at 3.2~4.9% of dry matter [22]. During the AD process, organic nitrogen components of waste are converted to ammonium-nitrogen (NH_4^+-N). The ammonia increases with temperature and pH. Furthermore, NH_3 is very toxic to microorganisms [26]. A low C/N ratio indicates that the substrate has high concentration of proteins which causes an increase in ammonia during AD process resulting in the inhibition of the methanogens. This results in decreased methane yield and accumulation of volatile fatty acids [27, 28].

A successful anaerobic treatment plant treating chicken manure introduced here was built in Shandong Province of China [29]. It has two clusters of biogas digesters constructed in two separate phases: the initial phase was built in 2009 and the second phase in 2015. The first cluster was designed and commissioned to produce biogas for generating electricity. It adopts continuously stirred tank reactors with a total working volume of 24,000 m^3 (8 × 3000 m^3). The first cluster biogas plant can treat raw chicken manure of an average amount of 300 tonnes d^{-1}. The TS content of raw chicken manure is approximately 20% but was diluted to between 8% and 9% before feeding into the digester. In 2015 the second cluster digester was built which uses similar technology and process to treat chicken manure. At the same time, biogas upgrading using membrane separation was also constructed. The biogas can be used for electricity and biomethane production. More flexible utilization of biogas makes the plant profitable. Moreover, the digestate is treated by separation and concentration technology. The concentrated fraction had upgraded value for use as a liquid fertilizer which is an innovative way for digestate utilization. In this example, the large-scale anaerobic digestion using chicken manure as a solo substrate is successful, which broadens the application of anaerobic technology application in animal manure treatment.

3.3.3 Strategies for Mitigating Ammonia Inhibition in Chicken Manure AD

3.3.3.1 Supplementation with Trace Elements

Trace elements play a critical role in the growth of anaerobic microorganisms [30]. Trace elements such as iron, nickel, cobalt, selenium, tungsten, and molybdenum can be used in various enzymatic reactions and are key co-enzyme or cofactors in enhancing the performance of AD [31]. These elements Co, Ni, Mo, Se, and W and other trace elements can significantly promote the growth of ammonia nitrogen-tolerant hydrogenotrophic methanogens, consequently, Se can rapidly increase the activity and quantity of hydrogenotrophic methanogens *Methanoculleus bourgensis* and improve the stability of the fermentation system and methane yields [32]. Table 3.4 shows how the trace elements can improve the methane yield and the degradation of volatile fatty acids (VFAs) in the AD of chicken manure.

The research team of Dr. QIAO Wei at China Agricultural University had developed trace element accelerants for pig manure, cow manure, and chicken manure, respectively, and achieved ideal results. Adding Fe^{2+} and Ni^{2+} to the anaerobic digestion reactor of chicken manure can simultaneously increase the acetoclastic and hydrogenotrophic methanogenic activity and increase the proportion of the methanogenic pathway of

Table 3.4 Positive effect of trace metals on chicken manure AD.

Organic loading rate g·VS/(L·d)	Trace element and concentration	Methane yields increase (%)	References
2.8	Se: 0.2	50	[33]
2.5	Ni: 1, Co: 1, Mo: 0.2, Se: 0.2, W: 0.2	54	[32]
2.8	Ni: 1, Co: 1, Mo: 0.2, Se: 0.2, W: 0.2, Fe: 5	34	[33]
3.6	Co: 1, Ni: 1, Mo: 0.2, Se: 0.2, W: 0.2	18	[34]
4.8	Fe: 280, Ni: 2	34	[35]

acetoclastic methanogen [35]. In addition, the research team verified the effectiveness of trace elements in the anaerobic treatment of chicken manure through a long-term (300 d) continuous operation of the pilot reactor (effective volume 500 m^3). After a third-party appraisal, the methane production performance increased by 19%.

3.3.3.2 In-situ Ammonia Stripping for Chicken Manure Digesters

In situ ammonia stripping and acid absorption can simultaneously reduce ammonium nitrogen concentration and recover nitrogen resources. In batch experiments, the in situ stripping process can reduce the ammonia nitrogen by 79% of the chicken manure in a high-temperature system [36]. Anaerobic digestion of chicken manure with 15% feed concentration by in situ ammonia stripping process at HRT of 20 d and OLR 5.3 g-VS L^{-1}·d^{-1}) showed a 20% reduction in ammonia-nitrogen and a 30% reduction in VFAs [37]. In addition, the in-situ ammonia stripping process can enhance the hydrolysis process of chicken manure for enhanced biogas production. In-situ ammonia stripping was integrated into the pretreatment reactor to remove ammonia nitrogen (18~31%) and to enhance the hydrolysis process (2.6~31.1%) and the performance of the reactor [38]. The in situ ammonia stripping process can effectively reduce the inhibition due to ammonia nitrogen concentration in the system and is suitable for the AD process of substrates with high concentration of nitrogen. So far, in situ stripping technology has not been used on an industrial scale in China. However ammonia stripping technology is rarely used at an industrial-scale in anaerobic digestion. Air ammonia stripping to recover nitrogen from digestate has been widely used, and it has been reported that at least 15 large-scale biogas projects in Germany used air stripping technology [39].

3.4 Food Waste Anaerobic Digestion

3.4.1 Challenges of Food Waste AD and the Solutions

3.4.1.1 VFAs Accumulation in Thermophilic AD of Food Waste

The C/N ratio of food waste is approximated to be between 12 and 18, which is associated with high biodegradability nature of the food waste, causing acidification of the reactors treating food waste. Food waste gets rapidly hydrolyzed to produce excess hydrogen that in the presence of carbondioxide and acetogenic bacteria leads to the acidification of the reactors, thus causing the inhibition of the methanogenic activity. Food waste contains high lipids that make the thermophilic process more suitable for these substrates, since the high temperature increases the solubility of the lipids and the contact between the microbes and substrates. However, thermophilic anaerobic microbes are more sensitive, and the process tends to fluctuate. In many studies and commercial thermophilic plants, the digester was normally operated under a low loading rate and thus greatly wasted their capacity.

3.4.1.2 AD Technologies for Food Waste

3.4.1.2.1 Trace Elements Addition for High Loading Rate Thermophilic Process

Concentrations of trace elements in food waste were reported to be 7.36–54 mg L^{-1} Fe, 0.015–0.38 mg L^{-1} Co, 0.44–6.72 mg L^{-1} Ni, 0.00–0.06 mg L^{-1} Se, and 0.00–0.26 mg L^{-1} Mo [40–42]. It was reported that the minimum requirement of trace elements per kilogram COD removed in thermophilic AD based on glucose was 450 mg Fe, 49 mg Ni, and 54 mg Co [43]. In contrast, the requirement of trace elements in mesophilic conditions was lower than that in thermophilic conditions. Moreover, in another report treating food waste, the requirement of trace elements was 276 mg Fe, 4.96 mg Ni, and 4.43 mg Co [43]. So far, there is still a lack of widely accepted methods for quantitatively estimating the optimal concentrations of trace elements required during the anaerobic digestion of food waste. Further studies on the effects of trace elements on thermophilic anaerobic treatment of food waste are necessary for evaluating the concentrations of the different TE content of food waste and the influence of operating conditions such as temperature. Dr. QIAO Wei research team at China Agricultural University piloted an study to determine the effect of switching the mesophilic food waste AD to a thermophilic process by treating the liquid fraction of food waste after lipids extraction. With the addition of trace elements, the pilot digester worked steadily under a loading rate of 4.3 g COD/Ld and a hydraulic retention time of 20 days. The concentration of the effluent volatile fatty acids was as low as 500 mg L^{-1}. The performance of the pilot study was much more than many industrial and commercial plants. The main reason for the successful pilot study was probably the addition of trace elements.

3.4.1.2.2 Co-digestion

In China, co-digestion technology has been attracting more and more interest. Ideally, in a co-digestion plant, different sources of waste such as sewage sludge, food waste, organic fraction of municipal solid waste, market waste, and night soil can be treated together in a reactor. The mixture of the different substrates can provide more balanced nutrients and make anaerobic digestion process have high efficiency. The co-digestion of food waste with sewage sludge was extensively studied. In this scenario, food waste could be treated in sewage plants. The introduction of food waste can significantly increase biogas production for electricity, which can compensate for the energy consumption for wastewater treatment. The co-digestion of food waste and straw was also reported to have many advantages [44]. Some reports proved that the co-digestion of food waste with activated sludge is an efficient method to adjust the C/N ratio to provide better buffer performance for the reactor [45], which was an efficient and environmental-friendly method.

3.4.1.3 Anaerobic Membrane Bioreactor Technology for Food Waste

Anaerobic membrane bioreactor technology was developed for retaining microorganisms in the bioreactor, thus enhancing the AD process's stability. The advantages of the AnMBR technologies have been widely reported and studied. A thermophilic AnMBR was operated for 75 days in the AD of food waste [46]. The results showed that methane yield was $0.30\,L\,kg{\cdot}COD^{-1}$ under an organic loading rate of $7.3\,kg\,COD\,m^{-3}\,d^{-1}$. Organic removal efficiencies, including COD, carbohydrates, and protein, were higher than 98%. Moreover, the concentration of VFAs was lower at $274\,mg\,L^{-1}$, indicating a well-balanced VFA production and conversion metabolism. It could be concluded that a thermophilic AD in AnMBR would be promising and practical technology for food waste treatment with high efficiency.

References

1. People's Republic of China (2020). *Ministry of agriculture and rural affairs of the People's Republic of China*, http://zdscxx.moa.gov.cn:8080/nyb/pc/index.jsp (accessed 8 February 2023).
2. People's Republic of China (2021). *National bureau of statistics*, http://www.stats.gov.cn/tjsj./ndsj/ (accessed 4 February 2022).
3. Food and Agriculture Organization of the United Nations. (2021). http://www.fao.org/home/en/ (accessed 4 February 2022).
4. Sun, C., Cao, W., Banks, C.J. et al. (2016). Biogas production from undiluted chicken manure and maize silage: a study of ammonia inhibition in high solids anaerobic digestion. *Bioresource Technology* 218: 1215–1223. https://doi.org/10.1016/j.biortech.2016.07.082.
5. Wandera, S.M., Qiao, W., Algapani, D.E. et al. (2018). Searching for possibilities to improve the performance of full scale agricultural biogas plants. *Renewable energy* 116: 720–727. https://doi.org/10.1016/j.renene.2017.09.087.
6. Li, K., Liu, R., and Sun, C. (2015). Comparison of anaerobic digestion characteristics and kinetics of four livestock manures with different substrate concentrations. *Bioresource Technology* 198: 133–140. https://doi.org/10.1016/j.biortech.2015.08.151.
7. Ileleji, E.K., Martin, C., and Jones, D. (2015). Basics of energy production through anaerobic digestion of livestock manure. *Bioenergy* 1: 287–295. https://doi.org/10.1016/B978-0-12-407909-0.00017-1.
8. Ramm, P., Abendroth, C., Latorre-Pérez, A. et al. (2020). Ammonia removal during leach-bed acidification leads to optimized organic acid production from chicken manure. *Renewable Energy* 146: 1021–1030. https://doi.org/10.1016/j.renene.2019.07.021.
9. Yu, Q., Sun, C., Liu, R. et al. (2020). Anaerobic co-digestion of corn stover and chicken manure using continuous stirred tank reactor: the effect of biochar addition and urea pretreatment. *Bioresource Technology* 319: 124197. https://doi.org/10.1016/j.biortech.2020.124197.
10. Bayrakdar, A., Sürmeli, R.Ö., and Çalli, B. (2018). Anaerobic digestion of chicken manure by a leach-bed process coupled with side-stream membrane ammonia separation. *Bioresource Technology* 258: 41–47. https://doi.org/10.1016/j.biortech.2018.02.117.
11. Chang, L., Wang, J., Ji, X. et al. (2016). The biomethane producing potential in China: a theoretical and practical estimation. *Chinese Journal of Chemical Engineering* 24: 920–928. https://doi.org/10.1016/j.cjche.2015.12.025.
12. Sun, H., Wang, E., Li, X. et al. (2021). Potential biomethane production from crop residues in China: contributions to carbon neutrality. *Renewable and Sustainable Energy Reviews* 148: 111360. https://doi.org/10.1016/j.rser.2021.111360.

13. Díaz, I., Donoso-Bravo, A., and Fdz-Polanco, M. (2011). Effect of microaerobic conditions on the degradation kinetics of cellulose. *Bioresource Technology* 102 (21): 10139–10142. https://doi.org/10.1016/j.biortech.2011.07.096.
14. Singh, R. and Kumar, S. (2019). A review on biomethane potential of paddy straw and diverse prospects to enhance its biodigestibility. *Journal of Cleaner Production* 217: 295–307. https://doi.org/10.1016/j.jclepro.2019.01.207.
15. Jin, Y., Chen, T., Chen, X. et al. (2015). Life-cycle assessment of energy consumption and environmental impact of an integrated food waste-based biogas plant. *Applied Energy* 151: 227–236. https://doi.org/10.1016/j.apenergy.2015.04.058.
16. Meng, Y., Li, S., Yuan, H. et al. (2015). Evaluating biomethane production from anaerobic mono- and co-digestion of food waste and floatable oil (FO) skimmed from food waste. *Bioresource Technology* 185: 7–13. https://doi.org/10.1016/j.biortech.2015.02.036.
17. Yang, Y., Bao, W., and Xie, G.H. (2019). Estimate of restaurant food waste and its biogas production potential in China. *Journal of Cleaner Production* 211: 309–320. https://doi.org/10.1016/j.jclepro.2018.11.160.
18. Hang, S. (2010). Anaerobic sludge digestion process from the perspective of carbon emission reduction. *Water & Wastewater Information (China)* 4: 15–16.
19. Mussoline, W., Esposito, G., Giordano, A. et al. (2013). The anaerobic digestion of rice straw: a review. *Critical Reviews in Environmental Science & Technology* 43 (9): 895–915. https://doi.org/10.1080/10643389.2011.627018.
20. Morales-Polo, C., Cledera-Castro, M.M., and Moratilla, B.Y. (2018). Reviewing the anaerobic digestion of food waste: from waste generation and anaerobic process to its perspectives. *Applied Sciences* 8 (10): 1804. https://doi.org/10.3390/app8101804.
21. Chae, K.-J., Jang, A., Yim, S.-K. et al. (2008). The effects of digestion temperature and temperature shock on the biogas yields from the mesophilic anaerobic digestion of swine manure. *Bioresource Technology* 99 (1): 1–6. https://doi.org/10.1016/j.biortech.2006.11.063.
22. Shapovalov, Y., Zhadan, S., Bochmann, G. et al. (2020). Dry anaerobic digestion of chicken manure: a review. *Applied Sciences* 10: 7825. https://doi.org/10.3390/app10217825.
23. Lili, D., Guangli, C., Xianzhang, G. et al. Efficient biogas production from cattle manure in a plug flow reactor: a large scale long term study. *Bioresource technology* 278: 450–455. https://doi.org/10.1016/j.biortech.2019.01.100.
24. Department of Agriculture and Rural Affairs of Jiangxi Province (2020). N2N regional circular agriculture model in China. http://nync.jiangxi.gov.cn. (accessed 15 January 2022).
25. Orhan, Y. and Burak, D. (2013). Ammonia inhibition in anaerobic digestion: a review. *Process Biochemistry* 48 (5-6): 901–911. https://doi.org/10.1016/j.procbio.2013.04.012.
26. Sprott, G.D. and Patel, G.B. (1986). Ammonia toxicity in pure cultures of methanogenic bacteria. *Systematic & Applied Microbiology* 7 (2): 358–363. https://doi.org/10.1016/S0723-2020(86)80034-0.
27. Park, S., Cui, F., Mo, K. et al. (2016). Mathematical models and bacterial communities for ammonia toxicity in mesophilic anaerobes not acclimated to high concentrations of ammonia. *Water Science & Technology* 74 (4): 935–942. https://doi.org/10.2166/wst.2016.274.
28. Niu, Q., Hojo, T., Qiao, W. et al. (2014). Characterization of methanogenesis, acidogenesis and hydrolysis in thermophilic methane fermentation of chicken manure. *Chemical Engineering Journal* 244: 587–596. https://doi.org/10.1016/j.cej.2013.11.074.
29. EA Bioenergy: Task 37: 06 2021, Case Story: Minhe Chicken Manure Biogas Plant, http://task37.ieabioenergy (accessed on 7 February 2022).
30. Fermoso, F.G., van Hullebusch, E., Collins, G. et al. (2019). *Trace Elements in Anaerobic Biotechnologies[M]*. IWA Publishing, Available online: https://www.iwapublishing.com/books/9781789060218/trace-elements-anaerobic-biotechnologies (accessed 15 January 2022).

31. Choong, Y.Y., Ismail, N., Abdullah, A.Z. et al. (2016). Impacts of trace element supplementation on the performance of anaerobic digestion process: a critical review. *Bioresource Technology* 209: 369–379. https://doi.org/10.1016/j.biortech.2016.03.028.
32. Molaey, R., Bayrakdar, A., Recep, S. et al. (2018). Anaerobic digestion of chicken manure: mitigating process inhibition at high ammonia concentrations by selenium supplementation. *Biomass and Bioenergy* 108: 439–446. https://doi.org/10.1016/j.biombioe.2017.10.050.
33. Molaey, R., Bayrakdar, A., Sürmeli, R.O. et al. (2018). Influence of trace element supplementation on anaerobic digestion of chicken manure: linking process stability to methanogenic population dynamics. *Journal of Cleaner Production* 181: 794–800. https://doi.org/10.1016/j.jclepro.2018.01.264.
34. Molaey, R., Bayrakdar, A., and Calli, B. (2018). Long-term inuence of trace element deciency on anaerobic mono-digestion of chicken manure. *Journal of Environmental Management* 223: 743–748. https://doi.org/10.1016/j.jenvman.2018.06.090.
35. Bi, S., Westerholm, M., Qiao, W. et al. (2019). Enhanced methanogenic performance and metabolic pathway of high solid anaerobic digestion of chicken manure by Fe^{2+} and Ni^{2+} supplementation. *Waste Management* 94: 10–17. https://doi.org/10.1016/j.wasman.2019.05.036.
36. Abouelenien, F., Fujiwara, W., Namba, Y. et al. (2010). Improved methane fermentation of chicken manure via ammonia removal by biogas recycle. *Bioresource Technology* 101: 6368–6373. https://doi.org/10.1016/j.biortech.2010.03.071.
37. Bi, S., Qiao, W., Xiong, L. et al. (2020). Improved high solid anaerobic digestion of chicken manure by moderate in situ ammonia stripping and its relation to metabolic pathway. *Renewable Energy* 146: 2380–2389. https://doi.org/10.1016/j.renene.2019.08.093.
38. Yin, D.M., Taherzadeh, M.J., Lin, M. et al. (2020). Upgrading the anaerobic membrane bioreactor treatment of chicken manure by introducing in-situ ammonia stripping and hyper-thermophilic pretreatment. *Bioresource Technology* 310: 123470. https://doi.org/10.1016/j.biortech.2020.123470.
39. Errico, M., Fjerbaek Sotoft, L., Kjærhuus Nielsen, A. et al. (2017). Treatment costs of ammonia recovery from biogas digestate by air stripping analyzed by process simulation. *Clean Technologies and Environmental Policy* 20 (7): 1479–1489. https://doi.org/10.1007/s10098-017-1468-0.
40. Zhang, W., Wu, S., Guo, J. et al. (2015). Performance and kinetic evaluation of semi-continuously fed anaerobic digesters treating food waste: role of trace elements. *Bioresource Technology* 178: 297–305. https://doi.org/10.1016/j.biortech.2014.08.046.
41. Zhang, W., Li, L., Wang, X. et al. (2020). Role of trace elements in anaerobic digestion of food waste: process stability, recovery from volatile fatty acid inhibition and microbial community dynamics. *Bioresource Technology* 315: 123796. https://doi.org/10.1016/j.biortech.2020.123796.
42. Voelklein, M.A., Shea, R.O., Jacob, A. et al. (2017). Role of trace elements in single and two-stage digestion of food waste at high organic loading rates. *Energy* 121: 185–192. https://doi.org/10.1016/j.energy.2017.01.009.
43. Niu, Q., Chi, Y., and Li, Y. (2013). Trace metals requirements for continuous thermophilic methane fermentation of high-solid food waste. *Chemical Engineering Journal* 222: 330–336. https://doi.org/10.1016/j.cej.2013.02.076.
44. Liu, Y.L., Qiao, W., Serena, C. et al. (2017). Continuous thermophilic anaerobic co-digestion of food waste and straw. *Zhongguo Huanjing Kexue/China Environmental Science (China)* 37 (6): 2194–2202.
45. Ma, Y., Yin, Y., and Liu, Y. (2017). New insights into co-digestion of activated sludge and food waste: biogas versus biofertilizer. *Bioresource Technology* 241: 448–453. https://doi.org/10.1016/j.biortech.2017.05.154.
46. Jiang, M., Wu, Z., Yao, J. et al. (2021). Enhancing the performance of thermophilic anaerobic digestion of food waste by introducing a hybrid anaerobic membrane bioreactor. *Bioresource Technology* 341: 125861.

4

Biogas Production from High-solid Anaerobic Digestion of Food Waste and Its Co-digestion with Other Organic Wastes

Le Zhang[1], To-Hung Tsui[2,3], Kai-Chee Loh[3,4], Yanjun Dai[3,5], Jingxin Zhang[3,6], and Yen Wah Tong[2,3,4]

[1]*Department of Resources and Environment, School of Agriculture and Biology, Shanghai Jiao Tong University, Shanghai, China*
[2]*NUS Environmental Research Institute, National University of Singapore, Singapore*
[3]*Energy and Environmental Sustainability for Megacities (E2S2) Phase II, Campus for Research Excellence and Technological Enterprise (CREATE), Singapore*
[4]*Department of Chemical and Biomolecular Engineering, National University of Singapore, Singapore*
[5]*School of Mechanical Engineering, Shanghai Jiao Tong University, Shanghai, China*
[6]*China-UK Low Carbon College, Shanghai Jiao Tong University, Shanghai, China*

4.1 Introduction

The sustainable management of enormous food waste amounts (generated from households, restaurants, cafeterias, retail stores, and industries) is currently a global issue [1–4]. The well-established technologies for food waste treatment include incineration, landfilling, and composting, and anaerobic digestion (AD). Among these options, AD has been regarded as an effective waste-to-energy technology to simultaneously mitigate two challenges, namely organic waste management and bioenergy recovery [5]. AD can be categorized into

Biogas Plants: Waste Management, Energy Production and Carbon Footprint Reduction,
First Edition. Edited by Wojciech Czekała

traditional wet AD (with a total solid of below 15%) and high-solid anaerobic digestion (HSAD) (with a total solid of above 15%) [6]. However, conventional AD operations frequently require relatively high energy requirements for the digester heating, a huge digester volume for operation, and costly downstream processing of a huge amount of digestate [7]. In contrast, the HSAD operations can reduce energy losses, the digester volume, the digestate volume, and the costs for organic waste transportation as well as increase the fertilizing potential of the discharged digestate [8, 9]. Hitherto, HSAD of food waste has already been established in some European countries and areas, where the food waste is often co-treated with other agricultural wastes (e.g. animal manure) [10]. Nevertheless, the efficiency of biogas production during the HSAD could be affected by several factors, including uneven substrates mixture, slow hydrolysis, and inhibitors (e.g. volatile fatty acids [11] and ammonia [12]). For instance, serious acidification phenomenon was observed during HSAD fed with food waste as a single substrate [13]. HSAD of nitrogen-rich feedstock like animal manures could generate intermediate ammonia [14]. The excess free ammonia in the HSAD systems can inhibit acetate metabolism of microbial communities, leading to considerable accumulation of propionic acid and other relatively long-chain volatile fatty acids and resultant process instability [15]. Meanwhile, HSAD of lignocellulosic substrates usually generated a low methane yield due to low biodegradation of recalcitrant cellulose and hemicellulose content [16].

The objective of this chapter is to facilitate implementation of HSAD technologies for effective food waste management. To this end, the most updated knowledge of fermentation technologies for HSAD was summarized in this chapter, which includes digester systems, process parameters, microbial communities, intensification strategies, and digestate management. Technical limitations and prospects of HSAD technologies for enhanced biogas production were also discussed. Critical analyses of HSAD strategies of food waste and its co-digestion with other organic wastes would facilitate the establishment of more efficient HSAD systems for better food waste management, renewable biofuel production, and sustainable environmental protection.

4.2 Reactor Systems for HSAD

4.2.1 High-solid Anaerobic Membrane Bioreactor

Reactor systems play a critical role in the HSAD of organic wastes for methane production [17]. Anaerobic membrane bioreactor (Figure 4.1) is one of the promising types among various reactor systems. However, a major bottleneck, namely membrane fouling or flux limitation, hinders the scale-up and wide application of anaerobic membrane bioreactors. Hence, novel approaches to enhance the flux at high-solid concentration are urgently needed to achieve a sustainable and highly efficient membrane reactor operation. In this regard, a high-solid anaerobic membrane bioreactor was recently optimized for enhancing flux during long-term continuous HSAD of food waste [18]. It was found that with an optimized pattern (i.e. 3-minutes filtration and varied relaxation time periods of 1–12 minutes) greatly enhanced the sustainable flux. Additionally, the anaerobic membrane bioreactor has also been investigated for high-solid thermophilic co-digestion of food waste and sewage sludge. The technical feasibility of this membrane bioreactor for HSAD was validated by the long-term stable operation [19]. More specifically, by increasing the sludge ratio from

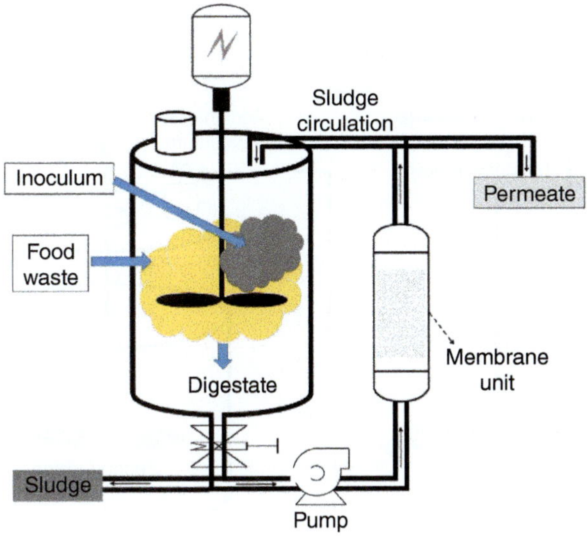

Figure 4.1 Schematic diagram of high-solid anaerobic membrane bioreactor.

0% to 100%, the biogas yield derived from this membrane bioreactor system ranged from 0.893 to 0.514 L g^{-1} of volatile solid (VS). Indeed, by coupling the membrane and filtration system, several drawbacks can be overcome by this membrane bioreactor, including unsatisfactory flocculation ability and slow microbial growth rate [20]. Subsequently, the same research team investigated the membrane fouling mechanism of the high-solid membrane bioreactor from several aspects, including the external forces analyses, particle size distribution of the digestate, and membrane filtration features, results of which greatly contributed to the further research and development of efficient high-solid anaerobic membrane bioreactors for treating food waste and sewage sludge [21]. More specifically, it was found that the soluble microbial byproducts played an important role in the membrane fouling process of HSAD membrane bioreactor [22]. During the HSAD process, a series of parameters were optimized including substrate to inoculum (S/I) ratio, co-substrate mixing ratio, inoculum acclimation period. To promote the practical engineering application of co-digestion of food waste and sludge through the anaerobic membrane bioreactor, large-scale tests deserve investment and investigation. Furthermore, Cheng et al. [23] experimentally validated the great potential of the high-solid anaerobic membrane bioreactor in the treatment of sewage sludge and food waste for supporting the design of smart cities. They also found that there was a close relationship between carbon-to-nitrogen ratio and net energy balance.

4.2.2 Two-stage HSAD Reactor System

The two-stage HSAD reactor system (Figure 4.2) includes the first stage (i.e. acidogenic fermentation) for hydrogen production and the second stage (i.e. methanogenesis) for methane production. The processes in both stages are catalyzed by corresponding microbes with different optimal working pH values. Hitherto, many studies have been reported

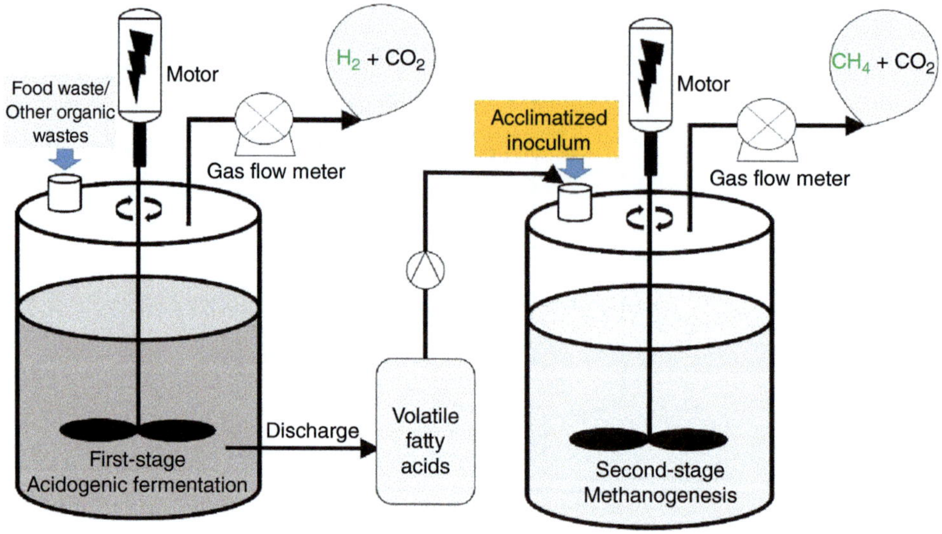

Figure 4.2 Schematic diagram of two-stage HSAD reactor system.

using two-stage HSAD reactor systems. For instance, two-stage HSAD of food waste and horticultural waste showed normal operation, while the HSAD of food waste failed due to the accumulation of high concentrations of volatile fatty acids [13]. By adjusting the mass ratio on basis of volatile solids among food waste, grass, and chicken manure, the HSAD was greatly improved for methane-rich biogas production. Compared to the single-stage HSAD, the two-stage HSAD exhibited much better performance in terms of the shorter AD time period and higher methane yields. The enhanced performance of the two-stage HSAD is mainly ascribed to the higher system stability so that organic particulates can be converted steadily and efficiently into methane. In another study on two-stage thermophilic fermentation of high-solid food waste, the long-term stability of the continuous operation was successfully achieved by recirculation of digester sludge [24]. In addition, the bioaugmentation strategies may further enhance the economic feasibility of the two-stage HSAD reactor systems (Figure 4.2) but requires more investigations on the reliability and sustainability of the systems [25].

4.2.3 High-solid Plug-flow Bioreactor

Plug-flow bioreactor (Figure 4.3) is another type of reactor system for HSAD of organic wastes. The plug-flow reactor is commonly adopted for bioenergy production due to its advantage of steady-state operation. Hitherto, pilot-scale plug-flow bioreactor has been utilized for co-digestion of food waste and rice husk due to the fact that high biodegradability and relatively low C/N ratio often caused rapid accumulation of volatile fatty acids and resultant process inhibition [26]. To solve this problem, the organic loading rates (OLRs) ranging from 5 to 9 kg VS m^{-3} d^{-1} were optimized for mesophilic (37 °C) co-digestion of food waste and rice husk. The results indicated that the highest volatile solids removal (82%) was obtained at an OLR of 5 kg VS m^{-3} d^{-1}. While at an OLR of 6 kg VS m^{-3} d^{-1},

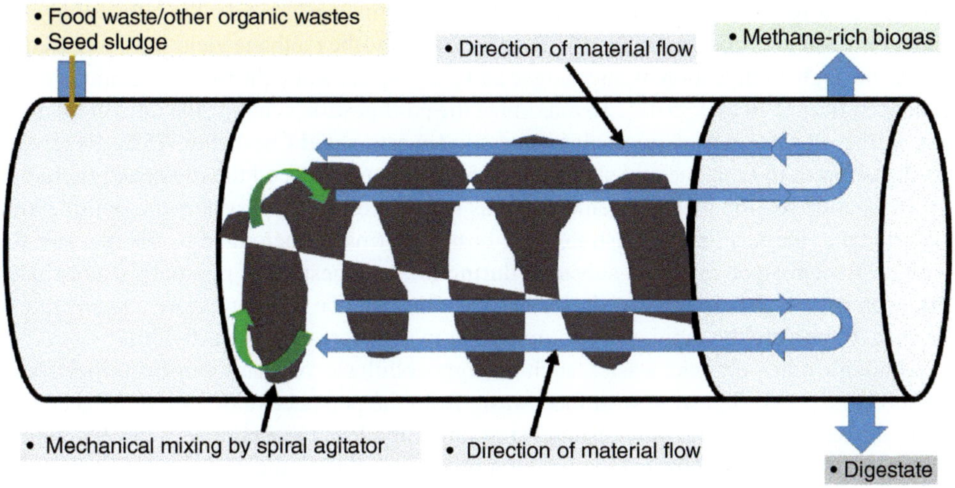

Figure 4.3 Schematic diagram of high-solid plug-flow bioreactor.

the daily biogas production was 196 L d^{-1} but decreased to 136 L d^{-1} at an OLR of 9 kg VS m^{-3} d^{-1}. The aforementioned results demonstrated that an OLR of 5–6 kg VS m^{-3} d^{-1} can be adopted to this plug-flow reactor for a stable performance. Recently, thermophilic and mesophilic high-solid digestion of food waste using industrial-scale plug-flow digesters were examined and compared [27]. Results showed that reasonable biogas yields obtained from HSAD of food waste were around 0.4–0.6 Nm^3 CH_4/kg volatile solids. Notably, severe process disturbance caused by organic acid accumulation (i.e. 6–14 g L^{-1}) and ammonia inhibition (i.e. 2 g NH_3-N/L) was observed under thermophilic conditions (i.e. 54 °C).

4.3 Intensification Strategies for HSAD

4.3.1 High-solid Anaerobic Co-digestion (HS-AcD)

Highly efficient AD of organic wastes relies on many conditions, including the proper composition of feedstocks and optimum carbon-to-nitrogen (C/N) ratio (e.g. 25–30). To adjust the C/N ratio to an optimum range, another suitable organic waste can be utilized as a co-substrate. In addition to the adjustment of C/N ratio, anaerobic co-digestion can also contribute to the dilution of toxic compounds [26]. High-solid anaerobic co-digestion (HS-AcD) can be beneficial for improving digestion stability by reducing the acidification degree, especially during the start-up period [28]. HS-AcD of food waste and sludge was reported to be beneficial to reduce power consumption [29]. In the anaerobic co-digestion of food waste and waste-activated sludge, the volatile fatty acids produced from hydrolysis and acidification of food waste can be neutralized by the ammonia produced from HSAD of wasted-activated sludge [14]. Successful HS-AcD of cattle slurry and vegetable processing waste for methane production was successfully achieved [30]. By reducing ammonia inhibition and Na^+ inhibition, the co-digestion of dewatered sludge and food waste greatly improved the system stability and enhanced volumetric biogas production [31]. The synergistic effect of high-solid co-digestion between microalgae (i.e. *Spirulina platensis*) and food waste was also observed [9]. More specifically, the methane yield was increased by 37.5% through co-digestion of microalgae and food waste, while the methanogenic rate was enhanced from 0.16 to 0.35 d^{-1}. To maximize the co-digestion synergy, the ratio of microalgae in the co-substrate of microalgae and food waste should be below 25%. Regarding co-digestion of microalgae and sludge, the co-digestion enhanced the methane production by 10%; however, the methanogenic rate was not improved. With systematic optimization experiments, the results demonstrated the best performance belonged to the digester fed with 66% microalgae in the co-substrate during the co-digestion of microalgae and sludge [9]. Moreover, a conclusion was drawn that rice husk can be an appropriate co-substrate to be used to mitigate the rapid acid accumulation in digester of food waste [26].

Although other organic wastes such as lignocellulosic biomass can be utilized as a co-substrate for co-digestion with food waste from the perspective of a more balance C/N ratio, the recalcitrant three-dimensional structure often leads to a relatively low biodegradability of the lignocellulosic biomass. For instance, during the high-solid co-digestion of food waste and landscape waste, the low biodegradability of landscape waste made it an unsuitable substrate to effectively help adjust the C/N ratio of the mixed waste [32]. Hence, pretreatments of lignocellulosic biomass wastes are highly recommended to help release the bioavailable carbon before subsequent co-digestion with food waste. Moreover, it is often difficult to determine a specific mixing ratio of different co-substrates due to inconsistent components, but previously determined blend ratios such as 50 : 50 between food waste and sewage sludge [33] corresponding to the highest performance can be used as a reference in the similar co-digestion scenarios.

4.3.2 Supplementation of Additives

To tackle process imbalance and enhance the process function, supplementation of additives such as biochar materials [34, 35] and trace elements [36, 37] has been explored in recent years. Biochar was found to be capable of enhancing HSAD performance of food waste by promoting volatile acid degradation, promoting direct interspecies electron transfer, and adjusting the methanogenic pathways [34, 38, 39]. Wang et al. [40] added Fe/C in a two-phase high-solid digester of food waste and obtained significantly increased methane yields. The experimental results also demonstrated that the supplemented F/C played an essential role in the acidification phase. In addition, the effect of trace metals (i.e. iron, cobalt, and nickel) on continuous thermophilic methane fermentation of high-solid food waste was investigated [36]. The results showed that the control digester showed gradually deteriorated reactor efficiency due to trace metals limitation caused by no supplementation of trace elements. In contrast, when the trace metals were supplemented to the reactor per 45 days, the decreased methane content, pH, and gas production were restored to normal levels. During the normal thermophilic methane fermentation, the removal efficiencies of carbohydrate, VS, and TS were 97%, 80%, and 77%, respectively. Similarly, the addition of trace elements (i.e. Fe, Co, and Ni) contributed to the recovery of methane production during a high-solid mesophilic (35 °C) methane fermentation of food waste [41]. These results proved the significance of trace metals to the HSAD of food waste. Indeed, the trace elements such as Fe, Co, and Ni play an important role in synthesizing essential cofactors and coenzymes in key enzymatic methanogenic pathways during the AD processes [42–44]. Moreover, the experimental results showed that the required values of Fe, Co, and Ni for an efficient high-solid thermophilic methane fermentation of food waste can be 276, 4.96, and 4.43 mg kg^{-1}-COD-removed, respectively [36]. Before the supplementation strategy of trace elements can be adopted in the industrial-scale anaerobic digesters, the cycle cost analysis and the economic feasibility analysis should be performed and considered.

4.3.3 Bioaugmentation Strategies for HSAD

HSAD of food waste or other organic wastes (e.g. animal manure) is a complicated process that is easily hampered by the high concentration of volatile fatty acids or excess free ammonia. On the one hand, in HSAD systems digesting food waste, manure, sludge, or other nitrogen-rich substrates, the accumulated high concentration of ammonia (e.g. >3 g L^{-1} NH_4^+-N or >0.15 g L^{-1} NH_3-N) can inhibit the normal growth of methanogenic microbes, leading to an inhibition of the methanogenesis step and a resultant decrease in methane yields. To mitigate the ammonia, the ammonia-tolerant microbial strains or consortia can be cultivated and added to these HSAD systems. The commonly used ammonia-tolerant microbes in anaerobic digesters contain several types, namely enriched ammonia-tolerant methanogenic cultures [45, 46] and various pure strains like syntrophic acetate-oxidizing bacteria as well as acetoclastic and hydrogenotrophic methanogens [47, 48]. To further improve the positive effects of bioaugmentation operations, the combinations of two or more strains can be further investigated, since the selection of appropriate strains holds great potential in offering synergistic effects and enhancing process efficiency.

On the other, bioaugmentation strategies can be used to mitigate the inhibition caused by the accumulation of volatile fatty acids. For instance, acid-tolerant methanogenic consortium or acid-degrading microbial consortium were supplemented into the overloaded anaerobic digesters, followed by relieving of high concentration of organic acids and recovery of the overloaded digesters [49, 50]. More specifically, acetate-degrading and propionate-catabolizing methanogenic consortia were the most representative acid-degrading microbes for bioaugmentation. Moreover, the high levels of salt from the feedstock like food waste can also inhibit methane production by affecting microbial activities. In this regard, *Methanosarcina* and *Methanosaeta* were found enriched in the anaerobic digester after acclimation to stepwise increased NaCl concentration from 1 to 16 g L^{-1} for around two months [51], indicating that the acclimatized microbial consortia can be potentially bioaugmented into the overloaded anaerobic digesters under high-salinity stress to enhance methane production.

4.3.4 Optimization of Process Parameters

Numerous previous studies have confirmed that the methane yields and productivity greatly relied on the process parameters, including mixing ratios of various co-substrates, pH, OLRs, temperature, the substrate-to-inoculum (S/I) ratio, mixing speed, C/N ratio, and concentration of inhibitors. Therefore, many investigations have been conducted to enhance the methane yield and productivity through parameter optimization. Yang et al.

[11] investigated the effect of pH adjustment on methane production performance in high-solid thermophilic anaerobic digesters fed with food waste and found that the digester with pH adjustment to 8 showed the highest methane content, cumulative methane yield, and cumulative biogas yields, which were 5-fold, 7.6-fold, and 1.5-fold higher than the equivalents in pH uncontrolled digester. Rajagopal et al. [52] investigated HS-AcD of food waste and dairy manure at low-to-moderate temperature (20–25 °C) conditions and observed that HS-AcD process at 25 °C was comparatively efficient in saving heat energy and producing methane compared to the operation at mesophilic conditions (35 °C). This finding could be beneficial to the HS-AcD operations in cold countries and areas. Even so, compared to the mesophilic HS-AcD, thermophilic HS-AcD of food waste and fats/oil was more effective for reduction of lipids and showing higher organic loading capacity [53]. Moreover, the HSAD performance can also be greatly affected by the S/I ratio. Specifically, a relatively low S/I ratio (e.g. 1.0 g total solids/g inoculum total solids) led to a relatively high cumulative methane yield, while a higher S/I ratio (e.g. 2.0–3.0) led to the accumulation of inhibitory intermediates (i.e. volatile fatty acids), mass transfer limitation, and resultant digester system failure [54]. In addition to the S/I ratio, the quality of the inoculum affected the HSAD performance. Reportedly, the acclimated granular sludge was a better inoculum than the conventional anaerobic sludge as the former contained higher archaea abundance [14]. Mixing strategies of HSAD is vital to achieving timely and efficient separation of gas, liquid, and solids in the anaerobic digester [29, 55]. Thus, it is necessary to optimize the mixing strategies of HSAD of food waste and co-digestion with other organic wastes. Recently, how the impeller width and mixing speed affects the relationships between hydrodynamics and energy consumption was simulated by playing the mixing model using CFD software [56]. Based on this study, the impeller width of 100 mm (10 rpm), 200 mm, and 300 mm (5 and 10 rpm) are recommended for achieving relatively high hydrodynamics and energy economics [56]. Moreover, the feedstock mixing ratios can greatly affect the fermentation performance. For instance, a higher ratio of food waste in co-mixture of food waste and sludge led to an increase in biogas production and volatile solids reduction [31]. Similarly, increasing the percentage of food waste in the mixed feedstock of food waste and green waste showed an increased methane yield. Interestingly, increasing the ratio of green waste in the mixture can shorten the retention time [57]. It may be difficult to determine the consistently optimal mixing ratio for all the co-digestion scenarios. Hence, the reported preferred ratios for the similar substrates and additional optimization experiments at the lab scale would be helpful to find a more appropriate mixing ratio for pilot-scale and industrial-scale co-digestion operations. Furthermore, it is crucial to select an appropriate OLR for a given HSAD system. Indeed, a too low organic loading could lead to a waste of treating capacity of the digester system, while a too high organic loading could cause the reactor failure due to overloading. For instance, the HSAD of food waste produced stably biogas of 650 $mL\ g^{-1}$ total solid (TS) at the OLR of 2–8 $g\text{-}TS\ L^{-1}\ d^{-1}$, but the HSAD performance significantly decreased when the OLR increased to 11 $g\text{-}TS\ L^{-1}\ d^{-1}$. These results demonstrated that the optimum OLR can be 8 $g\text{-}TS\ L^{-1}\ d^{-1}$ [6]. Moreover, when the HSAD was operated for 65 days, the concentration of volatile fatty acids, ammonia, and soluble chemical oxygen demand sharply increased to 16,366, 2711, and 41,024 $mg\ L^{-1}$, respectively [6]. These results indicated that inhibitors like volatile fatty acids and ammonia can be accumulated during the long-term operation (e.g. >65 days), thereby requiring more enhancing strategies

(e.g. co-digestion and supplementation of additives) for mitigating the inhibitory effects to maintain normal methane production.

Optimization of various process parameters can be accelerated through multi-parameters optimization via machine learning algorithms [58]. In recent years, several investigations have reported the application of machine learning in AD. For instance, Wang et al. [59] used four machine learning methods to model the AD process and identified the key process parameters through 17 samples. The results demonstrated that k-nearest neighbors (KNN) method was the best one and C/N ratio was a determinant operational parameter. Clercq et al. [60] applied several machine learning models to study an industrial-scale co-digestion system over 1398-day operation and found that the extreme gradient boosting (XGBoost) and random forest exhibited the highest methane prediction accuracy of 88%. Xu et al. [61] applied XGBoost to predict methane production and found that the dosages of zero-valent iron (ZVI), soluble chemical oxygen demand, and total solids are vital factors affecting methane yields from ZVI-amended anaerobic digesters. Additionally, Long et al. [62] adopted the machine learning model on optimization of operational parameters and obtained a prediction accuracy of 0.82. More recently, the application of machine learning in AD was critically summarized [63]. The aforementioned studies have proved the potential application of machine learning in the optimization of AD systems. However, the current investigations focused mainly on the prediction tasks and relatively simple feature importance analyses. To go a step further, more variety of organic wastes from different countries and regions and more data points during the AD processes should be collected and integrated to enlarge the size of datasets and further improve the established models. Such optimized machine learning models would be able to offer greater application potential of data-driven approaches in the HSAD systems for highly efficient waste-to-energy conversion when more and more experimental data were integrated.

4.4 Microbial Communities for HSAD

Investigation of microbial communities in the HSAD system with high HSAD performance can help explain the enhancing mechanisms of the adopted intensification strategies. It has been found that among various process parameters, temperature appeared to be a strong driver for diversified microbial community structures [27]. Yi et al. [64] characterized the microbial community structures in the HSAD system of food waste under mesophilic conditions (35 °C). They found that the predominant microbes responsible for organic matter catabolism during hydrolysis and acidogenesis stages belonged to *Clostridium* and *Symbiobacterium* organisms, while the predominant methanogen was *Methanosarcina*. The significance of *Methanosarcina* in maintaining efficient methane production was also confirmed by Capson-Tojo et al. [65]. Additionally, both hydrogenotrophic and acetoclastic pathways were supported in HSAD of food waste with syntrophism relationships [64]. Furthermore, it was found that the microbial community structure shifts correlated well with the process parameters during the HSAD process, indicating that it would be possible to diagnose the mesophilic HSAD of food waste by dynamically monitoring the microbial community shifts. In addition to the temperature, additional additives like Fe/C can also shift the microbial community structures. For instance, during high-solid digestion of food waste added by Fe/C, the predominant methanogenic species shifted from *Methanothrix* to *Methanospirillum* [40], which led to a change of methane production pathway from the use of acetic acid to hydrogen as the reactant.

4.5 Digestate Management for HSAD

Digestate, the byproduct of HSAD of organic wastes is a mixture of various microbial biomass, partially degraded organic materials and inorganic matters. As the solid-digestate-derived substrates contain substantial nutrition (e.g. N, P, and K) for plants, the digestate is often manufactured into biofertilizers [66, 67]. Prior to becoming a satisfactory fertilizer material, AD digestate should be pretreated to optimize the water content, nutrient content, physicochemical properties, and heavy metal concentration. As the solid content in HSAD digestate is above 15%, the rheological behaviors are significantly different from the digestate derived from the conventional AD with low-solid contents (i.e. <15%). Due to the special rheological behaviors, the posttreatments of the digestate from HSAD, including dewatering, conditioning, and land spreading, were greatly affected [7]. Therefore, novel solutions should be developed for the management of HSAD digestate. On the one hand, the HSAD digestate can be directly transferred to incinerators and landfills, which is an easy way to operate but will lead huge losses in energy and resource [68]. On the other, the HSAD can be converted into biochar through gasification or pyrolysis technologies [69]. The produced biochar can be partially re-utilized in the HSAD systems for microbial enhancement, while the remaining biochar can be utilized in soil improvement or as a solid fuel, or sold in a biochar market to make a profit. Furthermore, the discharged HSAD digestate can also be utilized to cultivate fungal or insects, which can be further utilized for the extraction of valuable compounds. From the perspective of circular economy, the integrated scenarios such as HSAD coupled with gasification are recommended as it can not only recover clean energy and improve the circular economy but can also minimize the environmental impact of HSAD digestate when applied in soil by killing the potential pathogens at high temperature (e.g. 600–800 °C). Hitherto, the technical readiness level of the integrated systems for the treatment of HSAD digestate is around lab-scale and pilot-scale; the full-scale application remains not appeared and should be explored.

4.6 Conclusions and Perspectives

HSAD has been confirmed as a promising technology for both organic waste (e.g. food waste) management and renewable energy recovery due to the fact that it can simultaneously treat multiple types of wastes, facilitate solving organic-wastes-related environmental issues, significantly reduce reactor volume, and produce digestate with high agronomic value. Nevertheless, HSAD of food waste and other organic wastes was often hindered by high concentrations of ammonia and/or volatile fatty. To mitigate the effects of these inhibitors, hitherto, anaerobic membrane bioreactor, two-stage reactor system, and plug-flow bioreactor have been developed and optimized for HSAD of food waste and other organic wastes. Several intensification approaches have been investigated, including HS-AcD, supplementation of additives like biochar and trace metals, bioaugmentation strategies for HSAD, and optimization of process parameters. More systematic microbial communities' data should be collected for the investigation of the enhancing mechanisms of the adopted intensification strategies.

Further studies should be carried out on basis of what has been discovered. HSAD with leachate recirculation deserves large-scale test bedding as it can increase the mass transfer of feedstock to the acclimated consortia, leading to a higher methane yield. The adoption of an acclimated inoculum can efficiently shorten the lag time period in the initial phase of the HSAD. Moreover, HSAD technologies can be improved in several aspects including how to achieve effective substrate mixing, how to better control the HSAD process to avoid process failure, and how to achieve more efficient separation of gas, liquid, and solid in the digester. The pretreatment of various substrates before feeding into the HSAD systems is also a key point as it has a close relationship with process inhibition. For instance, ultrasonication was beneficial to the solubilization of particulate organics and release soluble organic matters [70]. Therefore, the integrated system of feedstock pretreatment and HSAD of the pretreated waste mixture would be a technically viable approach to enhance methane production from the organic fraction of organic wastes. Additionally, automatic reactor systems for long-term continuous HSAD should be further investigated and developed in order to extend the application potential of this technology toward industrial installations. Furthermore, to tailor the optimal process parameters (including temperature, additive dosage, OLR, hydraulic retention time, etc., for a given HSAD treating a specific feedstock), machine learning as an emerging decision-making approach can be used to parameter optimization, leading to a reduction in digestion instability. The final determined parameter values would guide the experimental design and operations of HSAD for methane-rich biogas production, which can contribute to saving time and resources. Of course, further studies on the combination of more machine learning algorithms and practical application data related to HSAD are required.

Acknowledgments

This project was funded by the National Research Foundation, Prime Minister's Office, Singapore, under its Campus for Research Excellence and Technological Enterprise (CREATE) program.

References

1. Greses, S., Tomás-Pejó, E., and González-Fernández, C. (2021). Short-chain fatty acids and hydrogen production in one single anaerobic fermentation stage using carbohydrate-rich food waste. *Journal of Cleaner Production* 284: 124727. https://doi.org/10.1016/j.jclepro.2020.124727.
2. Huang, H., Qureshi, N., Chen, M.-H. et al. (2015). Ethanol production from food waste at high solids content with vacuum recovery technology. *Journal of Agricultural and Food Chemistry* 63 (10): 2760–2766. https://doi.org/10.1021/jf5054029.
3. Jones, R.J., Fernández-Feito, R., Massanet-Nicolau, J. et al. (2021). Continuous recovery and enhanced yields of volatile fatty acids from a continually-fed 100 L food waste bioreactor by filtration and electrodialysis. *Waste Management* 122: 81–88. https://doi.org/10.1016/j.wasman.2020.12.032.
4. Sugiarto, Y., Sunyoto, N.M.S., Zhu, M.M. et al. (2021). Effect of biochar addition on microbial community and methane production during anaerobic digestion of food wastes: the role of minerals in biochar. *Bioresource Technology* 323: 124585. https://doi.org/10.1016/j.biortech.2020.124585.

5. Anyaoku, C.C. and Baroutian, S. (2018). Decentralized anaerobic digestion systems for increased utilization of biogas from municipal solid waste. *Renewable & Sustainable Energy Reviews* 90: 982–991. https://doi.org/10.1016/j.rser.2018.03.009.
6. Zhao, M., Yang, L., Guo, M. et al. (2018). High-solids fermentation of food wastes for biogas recovery by using horizontal anaerobic reactor. *Journal of Renewable and Sustainable Energy* 10 (4): 043106. https://doi.org/10.1063/1.5032212.
7. Peng, W., Lü, F., Hao, L. et al. (2020). Digestate management for high-solid anaerobic digestion of organic wastes: a review. *Bioresource Technology* 297: 122485. https://doi.org/10.1016/j.biortech.2019.122485.
8. Di Capua, F., Spasiano, D., Giordano, A. et al. (2020). High-solid anaerobic digestion of sewage sludge: challenges and opportunities. *Applied Energy* 278: 115608. https://doi.org/10.1016/j.apenergy.2020.115608.
9. Du, X., Tao, Y., Li, H. et al. (2019). Synergistic methane production from the anaerobic co-digestion of *Spirulina platensis* with food waste and sewage sludge at high solid concentrations. *Renewable Energy* 142: 55–61. https://doi.org/10.1016/j.renene.2019.04.062.
10. Ebner, J.H., Labatut, R.A., Lodge, J.S. et al. (2016). Anaerobic co-digestion of commercial food waste and dairy manure: characterizing biochemical parameters and synergistic effects. *Waste Management* 52: 286–294. https://doi.org/10.1016/j.wasman.2016.03.046.
11. Yang, L., Huang, Y., Zhao, M. et al. (2015). Enhancing biogas generation performance from food wastes by high-solids thermophilic anaerobic digestion: effect of pH adjustment. *International Biodeterioration & Biodegradation* 105: 153–159. https://doi.org/10.1016/j.ibiod.2015.09.005.
12. Sun, C., Cao, W., Banks, C.J. et al. (2016). Biogas production from undiluted chicken manure and maize silage: a study of ammonia inhibition in high solids anaerobic digestion. *Bioresource Technology* 218: 1215–1223. https://doi.org/10.1016/j.biortech.2016.07.082.
13. Li, W., Loh, K.-C., Zhang, J. et al. (2018a). Two-stage anaerobic digestion of food waste and horticultural waste in high-solid system. *Applied Energy* 209: 400–408. https://doi.org/10.1016/j.apenergy.2017.05.042.
14. Mu, H., Li, Y., Zhao, Y. et al. (2018). Microbial and nutritional regulation of high-solids anaerobic mono-digestion of fruit and vegetable wastes. *Environmental Technology* 39 (4): 405–413. https://doi.org/10.1080/09593330.2017.1301571.
15. Peng, X., Zhang, S., Li, L. et al. (2018). Long-term high-solids anaerobic digestion of food waste: effects of ammonia on process performance and microbial community. *Bioresource Technology* 262: 148–158. https://doi.org/10.1016/j.biortech.2018.04.076.
16. Zhang, L., Loh, K.-C., and Zhang, J. (2018). Food waste enhanced anaerobic digestion of biologically pretreated yard waste: analysis of cellulose crystallinity and microbial communities. *Waste Management* 79: 109–119. https://doi.org/10.1016/j.wasman.2018.07.036.
17. Ellacuriaga, M., García-Cascallana, J., and Gómez, X. (2021). Biogas production from organic wastes: integrating concepts of circular economy. *Fuels* 2 (2): 144–167. https://doi.org/10.3390/fuels2020009.
18. Cheng, H., Li, Y., Kato, H., and Li, Y. (2020). Enhancement of sustainable flux by optimizing filtration mode of a high-solid anaerobic membrane bioreactor during long-term continuous treatment of food waste. *Water Research* 168: 115195.1–115195.12. https://doi.org/10.1016/j.watres.2019.115195.
19. Li, Y., Cheng, H., Guo, G. et al. (2020). High solid mono-digestion and co-digestion performance of food waste and sewage sludge by a thermophilic anaerobic membrane bioreactor. *Bioresource Technology* 310: 123433. https://doi.org/10.1016/j.biortech.2020.123433.
20. Duncan, J., Bokhary, A., Fatehi, P. et al. (2017). Thermophilic membrane bioreactors: a review. *Bioresource Technology* 243: 1180–1193. https://doi.org/10.1016/j.biortech.2017.07.059.
21. Cheng, H., Li, Y., Guo, G. et al. (2020). Advanced methanogenic performance and fouling mechanism investigation of a high-solid anaerobic membrane bioreactor (AnMBR) for the co-digestion of food waste and sewage sludge. *Water Research* 187: 116436. https://doi.org/10.1016/j.watres.2020.116436.

22. Cheng, H., Li, Y., Li, L. et al. (2020). Long-term operation performance and fouling behavior of a high-solid anaerobic membrane bioreactor in treating food waste. *Chemical Engineering Journal* 394: 124918. https://doi.org/10.1016/j.cej.2020.124918.
23. Cheng, H., Li, Y., Hu, Y. et al. (2021). Bioenergy recovery from methanogenic co-digestion of food waste and sewage sludge by a high-solid anaerobic membrane bioreactor (AnMBR): mass balance and energy potential. *Bioresource Technology* 326: 124754. https://doi.org/10.1016/j.biortech.2021.124754.
24. Lee, D.-Y., Ebie, Y., Xu, K.-Q. et al. (2010). Continuous H_2 and CH_4 production from high-solid food waste in the two-stage thermophilic fermentation process with the recirculation of digester sludge. *Bioresource Technology* 101: S42–S47. https://doi.org/10.1016/j.biortech.2009.03.037.
25. Tsui, T.H. and Wong, J.W. (2019). A critical review: emerging bioeconomy and waste-to-energy technologies for sustainable municipal solid waste management. *Waste Disposal & Sustainable Energy* 1 (3): 151–167. https://doi.org/10.1007/s42768-019-00013-z.
26. Jabeen, M., Zeshan, Y., S., Haider, M.R., and Malik, R.N. (2015). High-solids anaerobic co-digestion of food waste and rice husk at different organic loading rates. *International Biodeterioration & Biodegradation* 102: 149–153. https://doi.org/10.1016/j.ibiod.2015.03.023.
27. Westerholm, M., Liu, T., and Schnürer, A. (2020). Comparative study of industrial-scale high-solid biogas production from food waste: process operation and microbiology. *Bioresource Technology* 304: 122981. https://doi.org/10.1016/j.biortech.2020.122981.
28. Lee, E., Bittencourt, P., Casimir, L. et al. (2019). Biogas production from high solids anaerobic co-digestion of food waste, yard waste and waste activated sludge. *Waste Management* 95: 432–439. https://doi.org/10.1016/j.wasman.2019.06.033.
29. Latha, K., Velraj, R., Shanmugam, P., and Sivanesan, S. (2019). Mixing strategies of high solids anaerobic co-digestion using food waste with sewage sludge for enhanced biogas production. *Journal of Cleaner Production* 210: 388–400. https://doi.org/10.1016/j.jclepro.2018.10.219.
30. Yao, Y., Luo, Y., Yang, Y. et al. (2014). Water free anaerobic co-digestion of vegetable processing waste with cattle slurry for methane production at high total solid content. *Energy* 74: 309–313. https://doi.org/10.1016/j.energy.2014.06.014.
31. Dai, X., Duan, N., Dong, B., and Dai, L. (2013). High-solids anaerobic co-digestion of sewage sludge and food waste in comparison with mono digestions: stability and performance. *Waste Management* 33 (2): 308–316. https://doi.org/10.1016/j.wasman.2012.10.018.
32. Drennan, M.F. and DiStefano, T.D. (2014). High solids co-digestion of food and landscape waste and the potential for ammonia toxicity. *Waste Management* 34 (7): 1289–1298. https://doi.org/10.1016/j.wasman.2014.03.019.
33. Liu, C., Li, H., Zhang, Y., and Liu, C. (2016). Improve biogas production from low-organic-content sludge through high-solids anaerobic co-digestion with food waste. *Bioresource Technology* 219: 252–260. https://doi.org/10.1016/j.biortech.2016.07.130.
34. Cui, Y., Mao, F., Zhang, J. et al. (2021). Biochar enhanced high-solid mesophilic anaerobic digestion of food waste: cell viability and methanogenic pathways. *Chemosphere* 272: 129863. https://doi.org/10.1016/j.chemosphere.2021.129863.
35. Tsui, T.H., Zhang, L., Lim, E.Y. et al. (2021). Timing of biochar dosage for anaerobic digestion treating municipal leachate: altered conversion pathways of volatile fatty acids. *Bioresource Technology* 335: 125283. https://doi.org/10.1016/j.biortech.2021.125283.
36. Qiang, H., Niu, Q., Chi, Y., and Li, Y. (2013). Trace metals requirements for continuous thermophilic methane fermentation of high-solid food waste. *Chemical Engineering Journal* 222: 330–336. https://doi.org/10.1016/j.cej.2013.02.076.
37. Voelklein, M.A., O'Shea, R., Jacob, A., and Murphy, J.D. (2017). Role of trace elements in single and two-stage digestion of food waste at high organic loading rates. *Energy* 121: 185–192. https://doi.org/10.1016/j.energy.2017.01.009.

38. Sun, C., Liu, F., Song, Z. et al. (2019). Feasibility of dry anaerobic digestion of beer lees for methane production and biochar enhanced performance at mesophilic and thermophilic temperature. *Bioresource Technology* 276: 65–73. https://doi.org/10.1016/j.biortech.2018.12.105.
39. Zhang, L., Lim, E.Y., Loh, K.C. et al. (2020). Biochar enhanced thermophilic anaerobic digestion of food waste: focusing on biochar particle size, microbial community analysis and pilot-scale application. *Energy Conversion and Management* 209: 112654. https://doi.org/10.1016/j.enconman.2020.112654.
40. Wang, P., Ye, M., Cui, Y. et al. (2021). Enhancement of enzyme activities and VFA conversion by adding Fe/C in two-phase high-solid digestion of food waste: performance and microbial community structure. *Bioresource Technology* 331: 125004. https://doi.org/10.1016/j.biortech.2021.125004.
41. Qiang, H., Lang, D.-L., and Li, Y.-Y. (2012). High-solid mesophilic methane fermentation of food waste with an emphasis on iron, cobalt, and nickel requirements. *Bioresource Technology* 103 (1): 21–27. https://doi.org/10.1016/j.biortech.2011.09.036.
42. Molaey, R., Bayrakdar, A., Sürmeli, R.Ö., and Çalli, B. (2018). Influence of trace element supplementation on anaerobic digestion of chicken manure: linking process stability to methanogenic population dynamics. *Journal of Cleaner Production* 181: 794–800. https://doi.org/10.1016/j.jclepro.2018.01.264.
43. Zhang, L. and Loh, K.-C. (2019). Synergistic effect of activated carbon and encapsulated trace element additive on methane production from anaerobic digestion of food wastes – enhanced operation stability and balanced trace nutrition. *Bioresource Technology* 278: 108–115. https://doi.org/10.1016/j.biortech.2019.01.073.
44. Zhang, L., Zhang, J., and Loh, K.-C. (2019). Enhanced food waste anaerobic digestion: an encapsulated metal additive for shear stress-based controlled release. *Journal of Cleaner Production* 235: 85–95. https://doi.org/10.1016/j.jclepro.2019.06.301.
45. Fotidis, I.A., Treu, L., and Angelidaki, I. (2017). Enriched ammonia-tolerant methanogenic cultures as bioaugmentation inocula in continuous biomethanation processes. *Journal of Cleaner Production* 166: 1305–1313. https://doi.org/10.1016/j.jclepro.2017.08.151.
46. Mahdy, A., Fotidis, I.A., Mancini, E. et al. (2017). Ammonia tolerant inocula provide a good base for anaerobic digestion of microalgae in third generation biogas process. *Bioresource Technology* 225: 272–278. https://doi.org/10.1016/j.biortech.2016.11.086.
47. Fotidis, I.A., Wang, H., Fiedel, N.R. et al. (2014). Bioaugmentation as a solution to increase methane production from an ammonia-rich substrate. *Environmental Science & Technology* 48 (13): 7669–7676. https://doi.org/10.1021/es5017075.
48. Yang, Z., Wang, W., Liu, C. et al. (2019). Mitigation of ammonia inhibition through bioaugmentation with different microorganisms during anaerobic digestion: selection of strains and reactor performance evaluation. *Water Research* 155: 214–224. https://doi.org/10.1016/j.watres.2019.02.048.
49. Li, Y., Yang, G., Li, L., and Sun, Y. (2018). Bioaugmentation for overloaded anaerobic digestion recovery with acid-tolerant methanogenic enrichment. *Waste Management* 79: 744–751. https://doi.org/10.1016/j.wasman.2018.08.043.
50. Li, Y., Zhang, Y., Sun, Y. et al. (2017). The performance efficiency of bioaugmentation to prevent anaerobic digestion failure from ammonia and propionate inhibition. *Bioresource Technology* 231: 94–100. https://doi.org/10.1016/j.biortech.2017.01.068.
51. Zhang, J., Zhang, R., He, Q. et al. (2020a). Adaptation to salinity: response of biogas production and microbial communities in anaerobic digestion of kitchen waste to salinity stress. *Journal of Bioscience and Bioengineering* 130 (2): 173–178. https://doi.org/10.1016/j.jbiosc.2019.11.011.
52. Rajagopal, R.G., Bernard, G., and Hince, J.-F. (2019). High-solid anaerobic co-digestion of food waste and dairy manure: a pilot scale study at low-to-moderate temperature conditions. *Detritus* 5: 66–74. https://doi.org/10.31025/2611-4135/2019.13785.
53. Li, Y.Y., Sasaki, H., Yamashita, K. et al. (2002). High-rate methane fermentation of lipid-rich food wastes by a high-solids co-digestion process. *Water Science and Technology* 45 (12): 143–150. https://doi.org/10.2166/wst.2002.0420.

54. Dixon, P.J., Ergas, S.J., Mihelcic, J.R., and Hobbs, S.R. (2019). Effect of substrate to inoculum ratio on bioenergy recovery from food waste, yard waste, and biosolids by high solids anaerobic digestion. *Environmental Engineering Science* 36 (12): 1459–1465. https://doi.org/10.1089/ees.2019.0078.
55. Mao, L., Tsui, T.H., Zhang, J. et al. (2021). Mixing effects on decentralized high-solid digester for horticultural waste: startup, operation and sensitive microorganisms. *Bioresource Technology* 333: 125216. https://doi.org/10.1016/j.biortech.2021.125216.
56. Hu, F., Zhang, S., Wang, X. et al. (2022). Quantitative hydrodynamic characterization of high solid anaerobic digestion: correlation of "mixing-fluidity-energy" and scale-up effect. *Bioresource Technology* 344: 126237. https://doi.org/10.1016/j.biortech.2021.126237.
57. Chen, X., Yan, W., Sheng, K., and Sanati, M. (2014). Comparison of high-solids to liquid anaerobic co-digestion of food waste and green waste. *Bioresource Technology* 154: 215–221. https://doi.org/10.1016/j.biortech.2013.12.054.
58. Almomani, F. (2020). Prediction of biogas production from chemically treated co-digested agricultural waste using artificial neural network. *Fuel* 280: 118573. https://doi.org/10.1016/j.fuel.2020.118573.
59. Wang, L., Long, F., Liao, W., and Liu, H. (2020). Prediction of anaerobic digestion performance and identification of critical operational parameters using machine learning algorithms. *Bioresource Technology* 298: 122495. https://doi.org/10.1016/j.biortech.2019.122495.
60. Clercq, D.D., Wen, Z., Fei, F. et al. (2019). Interpretable machine learning for predicting biomethane production in industrial-scale anaerobic co-digestion. *Science of The Total Environment* 712: 134574. https://doi.org/10.1016/j.scitotenv.2019.134574.
61. Xu, W., Long, F., Zhao, H. et al. (2021). Performance prediction of ZVI-based anaerobic digestion reactor using machine learning algorithms. *Waste Management* 121: 59–66. https://doi.org/10.1016/j.wasman.2020.12.003.
62. Long, F., Wang, L., Cai, W. et al. (2021). Predicting the performance of anaerobic digestion using machine learning algorithms and genomic data. *Water Research* 199: 117182. https://doi.org/10.1016/j.watres.2021.117182.
63. Andrade Cruz, I., Chuenchart, W., Long, F. et al. (2022). Application of machine learning in anaerobic digestion: perspectives and challenges. *Bioresource Technology* 345: 126433. https://doi.org/10.1016/j.biortech.2021.126433.
64. Yi, J., Dong, B., Xue, Y. et al. (2014). Microbial community dynamics in batch high-solid anaerobic digestion of food waste under mesophilic conditions. *Journal of Microbiology and Biotechnology* 24 (2): 270–279. https://doi.org/10.4014/jmb.1306.06067.
65. Capson-Tojo, G., Trably, E., Rouez, M. et al. (2018). *Methanosarcina* plays a main role during methanogenesis of high-solids food waste and cardboard. *Waste Management* 76: 423–430. https://doi.org/10.1016/j.wasman.2018.04.004.
66. Fu, J., Ma, B., Xu, B. et al. (2020). Evaluation of the solid digestate from garage-type high solids anaerobic digestion of bundled rice straw and swine manure as a growth medium for seeding production. *Bioresources* 15: 3017–3028. https://doi.org/10.15376/biores.15.2.3017-3028.
67. Tsui, T.H., Zhang, L., Zhang, J. et al. (2022). Engineering interface between bioenergy recovery and biogas desulfurization: sustainability interplays of biochar application. *Renewable and Sustainable Energy Reviews* 157: 112053. https://doi.org/10.1016/j.rser.2021.112053.
68. Gao, W., Chen, Y., Zhan, L., and Bian, X. (2015). Engineering properties for high kitchen waste content municipal solid waste. *Journal of Rock Mechanics and Geotechnical Engineering* 7 (6): 646–658. https://doi.org/10.1016/j.jrmge.2015.08.007.
69. Chang, S., Zhang, Z., Cao, L. et al. (2020). Interaction and kinetics study of the co-gasification of high-solid anaerobic digestate and lignite. *Molecules* 25 (3): 459. https://doi.org/10.3390/molecules25030459.
70. Chowdhury, B., Magsi, S.B., Ting, H.N.J., and Dhar, B.R. (2020). High-solids anaerobic digestion followed by ultrasonication of digestate and wet-type anaerobic digestion for enhancing methane yield from OFMSW. *Processes* 8 (5): 555. https://doi.org/10.3390/pr8050555.

5

Biomethane – Production and Management

Wojciech Czekała, Aleksandra Łukomska, and Martyna Kulińska
Department of Biosystems Engineering, Poznań University of Life Sciences, ul. Wojska Polskiego 50, 60-627 Poznań, Poland

5.1 Introduction

The increase in the usage of biomethane has been observed in recent years, which, because of its properties and versatility, allows the decarbonization of many areas such as the electricity, gas, and transportation sectors. This fuel, which is also called renewable natural gas, is almost entirely composed of methane. Biomethane is produced from the processing of biomass through one of two processes. Biogas upgrading is first and the most popular method, accounting for 90% of its total production worldwide [1]. This method involves additional purification of the biogas produced from fermentation by the removal of carbon dioxide. This process is called upgrading and is closely related to biogas production [2].

Another method, however less commonly used, is biosyngas methanation. The gas is produced during the gasification of solid biomass, which, unlike to anaerobic digestion, is performed in the small presence of oxygen. Endothermic chemical reactions involving elemental carbon, carbon dioxide, carbon monoxide, hydrogen, water vapor, and methane are carried out [2]. The synthesized gas must be thoroughly purified to produce a pure biomethane stream. A catalyst stimulates a reaction between hydrogen and carbon monoxide or CO_2 [3]. The advantage of this method is that woody biomass can be converted to biomethane, which is not possible with anaerobic digestion due to the presence of lignin and cellulose.

The increased importance of biomethane production is linked to its ability to replace natural gas, a fossil fuel the combustion, which harms the environment. According to

Biogas Plants: Waste Management, Energy Production and Carbon Footprint Reduction,
First Edition. Edited by Wojciech Czekała

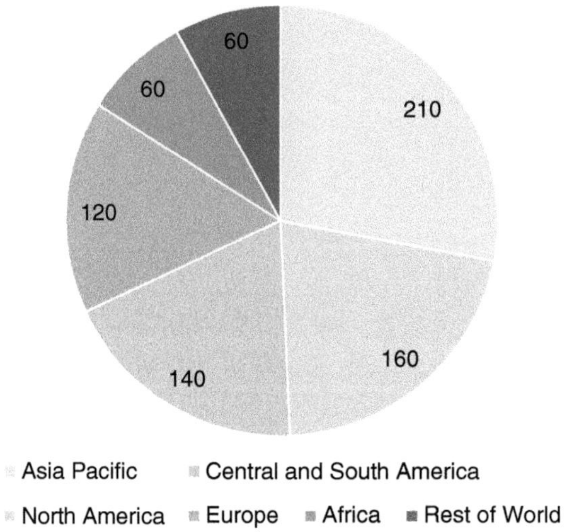

Figure 5.1 Production potential for biogas or biomethane by feedstock source (Mtoe) (own study based on IEA Report).

numerous announcements, natural gas is considered a transitional fuel, and within the subsequent few years, its consumption in various sectors is predicted to be reduced and eventually discontinued utterly. Due to its high methane content, biomethane can be used to produce heat and electricity. What is more it plays an essential role in the transportation industry. Similarly to biogas, biomethane is produced from a broad base of substrates, including waste. Thus, promoting a closed-loop economy involves reducing the waste production and using it better.

The development of the biomethane sector is uneven in different parts of the world, not only because of the availability of feedstock but also because of the policies of individual countries in this sector. Ninety percent of the world's biomethane production comes from Europe, China, and the United States [1]. In 2020, the total energy produced from Europe's biomethane was 32 TWh. It is an upward trend compared to previous years, and it is certain to continue. The most significant biomethane production – at 11 TWh – is in Germany where the energy market is mainly based on this fuel. Denmark is ranked second, with an annual production of more than 4 TWh, followed by countries such as France, the Netherlands, and Italy – more than 2 TWh [1]. Despite the differences in biomethane production, virtually every region has the potential for production (Figure 5.1).

It is assumed that the total production potential could be increased by 40% by 2040. Improved waste storage and management programs will expand the availability of waste and the ability to process it for energy. In addition, biogas and biomethane technologies are constantly being improved to boost the efficiency of the process and even allow for extensions of the feedstock base [1]. Since it is almost a clean source of methane, this gas can successfully replace high-carbon fossil fuels and thus be applied in the energy transition of high-carbon sectors such as electricity, gas, and transportation. Moreover, biomethane production, analogous to biogas, involves waste disposal from various industries and thus fits into a closed-loop economy. The aim of study to discuss the production process and the usage of biomethane, which, with the development of the biogas market, is gaining popularity worldwide.

Table 5.1 Composition of biogas from different generation sources.

Biogas components	Unit	Landfills	Industrial waste	Agriculture
Methane	%	50–80	50–70	50–80
Carbon dioxide	%	20–50	30–50	30–50
Oxygen	%	0–1	0–1	0–1
Nitrogen	%	0–3	0–1	0–1
Hydrogen disulfide	%	0.1	0.8	0.7

Source: Adapted from Chen et al. [5].

5.2 Purification and Usage of Biogas

Biogas, a high-calorie fuel for energy production, is produced from biomass, which can be either organic waste of various origins or specially grown vegetation. Depending on the substrate origin and chemical composition, the final composition of biogas can also vary. For example, agricultural biogas is characterized by a relatively high yield of methane and virtually trace amounts of oxygen (anaerobic digestion), while the opposite characteristic is for biogas from landfill degassing, where oxygen can occur as a result of partial contact between waste and air. Where the presence of hydrogen sulfide is noted, the highest values of this component can occur in biogas resulting from the anaerobic processing of food waste. To avoid this, some measures are used to bind hydrogen sulfide already at the fermentation stage [4]. Table 5.1 illustrates example values of biogas components depending on the origin of the substrates.

The produced biogas can be managed in numerous ways. Depending on the intended usage, these can include methods of direct implemention for energy purposes, which include combustion in thermal boilers, for example for heating purposes, and combustion in cogeneration engines. Cogeneration involves the simultaneous generation of two forms of energy: electricity and heat. In this system, a gas engine is connected to a generator, and exhaust gases are the heat source. When the exhaust gas comes into contact with water on the exchanger, the temperature of the water is raised, which can then be used for heating purposes.

Indirect methods include generating energy from biomethane, which is produced by purification and upgrading biogas. These methods consider transferring gaseous fuel through the network and converting it for energy at the destination. Thus, biomethane can be burned in thermal boilers or cogeneration engines at locations far from the generating unit, i.e. the biogas plant. Another method based on injecting purified biogas (with a composition similar to natural gas) into the network is further used for transportation purposes in the form of bioCNG. Additional compression of biomethane makes it possible to obtain another type of fuel, i.e. bioLNG, which can also be used, e.g. in wheeled transportation [6].

To obtain biomethane, several steps must be taken to remove impurities from the biogas (regardless of its origin). The composition of biogas, which is usually between 55% and 70% methane, contains compounds such as carbon dioxide, traces of water, hydrogen sulfide, ammonia, and even, in the case of some biogas plants, siloxanes [7]. The measures aforementioned include desulfurization, i.e. the removal of sulfur compounds, which have a very detrimental effect on gas processing equipment, leading to its corrosion [8, 9]. The subsequent step is so-called biogas upgrading, which involves separating carbon dioxide from methane, resulting in a gas with a composition similar to natural.

Different methods can be used to remove hydrogen sulfide depending on the H_2S content in the biogas and the requirements for an acceptable degree of purification. The choice of method can be conditioned primarily by the planned application and the size of the gas stream to be treated. In most agricultural biogas plants, a desulfurization method can be successfully applied based on microorganism actions. In situations where further gas purification is planned, such action can reduce the need for more expensive chemical methods.

5.2.1 Biological Desulfurization Within the Digester

Biological desulfurization relies on the ability of bacteria to oxidize sulfur compounds. The reaction occurs when air is supplied to the biogas tank, resulting in elemental sulfur and sulfite [10]. In this form, the separated contamination settles on the structural elements of the tank, while the gas can be collected and redirected to further stages of use. Biological oxidation activities are assumed to provide biogas purification efficiencies of 80–99% [11], and hydrogen sulfide concentrations after desulfurization can range from 20 to 100 ppm. Depending on the biogas purpose, biological desulfurization can be a preliminary treatment step for other methods. This method has many advantages, including low investment and operating costs, and due to its high efficiency, there is no need for additional chemicals within the tank. The installation mainly consists of a low-power air blower and tubes that supply air directly to the digester (Figure 5.2). The process is not complicated, so it can be carried out unattended [12].

5.2.2 Desulfurization by Adsorption on Iron Hydroxide

This type of desulfurization uses a reactor filled with a bed with adsorption properties (Figure 5.3). In this case, the sorbent is a highly porous granular material containing large amounts of iron hydroxide. The operation of such a desulfurization plant is based on the hydraulic distribution of biogas around the interior of the reactor and the simultaneous binding of hydrogen sulfide on the surface of the granules. Once the bed is saturated, further gas purification is impossible. To prolong the life of the bed, it is possible to subject it to regeneration, which involves supplying air (oxygen) to the reactor. The process of binding and regeneration of the bed follows the reactions:

$$2Fe(OH)_3 + 3H_2S \rightarrow Fe_2S_3 + 6H_2O$$

$$Fe_2S_3 + 1.5O_2 + 3H_2O \rightarrow 2Fe(OH)_3 + 3S$$

Regeneration can be carried out more frequently, which positively affects the life of the desulfurization material [13]. Moreover, when the oxygen content in the biogas is approximately 1–2%, the material regenerates itself.

5.2.3 Desulfurization by Adsorption on Activated Carbon

In addition to desulfurization plants using the adsorption properties of oxides and iron hydroxides, activated carbon desulfurization plants are considered more effective [14, 15]. The principle of operation is the same – the treated gas must pass through a filter column filled with a bed. When biogas comes into activated carbon, its adsorptive properties, i.e. its

Figure 5.2 Section of a biological desulfurization plant with a blower feeding air with oxygen to the digesters.

ability to absorb contaminants, are used. Reactors applied for hydrogen sulfide removal can be made either of plastic (HDPE material) or stainless steel. The shape of the reactor can be any; however, cylindrical vessels are used to maintain better contact between the bed and the gas (Figure 5.4).

When choosing activated carbon filters, it is crucial to remember that the moisture content of the injected biogas should be appropriate. Filter manufacturers stipulate that the injected gas should have a moisture content of 50–60%; otherwise, pores may become clogged, resulting in ineffective desulfurization. For this reason, at higher water contents in the biogas, preliminary drying is used, e.g. by lowering the temperature of the biogas on the heat exchanger and draining the condensate or raising the temperature by compressing the biogas with a blower (Figure 5.5).

5.3 Opportunities for Biogas Upgrading

5.3.1 CO_2 Separation Through Membranes

Membrane separation is based on gas solubility and diffusion, i.e. penetration, depending on the difference in concentration of its molecules through the separating material – the

Figure 5.3 Desulfurization plant consisting of two filters filled with iron hydroxide bed.

membrane. The filtration takes advantage of differences in the size of molecules of the gas components to be separated (Figure 5.6). Therefore, in the case of biogas, CO_2 penetrates membranes, while this is not possible for CH_4 molecules [17]. Gas transport through the filter walls is possible when different pressures, concentrations, or temperatures are on opposite sides. The most common membranes include polysulfone, polyimide, or polydimethylsiloxane [5].

To maintain the highest efficiency of the process, it is necessary to carry out preliminary purification of the biogas from hydrogen sulfide even before it enters the membranes. It is not advised to carry out the filtration process in the presence of H_2S in the gas mixture due to its harmfulness in contact with the membranes and a significant reduction in the service life of the membrane material [18]. It is necessary to perform the separation several times to achieve high process efficiency; nevertheless, it is possible to obtain high-quality methane – above 96% CH_4 (Figure 5.6) [19].

5.3.2 CO_2 Separation by Water Scrubbing

The separation process through water scrubbers takes advantage of the properties of gases, which relate to increasing solubility in water, depending on the pressure (Figure 5.7). The higher the pressure, the solubility of the gases separated in this case (methane and carbon

Figure 5.4 Filter made of HDPE plastic containing an activated carbon bed.

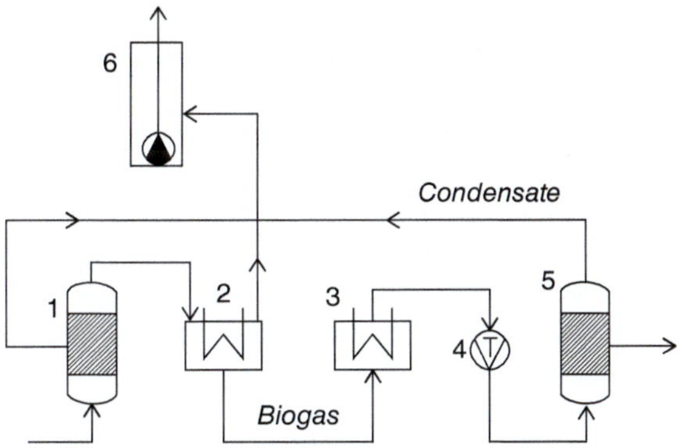

Figure 5.5 Schematic of biogas treatment plant including desulfurization plant and carbon filter. Legend: 1 – iron hydroxide filter, 2 – biogas dryer (heat exchanger), 3 – biogas heater (heat exchanger), 4 – compressor, 5 – activated carbon filter.

dioxide) increases. Hydrogen sulfide is also subject to this phenomenon, so using this method of biogas upgrading, there is no need for prior desulfurization (Figure 5.7) [16, 18].

The process uses adsorption columns in which compressed biogas at a pressure of approximately 10 bar flows from the bottom to the top of the column. At the same time,

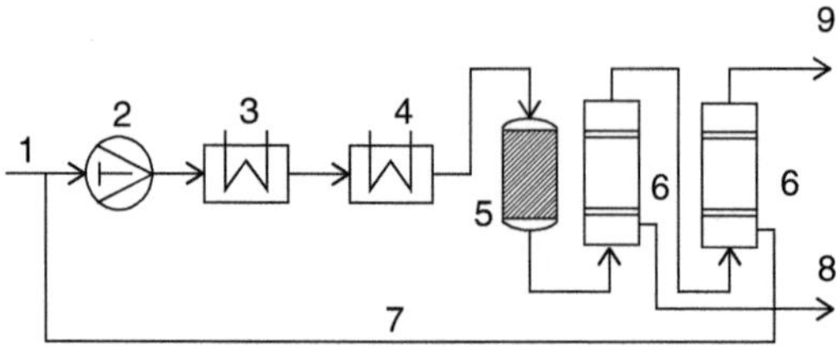

Figure 5.6 Membrane technology (own study based on [16]). Legend: 1 – raw biogas, 2 – compressor, 3 – biogas dryer (heat exchanger), 4 – biogas heater (heat exchanger), 5 – desulfurization filter, 6 – membrane columns, 7 – biogas recirculation, 8 – process gas containing CO_2, 9 – purified biomethane.

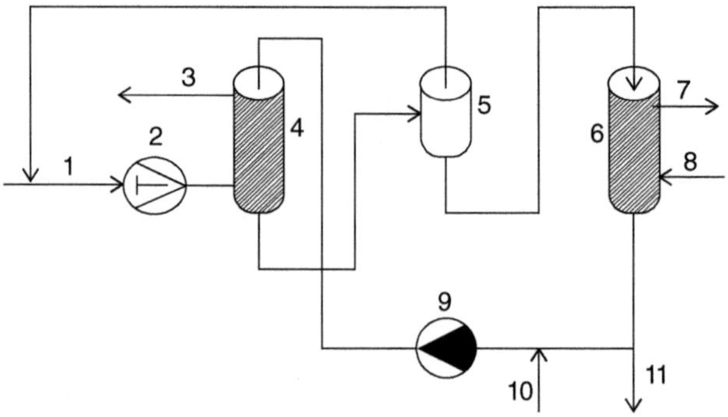

Figure 5.7 CO_2 separation by water scrubbers. Legend: 1 – raw biogas, 2 – compressor, 3 – treated biomethane, 4 – absorption column, 5 – storage tank, 6 – desorption column, 7 – CO_2-containing air, 8 – air, 9 – pump, 10 – post-process water, 11 – water bleed stream.

water flows through the reactor opposite to the gas, so contact between the agents occurs on the principle of countercurrent flow, resulting in the separated gases, namely CO_2 and H_2O, dissolving in the water. As a result, as the biogas flows through the water scrubber, the methane concentration increases and the concentration of carbon dioxide decreases. The efficiency of this process is relatively high, as the gas leaving the column contains up to 98% methane [20]. Some methane may also dissolve in the flowing water, so to avoid losses, the postprocess water is directed to a storage tank, where the released gas is directed back to the water scrubber and a new biogas stream. Water containing carbon dioxide and hydrogen sulfide can be reused by lowering the pressure in the desorption column and separating the gases [21, 22].

5.3.3 Chemical Separation of CO_2/Chemical Scrubbing

Like water scrubbers, chemical absorption also takes advantage of the higher solubility of carbon dioxide than methane. Scrubbers, in which chemical absorption occurs, ensure that the flowing biogas is in contact with the chemical agent. This method differs from a water scrubber in the pressure of the introduced biogas, which is approximately 1–2 bar, with a temperature between 20 and 65 °C, so compression is unnecessary [23]. As a result of biogas–liquid contact, CO_2 is absorbed, while methane remains within the flowing gas. However, removal of H_2S contained in the biogas does not occur, so pre-desulfurization

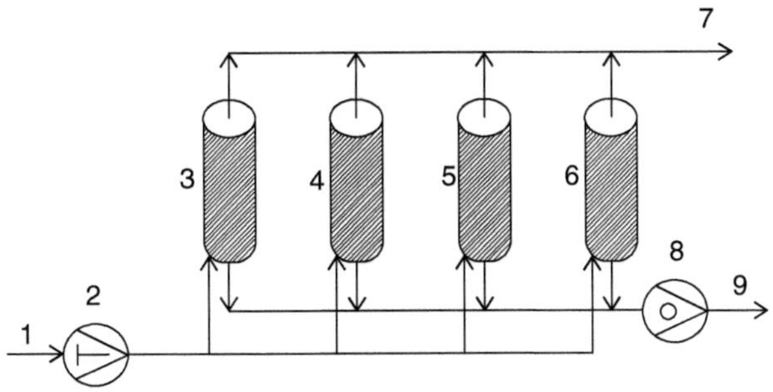

Figure 5.8 Pressure swing adsorption scheme (own study based on [18]). Legend: 1 – raw biogas, 2 – compressor, 3 – adsorption, 4 – pressure reduction (depressurize), 5 – desorption, 6 – pressure increase (pressurize), 7 – treated biogas, 8 – vacuum pump, 9 – waste gas (CO_2).

is recommended before biogas is fed to the scrubbers. After absorption, chemical solvent regeneration occurs in the desorption column by heating the stream to as high as 160 °C, releasing carbon dioxide [18]. The solvents in this method can be either alkali solutions, ammonia, or amines: diethanolamine, monoethanolamine, or methyl diethanolamine [5, 24]. With amines, it is possible to purify biogas from CO_2 very efficiently and obtain biomethane with CH_4 content even above 99% [23].

5.3.4 Pressure Separation of CO_2 (Pressure Swing Adsorption)

Pressure swing adsorption (PSA) is one of the most widespread practices for obtaining clean methane from biogas [25]. The process is based on the principle of adsorption, i.e. the binding of CO_2 molecules on the surface of reactor-filling material (Figure 5.8), such as activated carbon or zeolite molecular sieves, and since pressure also plays a vital role in this process, it is assumed that the higher the pressures the process is subjected to, the more gas molecules are bound [5]. Due to the perfect adsorption properties of the materials used, it is also possible to absorb H_2S and water vapor, but for removing primarily CO_2 other methods are recommended to pretreat biogas before using the PSA method. Otherwise, a particular deposit may deplete quickly (Figure 5.8) [17].

A PSA biogas upgrading plant is usually built with several reactors, called adsorption columns, the operation of which is divided into four phases and takes place in parallel. When the adsorption capacity of the bed in each column is fully utilized, pressure reduction occurs. This stage is gradual, and it allows both the regeneration of the bed and the release of carbon dioxide separated in the process. After regeneration, it is possible to carry out the process again, thus closing the cycle. It is possible that after the gas has passed through the cycle once, it is not thoroughly purified so that it can be returned to the system and reintroduced with raw biogas [20]. The most significant advantage of this method is the possibility of obtaining a gas with a very high methane concentration of 95–99%, which successfully meets the requirements for gas fed to the network [23].

5.3.5 Cryogenic CO_2 Separation

Cryogenic separation relies on the phenomena of liquefaction and sublimation of gases that are components of biogas (Figure 5.9). To obtain biomethane, high pressures (up to 80 bar)

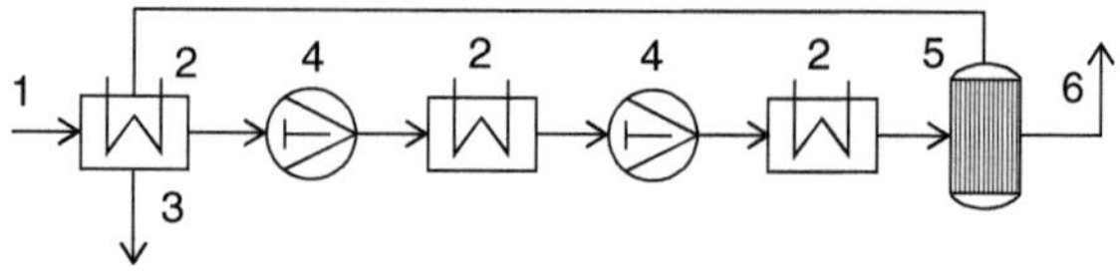

Figure 5.9 Cryogenic separation (own study based on [17]). Legend: 1 – raw biogas, 2 – biogas cooler, 3 – purified biogas, 4 – compressor, 5 – distillation column, 6 – waste stream.

and very low temperatures (−170 °C) are used in this method [26]. The entire process is carried out in stages to eliminate individual impurities gradually; e.g. when cooling biogas to a specific temperature, it is possible to liquefy carbon dioxide and separate it from methane, which is in gaseous form [5]. After cooling, the gas is then compressed to a very high pressure, which is done several times, to bring the biogas to the distillation column at the final stage, where the final purification of both carbon dioxide and hydrogen sulfide takes place. Such operations makes it possible to obtain up to 99% methane in the purified biogas (siloxanes are also removed during the process, along with water vapor). The high efficiency of cryogenic separation is very promising, but due to the complexity of the process and the amount of equipment needed, it turns out to be one of the most expensive methods for obtaining biomethane (Figure 5.9) [27].

5.4 Possibilities of Using Biomethane

Biomethane, unlike biogas, can be directly injected to the gas network, either as an addition to natural gas or as a replacement. In both cases, its implementation does not require modification of existing networks [28]. A biomethane plant can be connected to both the transmission and distribution networks. Due to the need of injecting biomethane at a higher pressure, transporting it through transmission networks is more expensive than through distribution networks [29]. Analogous to natural gas, biomethane can have a higher or lower methane content, directly affecting its calorific value. For biomethane to be fed into the network, it must have an appropriate methane content, as defined by law in each country. Due to the differences in legislation, there are significant disparities between the content of the injected gas in different countries. For example, a methane concentration of 85% in the Netherlands is sufficient, while countries such as Sweden require 97% [30]. There are also growing ideas that biomethane, if necessary, should be combined with other gaseous fuels such as propane, butane, or ethane [31]. As with natural gas, this could increase its heating value in some cases. Importantly, biomethane must be of high purity. The presence of impurities such as carbon dioxide, hydrogen sulfide, or water can cause the corrosion of gas pipelines [32]. To ensure that the injected gas does not adversely affect the gas infrastructure, it is required to check its parameters throughout the process, most often defined by the Network Operator [33]. Transmission of metering and dispatching data is crucial in the relationship between the biomethane plant and the distribution company. Information such as composition, instantaneous flow, pressure, and gas temperature in the metering system and at the outlet should be transmitted on an ongoing basis. Due to the symbiosis, any disruptions in operation or planned future investments for the operator and the biomethane plant should be mutually reported. Biogas that has been previously purified and upgraded is odorless. Following safety requirements, to detect dangerous concentrations in the air and ensure proper conditions for the use and distribution of the gas, odorization is used, i.e. the addition of a suitable odorizing agent with a characteristic scent. It is the final step in preparing biomethane before delivering it into the network [30].

5.4.1 Production of bioCNG and bioLNG Fuels

An alternative to the direct introduction of biomethane into the network is to use it in the production of transport biofuels. One method is to compress biomethane to a pressure of 200–250 bar using multistage compressors [34]. The fuel prepared in this way is called bioCNG. It has a relatively low density and is used in vehicles such as buses, passenger cars, and agricultural machinery. The second solution is to liquefy biomethane into bioLNG. This process occurs at temperatures below $-160\,^{\circ}\mathrm{C}$. Gas in this form has less pressure and is mainly used in long-haul vehicles and sometimes ships [35]. It can be transported over long distances using cryogenic tanks and regasified if necessary. The use of biomethane in the transportation sector is advantageous in vehicle engines that until now have been powered by natural gas, which do not require any modification and work just as well with both bioCNG and bioLNG. The same applies to the stations, where such vehicles are refueled. Replacing CNG and LNG with their renewable counterparts makes it possible to use existing infrastructure [36].

5.4.2 Production of Biohydrogen

Hydrogen is one of the essential elements, consisting of one proton and an electron. It has a high calorific value of $119.93\,\mathrm{MJ\,kg^{-1}}$ [37]. A source of energy is needed to produce hydrogen under laboratory or industrial conditions. Analogous to methane and biomethane production, hydrogen can be produced from renewable and nonrenewable sources. Because the produced compound in both cases has the same properties, color coding was introduced to identify its source. The gray color indicates hydrogen produced from fossil fuels, with high emissivity and environmental impact. The one with a blue marking was also created from nonrenewable sources, but the emissions of the entire process were reduced by using CO_2 capture equipment. The smallest, zero-emission class is marked in green, called biohydrogen. Currently, gray hydrogen is the most important, which is related to the fact that the most popular method, accounting for 48% of global production, is steam reforming of natural gas. Slightly less significant is petroleum/paraffin oil reforming, which accounts for 30% of global production and coal gasification at 18%. Water electrolysis is generally not used on an industrial scale, so only 4% of the world's hydrogen is produced by this method [38]. With the energy transition, there is an increasing demand for green hydrogen produced from technologies tempered solely by renewable energy sources [39]. There are many methods to produce green hydrogen, but in this chapter, it is worth mentioning the possibility of using biomethane as a feedstock for steam reforming. In this process, desulfurized and upgraded biogas is subjected to react with steam, separating it into a synthesis gas composed of hydrogen and carbon monoxide. The endothermic nature of the reaction requires an external heat input [40].

$$CH_4 + H_2O \rightarrow 3H_2 + CO$$

Steam reforming is the most economical of all the methods for producing green hydrogen. Compared to the gasification of wood residues or pyroreforming of glycerin, hydrogen obtained from biomethane has the highest energy content [41]. Due to their similar properties and potential for use, biomethane and biohydrogen have generally been considered competitive. Because of the enormous challenge of the energy transition, demand for both fuels has increased. It means that the development of one sector should not be an obstacle to the other – quite the opposite. Reducing the cost of biomethane production, if steam reforming is used, positively impacts the price of hydrogen.

5.5 Profitability of Biomethane Production and Recommended Support Systems

Due to its versatility, biomethane can be used in decarbonizing not one but several of the most emission-intensive areas. Eliminating natural gas in the transportation, power, or gas sectors would significantly reduce emissions of harmful gases and improve the profitability of individual areas [42]. A biomethane plant based on waste substrates also has a utilization role. When animal materials such as manure are used, not only their energy potential is fully utilized, but also uncontrolled methane emissions are prevented, which is essential for reducing the environmental impact of agriculture. Developing the biomethane market is not only an environmental benefit but also an economic one. For its production, which is analogous to biogas, industrial waste is increasingly used. Usually, such residues are given to specialized companies, which charge significant fees for their disposal. If it is a biomethane plant, and not a disposal plant, that processes the waste, it gains an additional source of income, thus raising the profitability of the entire investment.

Ensuring energy security is the cornerstone of any country's energy policy. It is defined as the energy system's resilience to exceptional and unpredictable events that, if they occur, could disrupt its functioning, either by limiting access to energy or by sudden increases in energy prices. A stable energy system should be characterized not only by a strong diversification of sources but also by the most significant possible independence from imported fuels, the prices of which are increasingly becoming a part of the political game. Natural gas production is closely linked to the availability of its deposits, which, because of geological conditions, are mainly distributed in unstable regions, with which relations are often difficult to predict. Domestic production of biomethane makes it possible to become independent of exported fuel and thus increases the country's energy security. Rising natural gas prices, while the cost of biomethane production is falling [43], are causing increasingly more countries to include green gas in their energy balance.

Despite favorable conditions for developing the biomethane market, there are still barriers that slow down such development or even halt it. A significant problem in many countries is the lack of appropriate legislation. Delivering the biomethane into the network must comply with the legal and technical regulations in each country. Developed countries with greater environmental awareness and with complete safety standards make provisions in such a way as to facilitate this. Unfortunately, there are also countries, where the process is deliberately blocked for various reasons. The most common practice is to raise technical requirements for gas parameters while failing to upgrade existing pipelines. An additional difficulty can also be insufficient absorption capacity of the network, preventing biomethane

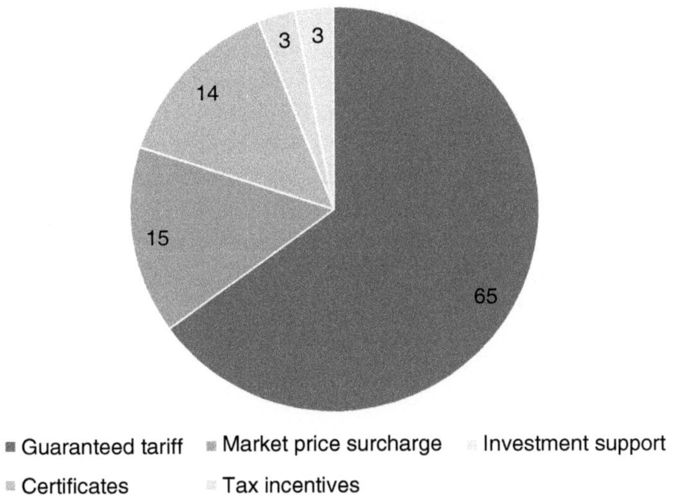

Figure 5.10 Percentage of biomethane support systems in Europe (own study based on IEA Report).

from being introduced into [32]. While this is of little importance in producing bioCNG and bioLNG, it is crucial in producing biomethane for gas purposes. Analogous to other renewable sources, the development of this sector involves new investments, the creation of which requires financial outlays and stable market conditions. Setting the framework for energy and climate policy, it is necessary to build appropriate support systems that would result in the development of the biomethane sector. Countries that have properly selected such regulations can boast of their significant production [44]. The most common are feed-in tariffs (FIT) and feed-in premiums (FIP), while tax incentives and investment support are the least popular (Figure 5.10).

Support systems become obsolete after some time and do not fulfill their intended role. In such cases, it is recommended to reform them, considering the actual needs and capabilities of the state. It should be born in mind that the biomethane sector has been undervalued and underfunded in many places for years, making it crucial to support its development in the context of increasing biomethane production.

5.6 Conclusion

A different composition characterizes biogas produced from biomass of different origins in terms of methane content and impurities. Depending on the installation where it is extracted, it is crucial to choose the appropriate technology for the purification and upgrading of biogas. The less methane the biogas contains, the more complicated the process of bringing it to a form of gas similar to natural gas, i.e. containing nearly 97–99% of methane. Each successive step in the process leading to the extraction of biomethane from biogas affects the complexity of the plant needed for this; therefore, it is also reflected in investment and operating costs.

Proper preparation of biomethane givens possibilities for long-distance transmission and usage for fuel or transportation purposes. Each country in the world has different conditions regarding the quality of biomethane that can be transmitted through the gas network, so the facilities used for this, or their complexity, may differ across national borders. Biomethane used in transportation comes in the form of bioCNG and bioLNG, an excellent alternative to the nonrenewable fuels that continue to play the first fiddle in the transportation industry worldwide. Another option for producing fuels from biomethane is obtaining green hydrogen, which, obtained by steam reforming, can bring many environmental and economic benefits.

To make the gas networks green, legislative barriers must first be removed, especially in terms of technical requirements, and the absorptive capacity of the networks must be increased through upgrades. Additionally, in countries where support systems for biomethane producers do not currently exist, they should be introduced as soon as possible.

References

1. Report International Energy Agency (2020). Outlook for biogas and biomethane: prospects for organic growth. https://iea.blob.core.windows.net/assets/03aeb10c-c38c-4d10-bcec-de92e9ab815f/Outlook_for_biogas_and_biomethane.pdf
2. Ardolino, F., Cardamone, G.F., Parrillo, F., and Arena, U. (2021). Biogas-to-biomethane upgrading: a comparative review and assessment in a life cycle perspective. *Renewable and Sustainable Energy Reviews* 139: https://doi.org/10.1016/j.rser.2020.110588.
3. Schiaroli, N., Battisti, M., Benito, P. et al. (2022). Catalytic upgrading of clean biogas to synthesis gas. *Catalysts* 12 (2): https://doi.org/10.3390/catal12020109.
4. Li, Y., Alaimo, C.P., Kim, M. et al. (2019). Composition and toxicity of biogas produced from different feedstocks in California. *Environmental Science & Technology* 53 (19): 11569–11579. https://doi.org/10.1021/acs.est.9b03003.
5. Chen, X.Y., Vinh-Thang, H., Ramirez, A.A. et al. (2015). Membrane gas separation technologies for biogas upgrading. *RSC Advances* 5 (31): 24399–24448. https://doi.org/10.1039/C5RA00666J.
6. Dzene, I., Romagnoli, F., Seile, G., and Blumberga, D. (2014). Comparison of different biogas use pathways for Latvia: biogas use in CHP vs. biogas upgrading. *The 9th International Conference "Environmental Engineering,"* Vilnius, Lithuania (22–23 May 2014). https://doi.org/10.3846/enviro.2014.017.
7. Ardolino, F., Parrillo, F., and Arena, U. (2018). Biowaste-to-biomethane or biowaste-to-energy? An LCA study on anaerobic digestion of organic waste. *Journal of Cleaner Production* 174: 462–476. https://doi.org/10.1016/j.jclepro.2017.10.320.
8. Budzianowski, W.M. (2016). A review of potential innovations for production, conditioning and utilization of biogas with multiple-criteria assessment. *Renewable and Sustainable Energy Reviews* 54: 1148–1171. https://doi.org/10.1016/j.rser.2015.10.054.
9. Xiao, C., Ma, Y., Ji, D., and Zang, L. (2017). Review of desulfurization process for biogas purification. *IOP Conference Series: Earth and Environmental Science* 100 (1): https://doi.org/10.1088/1755-1315/100/1/012177.
10. Miltner, M., Makaruk, A., and Harasek, M. (2017). Review on available biogas upgrading technologies and innovations toward advanced solutions. *Journal of Cleaner Production* 161: 1329–1337.
11. Ramos I., Fdz-Polanco M. 2014. Microaerobic control of biogas sulphide content during sewage sludge digestion by using biogas production and hydrogen sulphide concentration. *Chemical Engineering Journal* 250, 303–311. https://doi.org/10.1016/j.cej.2014.04.027.
12. Petersson, A. (2013). Biogas cleaning. *The Biogas Handbook* (eds. A. Wellinger, J. D. Murphy, and D. Baxter), 329–341. Woodhead Publishing. https://doi.org/10.1533/9780857097415.3.329.

13. Hernández, S.P., Scarpa, F., Fino, D., and Conti, R. (2011). Biogas purification for MCFC application. *International Journal of Hydrogen Energy* 36 (13): 8112–8118. https://doi.org/10.1016/j.ijhydene.2011.01.055.
14. De Arespacochaga, N., Valderrama, C., Mesa, C. et al. (2014). Biogas deep clean-up based on adsorption technologies for solid oxide fuel cell applications. *Chemical Engineering Journal* 255: 593–603. https://doi.org/10.1016/j.cej.2014.06.072.
15. Micoli L., Bagnasco G., Turco M. 2014. H_2S removal from biogas for fuelling MCFCs: new adsorbing materials. *International Journal of Hydrogen Energy* 39(4), 1783–1787. https://doi.org/10.1016/j.ijhydene.2013.10.126.
16. Mroczkowski, P. and Seiffert, M. (2011). Oczyszczanie i zatłaczanie biogazu na przykładzie Niemiec. Możliwości wdrożenia technologii w Polsce. EC BREC Instytut Energii Odnawialnej, Niemieckie Centrum Biomasy DBFZ. https://www.cire.pl/pliki/2/Mroczkowski_Seiffert_oczyszczanie_biomethane.pdf
17. Biernat, K., Gis, W., and Samson-Bręk, I. (2012). Review of technology for cleaning biogas to natural gas quality. *Combustion Engines* 51: 33–39.
18. Ryckebosch, E., Drouillon, M., and Vervaeren, H. (2011). Techniques for transformation of biogas to biomethane. *Biomass and Bioenergy* 35 (5): 1633–1645. https://doi.org/10.1016/j.biombioe.2011.02.033.
19. Huertas, J.I., Giraldo, N., and Izquierdo, S. (2011). Removal of H_2S and CO_2 from biogas by amine absorption. In: *Mass Transfer in Chemical Engineering Processes* (ed. J. Markoš), 133–135. IntechOpen https://doi.org/10.5772/20039.
20. Scholwin, F. (2010). Present state and development of technologies for treatment of biogas to the natural gas grade. *19 Annual Meeting of Fachverband Biogas e.V. Conference Proceedings*. https://www.cire.pl/pliki/2/Mroczkowski_Seiffert_oczyszczanie_biometanu.pdf
21. Rotunno, P., Lanzini, A., and Leone, P. (2017). Energy and economic analysis of a water scrubbing based biogas upgrading process for biomethane injection into the gas grid or use as transportation fuel. *Renewable Energy* 102: 417–432. https://doi.org/10.1016/j.renene.2016.10.062.
22. Tynell, A. (2007). Microbial growth on pall-rings: a problem when upgrading biogas with the technique absorption with water wash. *Applied Biochemistry and Biotechnology* 141 (2): 299–319. https://doi.org/10.1007/BF02729069.
23. Prussi, M., Padella, M., Conton, M. et al. (2019). Review of technologies for biomethane production and assessment of Eu transport share in 2030. *Journal of Cleaner Production* 222: 565–572. https://doi.org/10.1016/j.jclepro.2019.02.271.
24. Maile, O.I., Muzenda, E., and Tesfagiorgis, H. (2017). Chemical absorption of carbon dioxide in biogas purification. *Procedia Manufacturing* 7: 639–646. https://doi.org/10.1016/j.promfg.2016.12.095.
25. Zhao, Q., Leonhardt, E., MacConnell, C., Frear, C., and Chen, S. (2010). Purification technologies for biogas generated by anaerobic digestion. *Compressed Biomethane, CSANR, Ed.* p. 24.
26. Sahota, S., Shah, G., Ghosh, P. et al. (2018). Review of trends in biogas upgradation technologies and future perspectives. *Bioresource Technology* 1: 79–88. https://doi.org/10.1016/j.biteb.2018.01.002.
27. Wellinger, A. and Lindberg, A. (2000). Biogas upgrading and utilization. Task 24: Energy from biological conversion of organic waste, IEA Bioenergy.
28. Quintino, F.M., Nascimento, N., and Fernandes, E.C. (2021). Aspects of hydrogen and biomethane introduction in natural gas infrastructure and equipment. *Hydrogen* 301–318. https://doi.org/10.3390/hydrogen2030016.
29. Barczyński, A. (2021). Inserting biogas (biomethane) into the gas network – chances and dangers. *Wiadomości Naftowe i Gazownicze* 3 (268): 4–11. http://yadda.icm.edu.pl/yadda/element/bwmeta1.element.baztech-1e9db596-7d23-47bd-9608-616e8a7ecacd.
30. Savickis, J., Zemite, L., Zeltins, N., and Bode, I. (2020). The biomethane injection into the natural gas networks: the EU's gas synergy path. *Latvian Journal of Physics and Technical Sciences* 57: 34–50. https://doi.org/10.2478/lpts-2020-0020.

31. Wojtowicz, R. (2017). Co-firing of mixtures agricultural biogas with LNG or LPG as an alternative to injection of biogas to the grid. *Transactions of the Institute of Fluid-Flow Machinery* 137: 123–129.
32. Nevzorova, T. and Kutcherov, V. (2019). Barriers to the wider implementation of biogas as a source of energy: a state-of-the-art review. *Energy Strategy Reviews* 26: https://doi.org/10.1016/j.esr.2019.100414.
33. Report ACER (2020). ACER report on NRAs survey – hydrogen, biomethane, and related network adaptations. https://www.acer.europa.eu/Official_documents/Acts_of_the_Agency/Publication/ACER%20Report%20on%20NRAs%20Survey.%20Hydrogen%2C%20Biomethane%2C%20and%20Related%20Network%20Adaptations.docx.pdf
34. Yadav, K. and Sircar, A. (2021). Fundamentals and developments of compressed biogas in city gas distribution network in India: a review. *Petroleum Research* https://doi.org/10.1016/j.ptlrs.2021.12.003.
35. Report European Biogas Association (2020). BioLNG in transport: making climate neutrality a reality. https://www.europeanbiogas.eu/wp-content/uploads/2020/11/BioLNG-in-Transport_Making-Climate-Neutrality-a-Reality.pdf
36. Prussi, M., Julea, A., Lonza, L., and Thiel, C. (2021). Biomethane as alternative fuel for the EU road sector: analysis of existing and planned infrastructure. *Energy Strategy Reviews* 33 (17): https://doi.org/10.1016/j.esr.2020.100612.
37. Farias, C.B.B., Barreiros, R.C.S., da Silva, M.F. et al. (2022). Use of hydrogen as fuel: a trend of the 21st century. *Energies* 15 (1): https://doi.org/10.3390/en15010311.
38. Yusaf, T., Laimon, M., Alrefae, W. et al. (2022). Hydrogen energy demand growth prediction and assessment (2021–2050) using a system thinking and system dynamics approach. *Applied Sciences* 12 (2): https://doi.org/10.3390/app12020781.
39. Eljack, F. and Kazi, M. (2021). Prospects and challenges of green hydrogen economy via multi-sector global symbiosis in Qatar. *Frontiers in Sustainability* https://doi.org/10.3389/frsus.2020.612762.
40. Antonini, C., Treyer, K., Streb, A. et al. (2020). Hydrogen production from natural gas and biomethane with carbon capture and storage – a techno-environmental analysis. *Sustainable Energy & Fuels* 4: 2967–2986. https://doi.org/10.1039/D0SE00222D.
41. Berdechowski, K. (2019). Analysis of biohydrogen production methods in terms of GHG emission value. *Nafta-Gaz* 4: 230–235. https://doi.org/10.18668/NG.2019.04.05.
42. Pääkkönen, A., Aro, K., Aalto, P. et al. (2019). The potential of biomethane in replacing fossil fuels in heavy transport—a case study on Finland. *Sustainability* 11 (17): https://doi.org/10.3390/su11174750.
43. European Biogas Association (2022). A way out of the EU gas price crisis with biomethane. https://www.europeanbiogas.eu/a-way-out-of-the-eu-gas-price-crisis-with-biomethane (accessed 30 March 2022).
44. Repele, M., Udrene, L., and Bazbauers, G. (2017). Support mechanisms for biomethane production and supply. *Energy Procedia* 113: 304–310. https://doi.org/10.1016/j.egypro.2017.04.070.

6

The Biogas Use

Muhammad U. Khan[1], Abid Sarwar[2], Nalok Dutta[3], and Muhammad Arslan[1]

[1] *Department of Energy Systems Engineering, University of Agriculture, Faisalabad, Pakistan*
[2] *Department of Irrigation and Drainage, University of Agriculture, Faisalabad, Pakistan*
[3] *Bioproducts Sciences and Engineering Laboratory, Washington State University, USA*

6.1 Introduction

Energy is considered as a vital parameter for economic development, social development, human wellbeing, and raising living standards. Worldwide energy demands are rising because of industrialization and population growth. There is an increased reliance on conventional fuels, which are fossil fuels, because increased energy consumption is causing environmental problems. Concerns about the depletion of fossil fuels, greenhouse gas (GHG) emissions, and energy security have also sparked research into renewable energy sources. In the areas of solar, wind, tidal, geothermal, and biomass feedstock, there are numerous promising renewable energy sources; however, their adoption is constrained by cost and lack of technological expertise. One of the difficulties in building such energy systems is the choice and application of developing fuels and technologies. Biogas is a fuel that can be produced from a number of biodegradable materials, such as straw, animal waste, solid municipal trash, and biomass feedstocks. Livestock manure is the most prevalent source of feedstock in agricultural countries, followed by energy crops like beets and corn [1].

The switch to lower carbon emissions depends on biomass. One of the most promising biomass sources is the production of biogas from waste, which has the potential to replace conventional fuels and reduce our reliance on them. The production of biogas contributes to waste management and the development of a sustainable ecosystem because it is an energy

Biogas Plants: Waste Management, Energy Production and Carbon Footprint Reduction,
First Edition. Edited by Wojciech Czekała

source made from organic feedstock and biodegradable materials. Combustible gas created by the anaerobic digestion (AD) of organic wastes is known as biogas. Complex biological and physiochemical processes that are involved in the creation of biogas are regulated by a number of different variables, including substrate type, pH, temperature, and others. The primary byproducts of AD are biogas and waste slurry. Biogas and slurry are the main byproducts of AD. Biogas is made up of biomethane (CH_4) and carbon dioxide (CO_2), along with minor amounts of moisture, hydrogen sulfide (H_2S), and hydrogen (H_2) gas. A renewable energy source can be used in a variety of applications.

A 20% contribution of energy from sustainable sources was the target set by the EU Renewable Energy Directive (2009/28/EC) until 2020 [2]. Governmental policies, commercial operations, academic research, and industry studies are now focused on finding alternative, environmentally friendly, and energy-efficient technologies. Biogas is a rapidly growing option for producing an alternative energy source due to its relatively low startup and operating costs and the use of a variety of organic biomass residues as input raw materials (such as manures, agricultural residues, and municipal solid waste) that would otherwise be landfilled and added value to emissions of GHGs and leachate into the water table [3]. Furthermore, due to its greater energy output/input ratio, AD is regarded as the most efficient renewable energy technology when compared to other thermochemical and biological systems [2].

AD of biomass, such as municipal solid waste, agricultural wastes, sewage sludge, and energy crops, can yield biogas, a sustainable energy resource, which raises several concerns. About 50–60% of biogas is made up of methane (CH_4), 40–50% of carbon dioxide (CO_2), and some insignificant amounts of water and hydrogen sulfide (H_2S). Biogas might replace fossil fuels in the generation of energy, lowering emissions of GHGs and reduced reliance on imported energy [3]. Biomethane produced by AD of organic waste can be utilized for a variety of applications, namely energy generation, transportation, including heating, and injection into the natural gas infrastructure after the impurities have been removed [4]. Biogas energy use is predictable to twice in 10 years, from 14.5 GW in 2012 to 29.5 GW [5].

The question of what the best use for biogas should be raised when biogas collection and usage technique have advanced through time and biogas energy recovery has developed into one of the more established and productive wastes/residues-to-energy systems. Several factors, including the volume of gas produced, the cost of energy, how much energy the plant uses, and other incentives, influence this. Biogas is frequently used as a boiler fuel to supply the heat required to maintain an anaerobic digester at the proper temperature.

6.2 Biogas Utilization Technologies

Most of the industries need both heat and electricity to compensate the energy demand. The combined creation of productive heat and electricity generation from the combustion of a single fuel is known as cogeneration. Both the industrialized and tertiary sectors have used this method of producing combined heat and electricity. It has mostly been utilized due to its high level of general efficiency and the assurance of electricity with minimal environmental hazard [6]. When compared to other renewable energy sources, biogas has a wider range of applications. Biogas has traditionally used as energy source in boilers to generate heat or in combined power and heat (CHP) generation plants to cogenerate heat and electricity.

Electricity is created in CHP plants by burning the fuel (biogas or natural gas), which is then captured by a heat regaining unit from the burning system's from exhaust stream. This heat can be transformed into usable thermal energy, which is commonly steam or hot water. A diesel engine or four-stroke engine is typically used in these CHP systems. The organic Rankine cycle (ORC), the Cheng cycle, the steam piston engine, the steam turbine, and/or the steam screw engine are all instances of biogas being utilized in a boiler to harvest steam for turbines or driving motors. Stirling engines are another option for using biogas as a fuel.

Biogas is an energy-producing source and is a time-tested technique in several European countries. As a CHP system harvests dual-energy yields (heat and electricity) from a single process, they are more cost-effective and efficient. One of the most significant pros of using biogas in CHP is that the biogas does not need to be upgraded and can be used in the system immediately after generation. Since the use of biogas to CHP is on a smaller scale than that of biogas as a transportation fuel, the infrastructure required is less sophisticated. Discovery of a market for the heat can be problematic because there is no way of district heating, and even if district heating was commonly used, there may not be a nearby market available for an agricultural digester. The thermal energy could be used to heat pig housing units or to dry wood chips. Although finding a market for the electricity may be quite easier, the contract between the digester operator and the Electricity Supply Board Network in Ireland is signed. Another requirement is a price for connecting to the grid; this charge is figured out by the cost of improving the grid in the digestion proximity, such as grid transformers. The price is calculated depending on the cost of elevation of the grid in the digester's vicinity, such as grid transformers. Biogas is becoming widely used as a transportation fuel all over the world. Biogas could also be used in natural gas vehicles (NGVs) as fuel if the biogas is upgraded to natural gas purity. When biogas is utilized as a transportation fuel, it must be upgraded since high pollutant levels can damage a car engine and compromise the integrity of the national gas infrastructure if it is not decreased to an appropriate level.

The removal of contaminants from biogas is compulsory before its usage for different applications because these contaminants can deteriorate different equipment. Hydrogen sulfide (H_2S) in raw biogas has the potential to damage metal components in internal combustion (IC) engines, boilers, and gas pipes [7]. When biogas is utilized as energy source in IC engines, silicon compounds formed on the engine's walls can cause the exhaust pipe and catalytic converter to deteriorate. The presence of halogenated hydrocarbons and ammonia in biogas alters its ignition attributes and can result in corrosion in CHP engines and gas pipelines after burning [8]. Before connecting biogas to the natural gas grid, carbon dioxide should be eliminated because higher CO_2 concentrations diminish the heating value and Wobbe index of the gas. The principal impurities of biogas that are important for its usage in vehicle engines, CHP engines, boilers, and the natural gas grid are carbon dioxide, hydrogen sulfide, and water vapors, and they should all be eliminated before using the biogas [9]. Table 6.1 shows the requirements for treating biogas impurities for the utilization in different areas.

6.3 Use of Biogas as Trigeneration

Trigeneration is the phenomena in which biogas is used for heat, power, and cooling purposes. However, in this phenomenon, the energy is used in an efficient way as it is the

Table 6.1 Requirements for treating biogas impurities [4, 10].

	Boiler	Vehicle fuel	Natural gas grid	Stationary engine	Kitchen stove
H_2S (ppm)	<250	<5	<4	<1000	<10
CO_2 (vol%)	Removal not required [11–16]	Recommended (<4)	Removal required (≤3)	Removal not required [11–16]	Removal not required [11–16]
H_2S	Removal not needed	Removal needed	Removal needed	Removal not needed	Removal not needed

extended form of cogeneration. This technique helped to reduce the greenhouse emission and global warming [17].

6.4 Biogas as a Transportation Fuels

Biogas is composed of 55–60% methane, 30–40% of CO_2, and trace amounts of other gases. In biogas, methane is the major combustible gas, while carbon dioxide is an undesired gas that is incombustible. Water scrubbing technology, pressurized swing adsorption technology (PSA), cryogenic upgradation technology, and other biogas scrubbing technologies can be used to remove this undesirable CO_2 and other gases that are present in traces. After high-pressure compression at 200 bar in a cylinder, upgraded biogas containing 90% methane can be used as a source of transportation fuel. Bio-compressed natural gas (CNG) is a high-pressure methane gas that is created by an AD of organic wastes [4].

Using biomethane as a vehicle fuel is an efficient way to integrate it into the transportation industry. This application of biogas as an automobile fuel is enhancing all over the world, particularly in European countries such as Sweden and Switzerland. About 6500 vehicles are running on upgraded biogas in these two countries [18]. The use of biomethane as a vehicle fuel is simple to implement in most countries. Several rising countries such as Brazil, India, Pakistan, and Iran are already using CNG as a vehicular from the last two decades. Besides the countries, it is also used as a fuel in several European countries such as Germany, Austria, and Italy. These countries were using almost 27.4 million natural gas automobiles on the road worldwide in 2018, with Asia-Pacific leading the way with 19.8 million [19].

Vehicles may easily be converted to run on natural gas in dual-fuel mode. Commercially, equipment is available for converting petrol engines to natural gas. This equipment can also be utilized for improving the usage of biogas. As the conversion technique and kit for using upgraded biogas are identical to those for using natural gas, it can easily be replaced in current natural gas cars. Both light- and heavy-duty cars may run on biomethane. Light-duty vehicles can run on natural gas and biomethane with less modification, whereas heavy-duty vehicles may also need some major tweaking. Many national policies, procedures, and standards are developed in a number of countries, including Netherlands, Germany, Switzerland, and India, when using advanced biogas as a car fuel or injection into the natural gas system [19].

According to Tippayawong et al. [20], biogas has a great potential as a transportation vehicle fuel and can be used as a viable alternative to natural gas. Biogas is one of the renewable energy sources evaluated for supplying the required heat load in greenhouses. The results of the experiments demonstrate that it may be used as a heat source in a satisfactory manner. Impurities such as hydrogen sulfide (H_2S) and carbon dioxide (CO_2) are present in raw biogas, and its high heating value (HHV) level is lower than that of natural gas. For sweetening raw biogas, various techniques are proposed.

When biogas is utilized as a transportation fuel, it must meet a number of requirements in terms of heating value and contaminants, such as methane concentration of at least 97%. Upgrading is the term used to describe the process of purifying biogas and eliminating CO_2. The scale of production has a significant impact on the cost of refining biogas to transportation fuel. As a result, the $1000\,m^3\,h^{-1}$ biogas production scale chosen is economical based on an economic analysis. With a methane concentration of 55% in the biogas, this corresponds to a biogas production of $172\,TJ\,yr^{-1}$ (lower heating value [LHV]), the size of the biogas production plant utilized in the evaluation for all crops. Water scrubbers, chemical scrubbers, membranes, and PSA are some of the available upgrading technologies [21].

The type of feedstock, production location, climatic conditions, and technology used determine the composition of the biogas produced. The amount of methane in biogas ranges between 50% and 75%. Carbon dioxide (CO_2) constitutes 25–50% of the biogas, and hydrogen sulfide (H_2S) levels range from 100 to 10,000 ppm. The presence of these contaminants in biogas has a negative impact on engine performance. Biogas quality is improved by lowering CO_2 and H_2S levels. When biogas is upgraded to biomethane in a biogas treatment facility with around 98% methane, the biomethane has the same qualities as natural gas [22].

Transport sector in all over the world used the most of the fuel and also approximately shared 14% of the anthropogenic emission globally [23]. Diesel, petrol, and natural gas are almost the leading fuel used. Biogas can also be used as fuel by replacing the natural gas, which is renewable and sustainable. Moreover, different analyses show that the use of biogas as a fuel is having the low impact over the use of natural gas [24]. Compressed biomethane is having the same properties as CNG. To use biogas as fuel, the impurities removed from that to make the biomethane more than 97% pure and compressed at pressure level of 20–25 MPa.

Biomethane may also convert into liquefied petroleum gas (LPG) at 0.5–15 MPa. Methanol is also produced from biomethane by partial oxidation. Furthermore, biogas is converted into CNG, LPG, and syngas by Fischer–Tropsch process [25].

6.5 Use of Biogas in Reciprocating Engine

For power generation, reciprocating IC engines are commonly utilized. For a long time, biogas is mostly used as a fuel in IC engines. Biogas is utilized to power various applications such as fans and pumps, blowers, and air conditioners with shaft or electric power. IC engines come in a variety of sizes, ranging from several kW to MW. Engine size affects power generation, with efficiency ranging between 40% and 30% for large and small engines, respectively. Thermal efficiency ranges from 45% to 60%, resultant in overall efficacies of up to 90% [26]. In general, these engines need more than 30% by mol of CH_4 in the fuel to attain optimum combustion; however, a concentration of CH_4 as low as 21% by mol can be effectively accommodated. The acceptable H_2S limit in biogas for these engines is 500 ppm [27].

However, condensing water vapor is recommended to eluding condensation in pipes and the development of acids. Siloxanes must also be removed because of their coarse nature; thus, they must be removed. Since upgrading biogas and thus removing CO_2 are not necessary for heat and power production, raw biogas can be utilized in the engines. The biogas-based engine de-rated because of the less volumetric energy concentration of biogas when matched with natural gas and diesel. The engine needs regular oil changes and a significant overhaul for every five years, in addition to expensive running and maintenance expenditures. SI engines (spark ignition) and dual-fueled (gas and liquid fueled spark and stirling engines, compression ignition engines (CI), and micro-gas turbines) biogas IC engines are available [19].

In Japan, reciprocating engine is mainly used for gasification of the wood and other residues to fulfill the elevation in the energy demand gradually. In Japan, a gasification using a reciprocating engine is an efficient way to utilize these wasted trees and woods. The two most common methods for using biomass as a fuel to create electricity are direct combustion and gasification. Direct combustion is accomplished by the use of a steam turbine. The gasification process uses a reciprocating engines and a gas turbine [28]. A general-purpose reciprocating engine typically uses gaseous or liquid fuels [29]. The gasification process is studied in 2012 to explore using thermal decomposition to compute the production element ratio. According to the report,

(1) A liquid fuel with 75% of the energy input should be produced by fast thermal decomposition, and
(2) The liquid fuel contains excessively water to run a reciprocating engine.

On the other hand, in the scenario of gaseous fuel, the energy exchange ratio (cold gas efficiency) from wood to gas fuel might be close to 85%. Then, a gasifier and reciprocating engine having SI is superior to a power plant.

In order to reduce environmental pollution, power plants typically have regulatory restrictions on their emissions. The primary substances that regulated include CO_2, NO_x, HC, soot, and CO. Although wood would absorb CO_2 through growth, the CO_2 emission can be disregard by using wood chips. In order to run a reciprocating engine, Nadaleti and Przybyla [30] used bio-syngas (synthetic gas from wood). As a response, they noted that the exhaust included significantly more CO than when natural gas and fermentation gas used. When using a mixture of H_2, CO, CO_2, and CH_4 as fuel in a reciprocating engine, Arroyo et al. [11] found that the CO emission was higher than when using CH_4 or fermentation gas. The resultant cause is not made explicit in these studies. Since CO is flammable, this must be burned completely. Furthermore, due to CO toxicity, CO emission is severely constrained legally. As bio-syngas has a LHV than fossil fuels like natural gas or gasoline, such as bio-syngas, the ignition temperature in the chamber must be lower. NO_x density increases with higher LHV when methane or hydrogen is added. Because it is more economical, a straightforward reduction technique using urea will be utilized to remove NO_x. So the methane proportion in bio-syngas is lower than 2.0 vol%, and hydrogen addition is less than 20 ppm by vol. When Shah et al. [12] operated a reciprocating engine with a bio-syngas of 5.8 MJ Nm^{-3} derived from wood biomass, they found that the HC ratio in the exhale is much less than 40 ppm. The CO ratio in bio-syngas, on the other hand, is greater than 17.0 vol%. This means that "without perfect combustion," the CO in the fuel should be expelled (unburned). Modern car reciprocating engines have little gaps in the combustion chamber. Lowest emissions and greatest thermal efficiency are realized by these small gaps. A general-purpose reciprocating engine should, however, be inexpensive, simple to maintain, and long-lasting, and the royal regulation is not as strict as that for vehicle emissions. The gap present there could form the unburned gases that remain. The low heating value modified at the same excess air ratio in the reciprocating engine is powered by biogas, a gasifier that produces gas-using wood, to lessen the amount of incomplete combustion in the emissions (almost 1.3). To alter the LHV, hydrogen and a substitute for methane (city gas 13A) are introduced. Higher LHV case exhibits lower CO ratio in the exhaust as a result. Because methane's LHV is almost four times greater than that of hydrogen's, using less methane could result in a lower CO ratio in the output than in the hydrogen instance.

6.6 Spark Ignition Gas Engine

SI engines are the most prevalent biogas-powered engines. SI engines are easy to operate and maintain, with low to medium initial investment expenses. Biogas-based dual fuel or 100% biogas engines are available commercially for power generation, on medium to large scales (100 kW to MW). These engines have an electrical efficacy of up to 18–43%, depending on their size. Although these have a high tolerance for contaminants, they nevertheless require the removal of H_2S (below 100 ppm), siloxanes, and moisture [27]. The engine's exhaust is extremely hot, reaching temperatures of up to 650 °C. The leftover heat from the engine's exhaust gas can be effectively recuperated using a heat exchanger or water jacket. High contaminant discharges in the environment are a key drawback of these engines when related to fuel cells or gas turbines. Biogas can be combined with a small amount of biodiesel, bioethanol, or bio-dimethyl ether to create a co-fired fuel. Because the electrical efficiency is up to 43%, it is injected with plant or diesel oil [26]. Biogas requires a 20% diesel fuel augmentation for ignition due to its poor cetane rating and absence of ignition spark [19]. Gaseous fuels, in general, produce very little pollution and work effectively in both SI and CI engines. Wide ignition limits are seen in gaseous fuels, which mix readily with air to produce combustion efficiency. Lean combinations may also be used here. Additionally, gaseous biofuels have high hydrogen to carbon concentrations, which will result in minimal pollutants made up primarily of carbon [13]. Yoon and Lee [14] studied that by appropriately modifying the required engine, biogas may be utilized in IC engines. Due to its high octane rating, which boosts thermal performance, biogas is appropriate for engines with significant compression ratios. As because biogas has less carbon than traditional diesel fuel, it produces less pollutants [15]. According to Refs. [16, 31], the most crucial aspect of using biogas in CI engines is the lack of power de-rating, unlike with SI engines. The reality is due to the fact that SI engines exhibit substantial cycle-to-cycle variability due to their sensitivity to biogas concentration [31]. The CI engine operates through dual-fuel mode and utilizes biogas. Due to the high CO_2 component in biogas, it exhibits poor power density. Additionally, it requires less air per mass for burning. For biogas to work well, the air fuel ratio must be tightly controlled due to the biogas's low flammability restrictions. Biogas has a high self-ignition temperature due to its high CO_2 concentrations, and it prevents knocking, which is important in SI engines. Due to its high octane rating, biogas has a significant anti-knock index. Additionally, it has a minor amount of H_2S, which can corrode metal components like those in burners or engines [13]. Porpatham et al. [32] investigated that how a steady speed SI engine's efficiency, emission, and ignition are affected by a biogas's reduced CO_2 concentration. The tests run at a steady speed of 1500 rpm and a compression ratio of 13 : 1 for equivalent ratio in the range from rich to lean operational circumstances. Lean blends resulted in lower CO_2 and hydrocarbon emissions. The test findings demonstrated that the boost in thermal efficiency is caused by an increase in burning rates. 5%, 10%, and 15% hydrogen amounts were used in trials on a biogas-fueled SI engine on an adjusted basis. They discovered that by including hydrogen, biogas could minimize hydrocarbon emissions while increasing power and thermal efficiency. In order to prevent knock, the ignition time had to be delayed when the hydrogen level was above 15%. Lean mixture combustion also experienced a decrease in cycle-to-cycle fluctuations. Nagalingam et al. [33] proposed that the injection of hydrogen raised the lean limits of natural gas burning but reduced the energy due to the gas's poor volumetric thermal efficiency. The stated thermal efficiency decreased on H_2 addition because of a drop in the braking power to resistance power ratio and a rise in the fuel's calorific value. When operating with 100% hydrogen mixtures and natural gas, the ideal spark advancement could drop as low as 20°bTDC. Operating with pure hydrogen is observed to result in higher NO_x emissions standards. The findings of running a SI engine with two synthetic gases produced by catalytic biogas breakdown were compared in the comparison study [11]. Three distinct equivalency ratios and a diverse variety of velocities were used in the experimental experiments. It is observed that the synthetic gases' H_2 content enhanced the peak combustion pressures. When compared with other fuels, CO emissions increased due to CO and its amount contained in synthetic gases, while HC emissions reduced since only a small portion of CH_4 is not completely burned. The increased flame temperature caused by the syngas' high hydrogen content increased NO_x emissions.

The use of biogas for electricity generation is growing throughout many nations, and it has long-standing applications in combustion for lighting and cooking. A perfect energy source for stand-alone generating biogas electricity generation can be used for both off- and on-grid generation. According to studies, biogas electricity plants have a payback period of between five and eight years. Other research on the production of electricity from biogas in underdeveloped African nations revealed that small-scale grid-connected biogas power plants are not commercially viable.

6.7 Use of Biogas in Generator

In particular for grid power where rigorous frequency control is necessary, generators are frequently synchronous devices [34]. From a thermodynamic perspective, biogas fuel may be utilized as a fuel in any combustion engine to turn an electric generator that is linked to the engine [35]. Biogas may be used as a fuel in the primary movers of several engines, including internal and external combustion engines. The engines can function as solely biogas-fueled dedicated biogas engines, dual-fuel engines employing conventional fuels as pilot fuels, or solely biogas-fueled dedicated biogas engines [36]. Varied biogas or biomethane using engines will have different performance characteristics. Consequently, different engines require various adaptations in order to use biogas as fuel [37].

6.8 Use of Biogas in Gas Turbines

Large energy plants often employ biogas-powered turbines with a capacity of 3–5 MW or more. Small biogas power plants range in size from 30 to 75 kW [38], which are the most prevalent sizes on the market. Relatively inexpensive requirements and better efficiency, particularly in larger size plants, are two operational advantages of gas turbines [39]. Even though they are tiny, micro-turbines can be linked together to produce the desired sizes based on demand. Gas turbines are very tolerable and can run on a variety of fuels, even unprocessed biogas [40]. The difficulty with gas turbines is that their high rotational speeds and high operating temperatures necessitate the use of costly technical materials [41, 42].

6.9 Usage of Biogas in Fuel Cell

As long as the fuel utilized is renewable and sustainable, fuel cells provide a green approach for producing electricity [43]. Direct biogas to electricity conversion in fuel cells requires extremely clean biogas, but prices are still expensive at the moment, despite constant research and development to lower them [44]. Herein, fuel cell utilized the biogas, as hydrogen from that reacts with the atmospheric oxygen to generate the electricity. Although the process occurred in the fuel cell is electrochemical, this is used to produce the clean energy [45–48].

6.10 Hydrogen Production from Biogas

The use of the biogas is still under discussion in the fuel cell. Ultimately hydrogen used in the fuel is also an attractive technology [49]. Hydrogen used in the fuel cell used the biomethane as a feedstock, while biomethane is produced from the biogas. Hydrogen has a potential to replace the coal and other fuels due to its high calorific value. Hydrogen is produced from the biogas by autothermal reforming process (ATR), steam methane reforming (SMR), and electrolysis [49, 50].

6.11 Biogas Cleaning for its Utilization

The biogas consists of different other gases in traces. The major part of the biogas consists of biomethane that is combustible, CO_2, water, H_2S, oxygen, nitrogen, ammonia, and volatile compounds that also have siloxanes and other particles. It is necessary to use the biogas to clean before its use. Petersson [51] removed the impurities present in the bio-gas using different techniques.

6.11.1 Carbon Dioxide

Carbon dioxide is the primary component of biogas after biomethane, excluding methane. It develops during the breakdown of different kinds of substrates used to produce biogas. The conversion of substrates into biogas is a multi-step, multi-microbial process that takes several different phases. The various processes result in the formation of carbon dioxide, which the methanogenic bacteria is used as an electrophile. The biogas's volumetric energy content will decline due to the carbon dioxide. The carbon dioxide might be viewed as a contaminant that needs to be eliminated if high volumetric energy content is crucial (for instance, when the gas is utilized as a fuel in vehicles or fed into the gas grid). Other applications, such as the production of heat and power, are typically unaffected. However, carbon dioxide and condensed water will combine to make carbonic acid.

6.11.2 Water

Since water exists in AD at all times, part of it will vaporize in the digester and end up in the biogas that created. As a result, water is always present in the biogas as it exits the digester. How much water may be present in biogas before it becomes water-saturated determines how much water is in the gas. In biogas, the water concentration that corresponds to saturation depends on the temperature and pressure inside the digester. Problems can arise downstream from the digester because of water in raw biogas. For instance, the creation of carbonic acid can lead to pipeline corrosion when other molecules like carbon dioxide are present. Water also decreases the energy content of the gas, which can have a detrimental impact on how well the gas is used for energy. Water is produced when biogas is burned, even when the raw biogas is purified before use. Water may condense and cause issues in downstream heat exchangers and exhaust systems based on the temperature and pressure after the combustion stage.

6.11.3 Hydrogen Sulfide

Hydrogen sulfide is another typical contaminant found in landfill and biogas emissions. Raw biogas may also contain other sulfur-containing contaminants, but hydrogen sulfide is the most typical. Bacteria that have the capacity to convert sulfate present in the digester to hydrogen sulfide can produce this gas. These bacteria exist in the digester and compete with bacteria that produce methane for the same substrate, but they produce hydrogen sulfide instead of biogas. Stillage, macro-algae, and various substrates from the paper industry are examples of sulfate-rich substrates. When sulfur-containing proteins, including cysteine and methionine, are digested, hydrogen sulfide is also created during the digestive process. Hydrogen sulfide may also come from gypsum in landfill gas. Hydrogen sulfide, which is present when biogas is used, can cause corrosion because it reacts with water to generate sulfuric acid. When biogas-carrying hydrogen sulfide is burned, sulfuric acid is produced during combustion, which results in emissions. It is also critical to remember that hydrogen sulfide is extremely poisonous and poses major health hazards.

6.11.4 Oxygen and Nitrogen

Since biogas is produced anaerobically, oxygen and nitrogen are often absent, but they can be detected if air is permitted (or manages) to penetrate the system from somewhere. Nitrogen in the raw biogas may indicate leakages into the digester because oxygen will be used if it is present. In an effort to purify the biogas, air is occasionally injected to the process by reducing hydrogen sulfide. Because of the fact that the harvesting of landfill gas could result in a reduced pressure in the landfill, that will cause air to be pulled in, nitrogen is more frequently present in landfill gas; however, there may also be trace amounts of oxygen. Because oxygen can cause combustible mixes to form with the methane in biogas, the amount of oxygen must be carefully regulated.

6.11.5 Ammonia

A common contaminant in raw biogas is ammonia, which is produced in the digester when materials with proteins, including animal abattoir waste, are hydrolyzed. High quantities of ammonia in the digester may prevent the synthesis of methane.

6.11.6 Volatile Organic Compounds

Various types of organic chemicals known as volatile organic compounds can be detected in biogas in varying amounts. The substrate was used to produce biogas effects of the types of chemicals and their concentrations, and alkanes, siloxanes, and halogenated hydrocarbons are some examples. Siloxanes are substances that are utilized in items like fire retardants, and both siloxanes deodorants and shampoos are organic compounds made up of units of R_2SiO, where R belongs to the specific class, Si is silicon, and O is oxygen. As certain siloxanes evaporate, if they are included in the substrate going into the digester, they will also be present in trace levels in the biogas that released. As a result, siloxanes are present in biogas made from sewage sludge. How much water evaporates into the biogas will depend on the temperature within the digester. Siloxanes with a low molecular weight will vaporize more quickly with others. Due to siloxanes evaporating in the landfill, siloxanes can also be discovered in landfill gas. Engine corrosion can result from the combustion of siloxanes, which produces siloxane oxide. The formed siloxane oxide is resistant toward combustion and will also produce undesirable depositions in the equipment. Hydrocarbon molecules with halogens are known as halogenated hydrocarbons such as bromine, fluorine, or chlorine. Owing to the volatilization of halogen-containing materials in landfills, for instance, they may be found in raw biogas. When hydrogenated hydrocarbons are burned, acids are produced that can lead to corrosion and acidification (e.g. hydrochloric acid).

6.11.7 Particles

Raw biogas frequently contains particles. Particulates frequently act as the centers on which water droplets condense. Particles' abrasive characteristics might lead to equipment degradation. The most significant problem in using biogas as a fuel is the necessity of cleaning the gas to guarantee that gas fulfills quality criteria for equipment. Biogas scrubbing is a multistage, capital-intensive process that can also be expensive to maintain due to media alternates and/or power expenditures. If the pollutants in the gas are not removed, they might raise the maintenance costs of gas-fueled equipment and shorten the life of the equipment as well as increase the environmental risk. As a result, successful gas use necessitates biogas scrubbing to prevent condensation, reduce the level of H_2S, and remove siloxanes.

6.11.8 Foams and Solid Particles

Filters and/or cyclones are often used in all biogas facilities to remove particles from the biogas. Solid particles and oil-like components in biogas are filtered out using standard

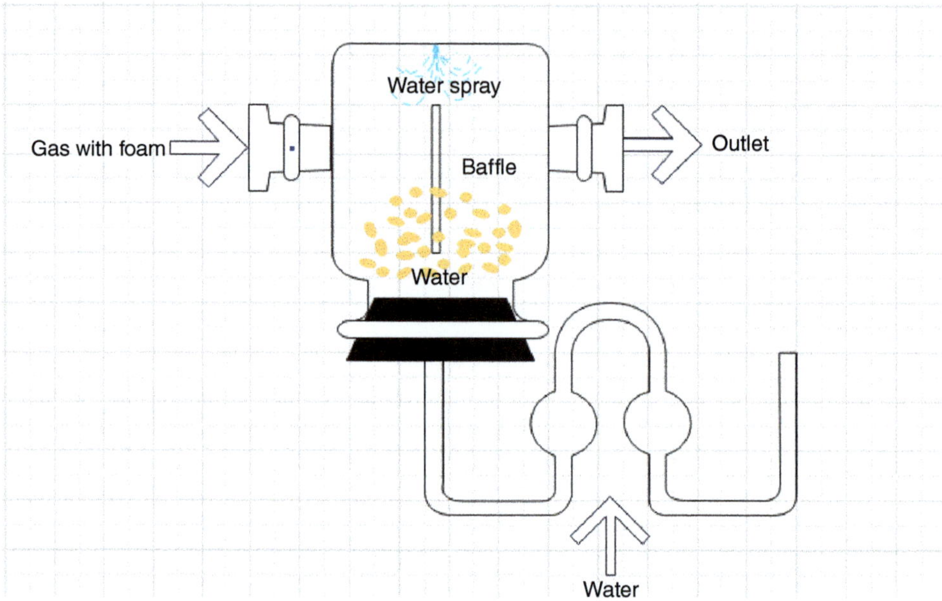

Figure 6.1 Schematic of cleaning of biogas.

dust collectors, while sludge and foam are separated using cyclones. Filters having a mesh size of 2–5 μm are typically considered suitable for most downstream applications [52]. A foam extractor in the digester gas piping separates any foam and sediments entrained in gas stream. The foam separator is a big tank with a baffle wall running along the center. Water nozzles are installed on the vessel's roof to give a continual spray wash. The gas, which packed with froth and particles, enters the vessel at the top. The gas goes down the spray wash under the baffle wall and then backs up to the discharge nozzle via a second spray wash. The gas that emerges from the foam separator will be substantially free of foam and debris. The cross section of a foam separator is shown in Figure 6.1 [53].

6.12 Different Approaches for H_2S Removal

The presence of H_2S in the biogas is not a good approach. The presence of H_2S also increases the environmental impact. H_2S act as the corrosive in the fuel pipelines and engines metallic parts. The incomplete burning of H_2S results in the formation of SO_2 that produces environmental hazards. Scrubbing H_2S from biogas can be done in a number of ways. The use of chemical scrubbers or iron sponge and the addition of ferric (Fe^{3+}) salts to the input are the most prevalent approaches.

6.12.1 Iron Sponge

A permeable bed of iron sponge (hydrated ferric oxide) filters the digester gas. The hydrated ferric oxide medium employed is iron-dipped wood chips that have been soaked in water.

An exothermic reaction occurs as H_2S passes through the iron medium, converting H_2S to ferric sulfide (black solid) and water:

$$Fe_2O_3 \cdot +H_2O + 3H_2S \rightarrow Fe_2S_3 + 4H_2O \quad (6.1)$$

The iron sponge can be revived and was used repeatedly before being replaced. To remove sulfur from the iron and reconstruct the hydrated ferric hydroxide, an exothermic reaction called iron sponge regeneration employs water and air. The iron chips can be overheated and catch fire if they are not placed in a bath of moving water and adequately managed. If the vessel has not been completely cleared of digester gas, this could be much riskier. Typically, wood chips and ferric sulfide are used as the exhaust media. This can be disposed of at any landfill because it is not a toxic waste.

6.12.2 Proprietary Scrubber Systems

The H_2S-laden digester gas is routed over a bed of media in proprietary scrubber systems, which selectively interacts with the H_2S in the gas. The proprietary methods often employ a free-flow of granular media that does not harden like iron sponge media but cannot be regenerated. Multiple vessels are settled in a lead/lag configuration in these systems. The biogas containing sulfides initially flows through the leading vessel, where the majority of the sulfide interacts with the media. The gas next passes through the lag vessel, which removes nearly all of the H_2S from the biogas. When the lead vessel's media is depleted, the vessel is removed from the line and the media is switched. The lag vessel takes over as the primary scrubber while the lead vessel is offline. The original bowl is returned to service as a backup vessel after the media is replaced.

6.12.3 Ferric Chloride Injection

Unlike the other approaches, ferric chloride injection does not eliminate H_2S from the digestive gas stream, but it does reduce the quantity of H_2S created in the digesters. Injecting ferric chloride into a sludge stream causes H_2S to oxidize and create insoluble ferrous sulfide which is removed out of solution and removed during dewatering:

$$3H_2S + 2FeCl_3 \rightarrow S + 2FeS + 6HCl \quad (6.2)$$

$$Fe^{2+} + S^{2-} \rightarrow FeS \quad (6.3)$$

$$2Fe^{3+} + 3S^{2-} \rightarrow 2FeS + S \quad (6.4)$$

Using ferric chloride has two additional advantages. The odors created by the presence of H_2S will be decreased at the digesters because the ferric salts applied interact with the H_2S before it is delivered into the gas stream. Second, struvite (magnesium ammonium phosphate, $MgNH_4PO_4 \cdot 6H_2O$) development on process pump impellers, mixers, and heat exchangers downstream of anaerobic digesters has been demonstrated to be reduced by ferric chloride. Herein, also NaOH and $Fe(OH)_3$ are used in the term of physical method for the removal of H_2S.

6.12.4 Biological Method

If air or oxygen is added to the digestion, hydrogen sulfide will oxidize biologically with oxygen to produce elementary sulfur. Sulfate dioxide may also develop in small amounts. *Thiobacillus* bacteria, which are often contained in the digester, catalyze the oxidation. For these bacteria to thrive on, a physical structure typically is built in the digester's mouth. This method has the disadvantage that if excessive oxygen is introduced, then the digestion mechanism may be adversely affected. Additionally, caution must be exercised to prevent the formation of explosive methane and oxygen mixes.

6.13 Different Approaches for Moisture Reduction

Moisture plays a key factor in the formation of acid with other substances like H_2S, NH_3, and CO_2 that cause rusting in bulk compressors, gas storage tanks, motors, and pipelines. Moisture production of condensation that damages instruments and freezing of stored water at low temperatures and high pressures result in rust, corrosion, lubrication wipes, and pipe clogs and absorption/accumulation of other pollutants. Moisture removal is usually performed by reducing flow velocity by creating a broad spot in the gas collection line. The gas flow ratio via the pipe should not surpass $3.7\ m\,s^{-1}$ ($12\ ft\,s^{-1}$) to successfully remove the moisture. Sediment and moisture can be settled out of the flow at this lower velocity and can be eliminated by installing drip taps at all low points throughout the gas gathering system. It is advised that the gas piping should be sloped at least 1% toward the collecting site ($10.4\ mm\,m^{-1}$ [$0.125\ in\,ft^{-1}$]) (WEF Manual of Practice 81,998). Using a refrigerant-type drier, further reduction in the moisture can be obtained by cooling the gas to roughly 4 °C (40 °F). To prevent H_2S corrosion, the dryer is usually composed of stainless steel or other corrosion-resistant materials. By eliminating H_2S from the gas prior to drying, corrosion from condensing acid can be reduced. The three basic techniques are used in the following sections.

6.13.1 Compression or Condensation

The biogas pipeline can be cooled by laying it in the underground and outfitting it with flowing water or condensed traps. Cyclones and demisters are additional technical solutions for compressed water cooling.

6.13.2 Adsorption

Aluminum oxide, magnesium oxide, silica, zeolites, and activated carbon are among materials that can be used for adsorption. Normal methods of regeneration involve raising the temperature and/or lowering the pressure. Two columns often operate in parallel, with one trapping while the other regenerates.

6.13.3 Absorption

Glycol solutions, such as ethylene glycol, tri-ethylene glycol, and diethylene glycol, can be absorbed and then heated to reproduce them, or hygroscopic salts can be used instead. Whenever salt gets wet with moisture, it dissolves in and typically not reproduced.

6.14 Siloxane Removal

Siloxane presence mainly affects the efficiency of the fuel and reduces the life time of equipment in which biogas is used as a fuel. SiO_2 formed when siloxane is burned in the fuel; as a result, deposition occurred at the different parts of the engines that affect the engine efficiency and accumulation of SiO_2 also affect the fuel efficiency. Graphite molecular sieve scrubbers and low-temperature drying systems are the two most frequent forms of siloxane removal systems.

6.14.1 Gas Drying

A sizeable portion of the siloxanes can be removed along with the moisture when the biogas is dried because siloxanes are highly heavy and have a tendency to adhere to water vapor in the biogas stream. The digester gas is chilled to 23 °C (73.4 °F) in low-temperature gas drying systems utilizing a refrigerant-type dryer, which causes the siloxanes to separate out with the condensed moisture. According to dryer manufacturers, low-temperature drying eliminates 90–95% of siloxanes. The disadvantage of employing gas drying systems at temperatures of 23 °C or lower, despite their relatively straightforward design, is that they tend to accumulate a lot of ice. The refrigeration system may be damaged if the ice is not removed.

As a result, if the quantities of H_2S and other volatile organic compounds (VOCs) in the biogas are not lowered before entering the siloxane scrubbers, they will be adsorbed on to the media, necessitating more frequent media replacement and greater running expenses. Activated carbon or graphite media scrubbers, when used in conjunction with a gas dryer, are a cost-effective way to remove siloxanes from digester gas. With a gas dryer, chilling the gas to 4 °C (39.2 °F) would remove some of the water and about 30–40% of the siloxanes. Scrubbers can be used to remove the leftover siloxanes.

A complete digester of biogas treatment module is shown in Figure 6.2. It consists of an H_2S scrubber to lower H_2S concentration levels in the gas upstream of the siloxane

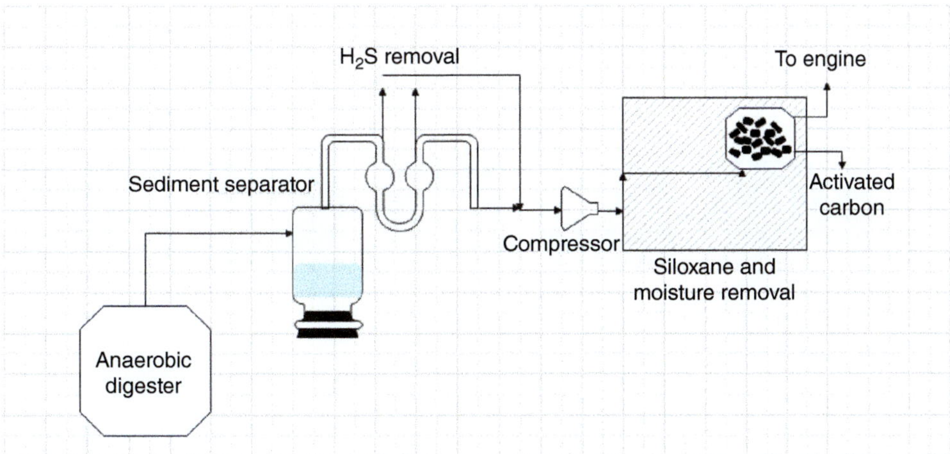

Figure 6.2 Schematic of biogas digester module.

scrubber. While gas dryer is to lower the temperature of the gas to 4 °C (39.2 °F), a foam splitter is to remove any absorbed foam and sediment from the gas, and a siloxane scrubber is to remove contaminants, heavy organics, and about 30–40% of the siloxanes. Depending on the gas pressures obtained from the digesters or storage equipment, a compressor may be required to generate the pressure required to transfer the digester gas through the dryer and the siloxane scrubber. Different gas cleaning standards may be the main tasks on the use and the levels of different pollutants in the gas.

6.15 CO_2 Separation

Removal of CO_2 is very necessary to eliminate from the biogas. As in the biogas, the composition of carbon dioxide is almost 25–30% when produced. It is necessary to extract the CO_2 portion from the biogas for its valuable utilization. Removal of CO_2 is very necessary as it decreases the calorific value of the fuel and decreases the fuel efficiency.

6.15.1 Cryogenic Technique

Different techniques have been utilized for the removal of CO_2, and cryogenic technique is one of the most used techniques. Elevated pressure (40 bar) and very low temperatures (approximately −100 °C) are the conditions in this process. The CO_2 is concentrated or sublimated from the raw biogas at a temperature where it can be segregated from biogas in a solid or liquid portion, while the methane builds up in the gaseous state. Process involved in the cryogenic is shown in Figure 6.3. In cryogenic process, in order to remove several unwanted chemicals from the biogas, including water vapor and siloxanes, and to maximize energy recovery, cooling is typically done in a series of phases. Its first heat exchanger cools down the raw biogas stream up to −70 °C as it comes through it. After that, this compressed gas is moved through the 40 bar pressure where it undergoes the process under different compressors and heat exchangers. After that, CH_4 is separated from the impurities such as H_2S and CO_2.

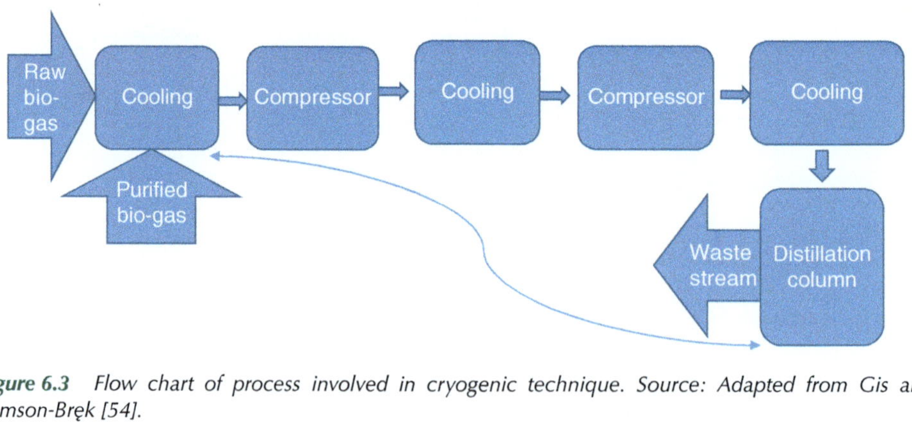

Figure 6.3 Flow chart of process involved in cryogenic technique. Source: Adapted from Cis and Samson-Bręk [54].

6.15.2 Water Scrubber

Water scrubber is one of the very finest techniques for the removal of CO_2 from the biogas. The increase in the solubility of the CO_2 in the water increases the feasibility of this technique. Different studies have shown that the solubility of CO_2 in the water at 25 °C is almost 26% higher than the solubility of CH_4. This method of purification could yield biomethane with a CH_4 content of 95–99% (v/v) [55]. One of the most popular techniques for biogas purification is water scrubbing [56]. Herein, freshwater is also delivered from the top of the columns at the same instant as biogas is pressurized (6–10 bar) from the lower side and sent to the absorber column. In order for the absorber column to function better efficiently, it is loaded with unorganized packaging material [55]. In order to reduce methane leakage, the saturated water is moved to the flash vessel, in which the pressure is reduced to about 3 bar. The desorption column receives the water after it exits the flash vessel. As the operating power for CO_2 elution is increased by lowering its partial pressure, air is added to the desorption container at 200 $m^3\ h^{-1}$ of water at 8 bar of approximate pressure at 20 °C and must be circulated by a 1000 $Nm^3\ h^{-1}$ raw biogas water scrubber for the improvement in the system [57, 58]. Because H_2S dissolves more readily in water than CO_2, the water scrubber has the advantage of removing H_2S along with CO_2. It is advised to use freshwater because the quality of water can quickly deteriorate if H_2S and CO_2 are removed simultaneously. A few benefits of the upgrading system are its affordability, great efficiency, lack of need for chemical reagents, and ability to collect methane at levels around 97% [10]. The schematic of water scrubber and steps also involved in the cleaning of biogas is shown in Figure 6.4.

6.15.3 Adsorption

The adsorbent of air molecules to a solid substrate is a different technique for separating CO_2 and CH_4. Adsorptive solid surfaces are porous materials with specified surface areas that are used in this technique. This method is also known as pressure swing adsorption (PSA). This method makes it possible to separate biogas in a certain manner using various

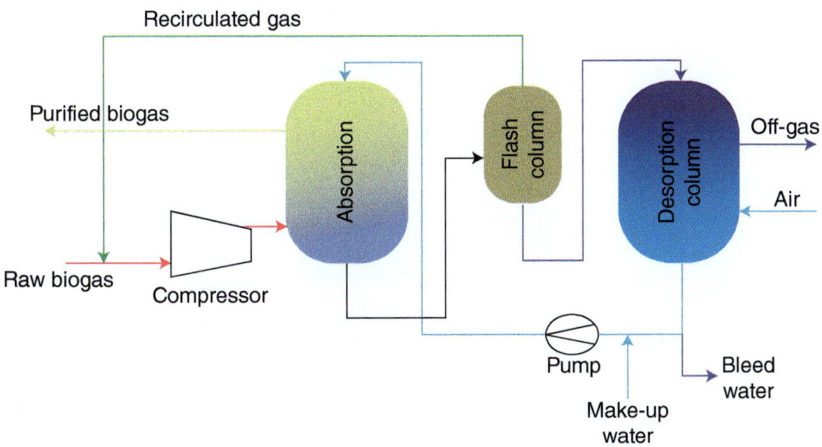

Figure 6.4 Schematic of water scrubber [57, 58].

adsorption equilibriums that can adsorb more CO_2 or various adsorption kinetics that can adsorb CH_4 more quickly than CO_2. Silica gels, activated carbon, titanosilicates, zeolites, and carbon molecular sieves are the adsorption materials utilized in this method [59, 60]. The vapors in the biogas should be preseparated in the adsorption process to the solid surface in order to avoid the possibility of harming the adsorbent [27]. To provide adequate dynamic force for this adsorption mechanism, the pressure of the biogas feed must also be 10 bar along with CO_2; this process can simultaneously adsorb N_2 and O_2. This technique has advanced to the point that it is now readily accessible, with ranging capabilities from 10 to 10,000 $m^3\ h^{-1}$ [10]. Moreover, the contaminants in raw biogas can have an impact on the process's effectiveness. The outputs of gases should be recirculated into the PSA systems because the process also loses 2–4% of the CH_4 that is produced [59]. The two another adsorption techniques are temperature swing adsorption (TSA) [61] and electrical swing adsorption (ESA) [62].

6.15.4 Membrane Separation

Another CO_2 removal process is membrane separation process. This technique has gained very crucial importance during the recent years [63]. Herein, pressures ranging from 5 to 30 bar are used to feed biogas into the membrane pores [64]. High-permeability gas types in biogas are more likely to pass through the membrane to the low-pressure permeable end than lower-permeability gas types with high-pressure side and leave the membrane module as a permeate side when they have lower permeability [63]. Energy consumption to maintain pressure for separation is one of the difficulties with membrane systems; more research should be done to lower pressure consumption. The three membrane materials that most frequently employed in this method are polysulphone, cellulose acetate (CA),

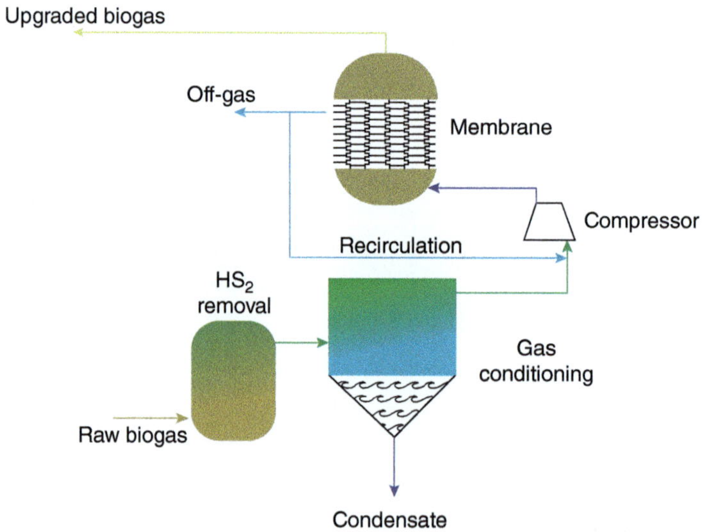

Figure 6.5 Schematic of membrane-based CO_2 removal technique. Source: Adapted from Bauer et al. [57].

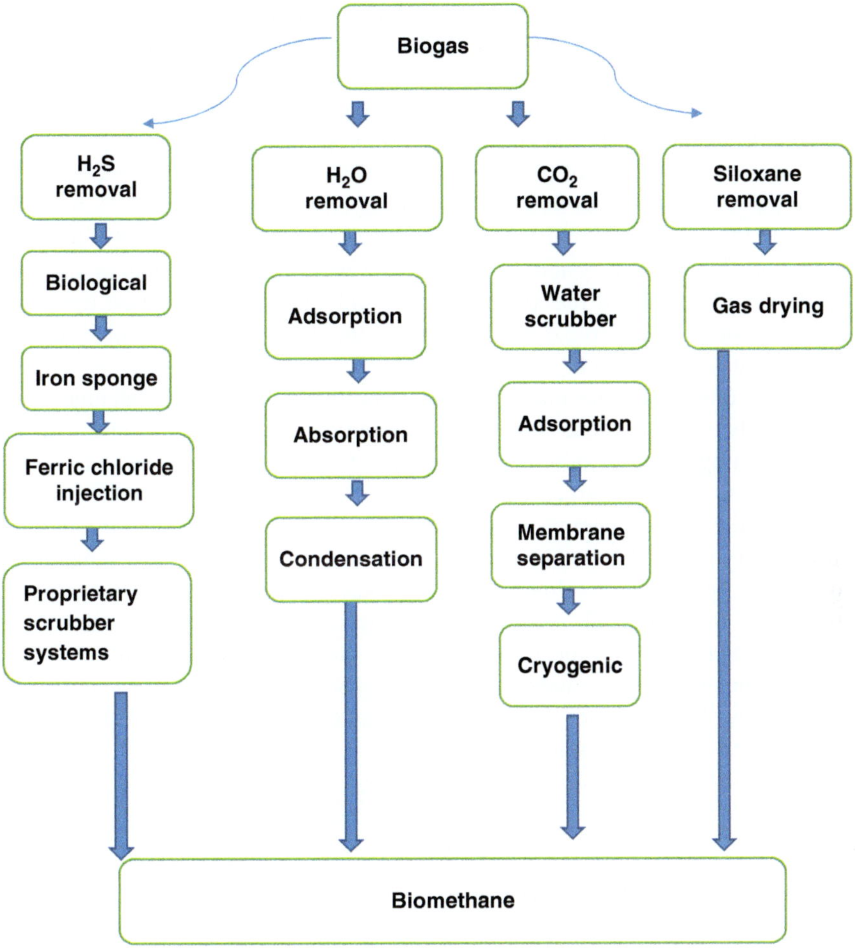

Figure 6.6 Overview of biogas cleaning techniques.

and polyimide. A CA membranes, on the other hand, does have a low softening pressure of about 8 bar as CO_2 can absorb in the membrane matrix due to its OH^--rich nature [65]. Membrane separation technique is shown in Figure 6.5. When membrane modules are utilized successively, the separation of the membranes offers up to 99.5% CH_4 collection and biomethane purity having at least 99% CH_4 [66]. Figure 6.6 also evaluates the steps involved in the cleaning of the major contents of the biogas that enhance the quality of biomethane.

6.16 Conclusion

The process of establishing a sound scientific foundation for the use of biogas is discussed in this chapter. The findings demonstrated that the availability of feedstock in the world is sufficient to ensure a significant development of biogas plants for CHP production that also uses animal manure as input resource. Moreover, it is also worth mentioning that the biogas can be used as transport fuel and CHP system. However, herein, it is demonstrated the production of the biogas and the composition of the biogas that produced from the different types of feedstock. The usage of the biogas for cogeneration is standardised as the best due to its dual part. The cleaning of the biogas is the major prospect as approximately biogas contains 60% of biomethane and 40% of CO_2. While in the world, energy efficiency ranged from 8% to 54% for producing electricity, 16% to 83% for producing heat, 18% to 90% for producing both electricity and heat, and 4% to 18% for producing transportation. The energy system adapts various ways to the demands of sustainability, even if the results indicated that paths via CHP units are indeed the best option for biogas consumption. The complete biogas usage system has been the subject of a system engineering study in this work.

Once core components are interconnected, optimization solutions become accessible. Thus, this study offers a crucial foundation that can be utilized to concentrate on a certain approach that might prove to be of significance. Once a route is found, it can be additionally examined using enlarged boundary conditions. However, other factors should be considered while making energy investments outside energy efficiency. It is advised that interdisciplinary research have been done to support system-level energy efficiency efforts and to provide a complete baseline model that takes into account not only energy but also environmental, technological, and socioeconomic concerns. The studies that have been done for this work serve as the foundation for a flexible framework that can help market and government decision-makers create a market to meet renewable energy goals in a sustainable way. This provides the proper pathway to the utilization of the biogas and to compare the best uses in a sustainable way.

References

1. Korberg, A.D., Skov, I.R., and Mathiesen, B.V. (2020). The role of biogas and biogas-derived fuels in a 100% renewable energy system in Denmark. *Energy* 199: https://doi.org/10.1016/j.energy.2020.117426.
2. Wang, Z., Wang, T., Si, B. et al. (2021). Accelerating anaerobic digestion for methane production: potential role of direct interspecies electron transfer. *Renewable and Sustainable Energy Reviews* 145: https://doi.org/10.1016/j.rser.2021.111069.
3. Abdeshahian, P., Lim, J.S., Ho, W.S. et al. (2016). Potential of biogas production from farm animal waste in Malaysia. *Renewable and Sustainable Energy Reviews* 60: 714–723. https://doi.org/10.1016/j.rser.2016.01.117.
4. Khan, M.U., Lee, J.T.E., Bashir, M.A. et al. (2021). Current status of biogas upgrading for direct biomethane use: a review. *Renewable and Sustainable Energy Reviews* 149: https://doi.org/10.1016/j.rser.2021.111343.
5. Adnan, A.I., Ong, M.Y., Nomanbhay, S. et al. (2019). Technologies for biogas upgrading to biomethane: a review. *Bioengineering* 6: https://doi.org/10.3390/bioengineering6040092.
6. Brizi, F., Silveira, J.L., Desideri, U. et al. (2014). Energetic and economic analysis of a Brazilian compact cogeneration system: comparison between natural gas and biogas. *Renewable and Sustainable Energy Reviews* 38: 193–211. https://doi.org/10.1016/j.rser.2014.05.088.
7. Esen, M. and Yuksel, T. (2013). Experimental evaluation of using various renewable energy sources for heating a greenhouse. *Energy and Buildings* 65: 340–351. https://doi.org/10.1016/j.enbuild.2013.06.018.

8. Anneli, P. and Arthur, W. (2014). Biogas upgrading technologies – developments and innovations, IEA bioenergy. IEA Bioenergy. 21. https://doi.org/10.1533/9780857097415.3.329.
9. Song, C., Liu, Q., Deng, S. et al. (2019). Cryogenic-based CO_2 capture technologies: state-of-the-art developments and current challenges. *Renewable and Sustainable Energy Reviews* 101: 265–278. https://doi.org/10.1016/j.rser.2018.11.018.
10. Khan, I.U., Othman, M.H.D., Hashim, H. et al. (2017). Biogas as a renewable energy fuel – a review of biogas upgrading, utilisation and storage. *Energy Conversion and Management.* 150: 277–294. https://doi.org/10.1016/j.enconman.2017.08.035.
11. Arroyo, J., Moreno, F., Muñoz, M. et al. (2014). Combustion behavior of a spark ignition engine fueled with synthetic gases derived from biogas. *Fuel* 117 (Part A): 50–58. https://doi.org/10.1016/j.fuel.2013.09.055.
12. Shah, A., Srinivasan, R., To, S.D.F., and Columbus, E.P. (2010). Performance and emissions of a spark-ignited engine driven generator on biomass based syngas. *Bioresource Technology* 101 (12): 4656–4661. https://doi.org/10.1016/j.biortech.2010.01.049.
13. Ravi, K., Mathew, S., Bhasker, J.P., and Porpatham, E. (2017). Gaseous alternative fuels for spark ignition engines – a technical review. *Journal of Chemical and Pharmaceutical Sciences* 10 (1): 93–99. www.jchps.com.
14. Yoon, S.H. and Lee, C.S. (2011). Experimental investigation on the combustion and exhaust emission characteristics of biogas-biodiesel dual-fuel combustion in a CI engine. *Fuel Processing Technology* 92 (5): 992–1000. https://doi.org/10.1016/j.fuproc.2010.12.021.
15. Walsh, J.L., Ross, C.C., Smith, M.S., and Harper, S.R. (1989). Utilization of biogas. *Biomass* 20 (3–4): 277–290. https://doi.org/10.1016/0144-4565(89)90067-X.
16. Bari, S. (1996). Effect of carbon dioxide on the performance of biogas/diesel duel-fuel engine. *Renewable Energy* 9 (1–4): 1007–1010. https://doi.org/10.1016/0960-1481(96)88450-3.
17. Gholizadeh, T., Vajdi, M., and Rostamzadeh, H. (2020). Exergoeconomic optimization of a new tri-generation system driven by biogas for power, cooling, and freshwater production. *Energy Conversion and Management* 205. https://doi.org/10.1016/j.enconman.2019.112417.
18. Larsson M. *Global Energy Transformation: Four Necessary Steps to Make Clean Energy the Next Success Story*. 2009. 1–306. Palgrave Macmillan London. https://doi.org/10.1057/9780230244092.
19. Kapoor, R., Ghosh, P., Tyagi, B. et al. (2020). Advances in biogas valorization and utilization systems: a comprehensive review. *Journal of Cleaner Production* 273: https://doi.org/10.1016/j.jclepro.2020.123052.
20. Tippayawong, N., Promwungkwa, A., and Rerkkriangkrai, P. (2007). Long-term operation of a small biogas/diesel dual-fuel engine for on-farm electricity generation. *Biosystems Engineering* 98 (1): 26–32. https://doi.org/10.1016/j.biosystemseng.2007.06.013.
21. Lantz, M. (2013). Biogas in Sweden – opportunities and challenges from a systems perspective. Doctoral thesis. Lund University. p. 93. https://doi.org/9789174734690.
22. Hosseinipour, S.A. and Mehrpooya, M. (2019). Comparison of the biogas upgrading methods as a transportation fuel. *Renewable Energy* 130: 641–655. https://doi.org/10.1016/j.renene.2018.06.089.
23. Scarlat, N., Dallemand, J.F., and Fahl, F. (2018). Biogas: developments and perspectives in Europe. *Renewable Energy* 129: 457–472. https://doi.org/10.1016/j.renene.2018.03.006.
24. Dahlgren, S. (2022). Biogas-based fuels as renewable energy in the transport sector: an overview of the potential of using CBG, LBG and other vehicle fuels produced from biogas. *Biofuels* 13 (5): 587–599. https://doi.org/10.1080/17597269.2020.1821571.
25. Brauer, H.B. and Khan, J. (2021). Diffusion of biogas for freight transport in Sweden: a user perspective. *Journal of Cleaner Production* 312: https://doi.org/10.1016/j.jclepro.2021.127738.
26. Allegue, L. B. and Hinge, J. (2012). Biogas and bio-syngas upgrading. Danish Technological Institute (December): pp. 1–97. http://www.teknologisk.dk/_root/media/52679_Report-Biogas and syngas upgrading.pdf.

27. Sun, Q., Li, H., Yan, J. et al. (2015). Selection of appropriate biogas upgrading technology – a review of biogas cleaning, upgrading and utilisation. *Renewable and Sustainable Energy Reviews* 51: 521–532. https://doi.org/10.1016/j.rser.2015.06.029.
28. Pellegrini, L.F., de Oliveira Júnior, S., and Burbano, J.C. (2010). Supercritical steam cycles and biomass integrated gasification combined cycles for sugarcane mills. *Energy* 35 (2): 1172–1180. https://doi.org/10.1016/j.energy.2009.06.011.
29. Bridgwater, A.V. (2012). Review of fast pyrolysis of biomass and product upgrading. *Biomass and Bioenergy* 38: 68–94. https://doi.org/10.1016/j.biombioe.2011.01.048.
30. Nadaleti, W.C. and Przybyla, G. (2018). SI engine assessment using biogas, natural gas and syngas with different content of hydrogen for application in Brazilian rice industries: efficiency and pollutant emissions. *International Journal of Hydrogen Energy* 43 (21): 10141–10154. https://doi.org/10.1016/j.ijhydene.2018.04.073.
31. Bedoya, I.D., Saxena, S., Cadavid, F.J. et al. (2012). Experimental study of biogas combustion in an HCCI engine for power generation with high indicated efficiency and ultra-low NO_x emissions. *Energy Conversion and Management* 53 (1): 154–162. https://doi.org/10.1016/j.enconman.2011.08.016.
32. Porpatham, E., Ramesh, A., and Nagalingam, B. (2012). Effect of compression ratio on the performance and combustion of a biogas fuelled spark ignition engine. *Fuel* 95: 247–256. https://doi.org/10.1016/j.fuel.2011.10.059.
33. Nagalingam, B., Duebel, F., and Schmillen, K. (1983). Performance study using natural gas, hydrogen-supplemented natural gas and hydrogen in AVL research engine. *International Journal of Hydrogen Energy* 8 (9): 715–720. https://doi.org/10.1016/0360-3199(83)90181-7.
34. Dev, S., Stevenson, D., Yousefi, A., Guo, H., and Butler, J. (2021). An experimental study on a dual-fuel generator fueled with diesel and simulated biogas. *Proceedings of ASME 2021 Internal Combustion Engine Division Fall Technical Conference, ICEF 2021*, virtual (13–15 October 2021). https://doi.org/10.1115/ICEF2021-67429.
35. Dev, S., Stevenson, D., Butler, J., Tartakovsky, B., Guo, H., and Hewko, R. (2021). Replacement of diesel by biogas generated from wastewater treatment in a small diesel generator by dual fuel technology. *ASME 2020 Internal Combustion Engine Division Fall Technical Conference, ICEF 2020*, virtual (4–6 November 2020). https://doi.org/10.1115/ICEF2020-3033.
36. Saeed, M., Fawzy, S., and El-Saadawi, M. (2019). Modeling and simulation of biogas-fueled power system. *International Journal of Green Energy* 16 (2): 125–151. https://doi.org/10.1080/15435075.2018.1549997.
37. Miao, C., Teng, K., Wang, Y., and Jiang, L. (2020). Technoeconomic analysis on a hybrid power system for the UK household using renewable energy: a case study. *Energies* 13 (12): https://doi.org/10.3390/en13123231.
38. Cao, Y., Dhahad, H.A., Hussen, H.M., and Parikhani, T. (2022). Proposal and evaluation of two innovative combined gas turbine and ejector refrigeration cycles fueled by biogas: thermodynamic and optimization analysis. *Renewable Energy* 181: 749–764. https://doi.org/10.1016/j.renene.2021.09.043.
39. Benaissa, S., Adouane, B., Ali, S.M. et al. (2022). Investigation on combustion characteristics and emissions of biogas/hydrogen blends in gas turbine combustors. *Thermal Science and Engineering Progress* 27: https://doi.org/10.1016/j.tsep.2021.101178.
40. Liu, A., Yang, Y., Chen, L. et al. (2020). Experimental study of biogas combustion and emissions for a micro gas turbine. *Fuel* 267: https://doi.org/10.1016/j.fuel.2020.117312.
41. Su, B., Han, W., He, H. et al. (2020). A biogas-fired cogeneration system based on chemically recuperated gas turbine cycle. *Energy Conversion and Management* 205: https://doi.org/10.1016/j.enconman.2019.112394.
42. Alshahrani, S. and Engeda, A. (2021). Performance analysis of a solar-biogas hybrid micro gas turbine for power generation. *Journal of Solar Energy Engineering ASME* 143 (2): https://doi.org/10.1115/1.4048157.

43. Thiruselvi, D., Kumar, P.S., Kumar, M.A. et al. (2021). A critical review on global trends in biogas scenario with its up-gradation techniques for fuel cell and future perspectives. *International Journal of Hydrogen Energy* 46 (31): 16734–16750. https://doi.org/10.1016/j.ijhydene.2020.10.023.
44. Saadabadi, S.A., Thattai, A.T., Fan, L. et al. (2019). Solid oxide fuel cells fuelled with biogas: potential and constraints. *Renewable Energy* 134: 194–214. https://doi.org/10.1016/j.renene.2018.11.028.
45. Wasajja, H., Lindeboom, R.E.F., van Lier, J.B., and Aravind, P.V. (2020). Techno-economic review of biogas cleaning technologies for small scale off-grid solid oxide fuel cell applications. *Fuel Processing Technology* 197: https://doi.org/10.1016/j.fuproc.2019.106215.
46. Kamalimeera, N. and Kirubakaran, V. (2021). Prospects and restraints in biogas fed SOFC for rural energization: a critical review in Indian perspective. *Renewable and Sustainable Energy Reviews* 143: https://doi.org/10.1016/j.rser.2021.110914.
47. Santoro, M., Di Bartolomeo, E., Luisetto, I. et al. (2022). Insights on the electrochemical performance of indirect internal reforming of biogas into a solid oxide fuel cell. *Electrochimica Acta* 409: https://doi.org/10.1016/j.electacta.2022.139940.
48. Minutillo, M., Perna, A., and Sorce, A. (2020). Green hydrogen production plants via biogas steam and autothermal reforming processes: energy and exergy analyses. *Applied Energy* 277: https://doi.org/10.1016/j.apenergy.2020.115452.
49. Deheri, C., Acharya, S.K., Thatoi, D.N., and Mohanty, A.P. (2020). A review on performance of biogas and hydrogen on diesel engine in dual fuel mode. *Fuel* 260: https://doi.org/10.1016/j.fuel.2019.116337.
50. Khan, I. (2020). Waste to biogas through anaerobic digestion: hydrogen production potential in the developing world – a case of Bangladesh. *International Journal of Hydrogen Energy* 45 (32): 15951–15962. https://doi.org/10.1016/j.ijhydene.2020.04.038.
51. Petersson, A. (2013). Biogas cleaning. In: *The Biogas Handbook: Science, Production and Applications*, 329–341. Woodhead Publishing https://doi.org/10.1533/9780857097415.3.329.
52. Persson, M., Jonsson, O., and Wellinger, A. (2007). Biogas upgrading to vehicle fuel standards and grid. IEA Bioenergy. pp. 1–32. https://doi.org/10.1002/bbb.1423.
53. Harikishan, S. (2009). Biogas processing and utilization as an energy source. In: *Anaerobic Biotechnology for Bioenergy Production: Principles and Applications*, 267–291. Wiley https://doi.org/10.1002/9780813804545.ch12.
54. Gis, W. and Samson-Bręk, I. (2012). Review of technology for cleaning biogas to natural gas quality. *Automotive Industry Institute (PIMOT)* 2012 (5): 33–39.
55. Ryckebosch, E., Drouillon, M., and Vervaeren, H. (2011). Techniques for transformation of biogas to biomethane. *Biomass and Bioenergy* 35: 1633–1645. https://doi.org/10.1016/j.biombioe.2011.02.033.
56. Angelidaki, I., Treu, L., Tsapekos, P. et al. (2018). Biogas upgrading and utilization: current status and perspectives. *Biotechnology Advances* 36: 452–466. https://doi.org/10.1016/j.biotechadv.2018.01.011.
57. Bauer, F., Persson, T., Hulteberg, C., and Tamm, D. (2013). Biogas upgrading – technology overview, comparison and perspectives for the future. *Biofuels, Bioproducts and Biorefining* 7 (5): 499–511. https://doi.org/10.1002/bbb.1423.
58. Mai-Moulin, T., Visser, L., Fingerman, K.R. et al. (2019). Sourcing overseas biomass for EU ambitions: assessing net sustainable export potential from various sourcing countries. *Biofuels, Bioproducts and Biorefining* 13 (2): 293–324. https://doi.org/10.1002/bbb.1853.
59. Awe, O.W., Zhao, Y., Nzihou, A. et al. (2017). A review of biogas utilisation, purification and upgrading technologies. *Waste and Biomass Valorization* 8: 267–283. https://doi.org/10.1007/s12649-016-9826-4.
60. Atelge, M.R., Atabani, A.E., Banu, J.R. et al. (2020). A critical review of pretreatment technologies to enhance anaerobic digestion and energy recovery. *Fuel* 270: https://doi.org/10.1016/j.fuel.2020.117494.
61. Mason, J.A., Sumida, K., Herm, Z.R. et al. (2011). Evaluating metal-organic frameworks for post-combustion carbon dioxide capture via temperature swing adsorption. *Energy & Environmental Science* 4 (8): 3030–3040. https://doi.org/10.1039/c1ee01720a.

62. Quan, W., Holmes, H.E., Zhang, F. et al. (2022). Scalable formation of diamine-appended metal-organic framework hollow fiber sorbents for postcombustion CO_2 capture. *Journal of the American Chemical Society* https://doi.org/10.1021/jacsau.2c00029.
63. Scholz, M., Melin, T., and Wessling, M. (2013). Transforming biogas into biomethane using membrane technology. *Renewable and Sustainable Energy Reviews* 17: 199–212. https://doi.org/10.1016/j.rser.2012.08.009.
64. Haider, S., Lindbråthen, A., and Hägg, M.B. (2016). Techno-economical evaluation of membrane based biogas upgrading system: a comparison between polymeric membrane and carbon membrane technology. *Green Energy & Environment* 1 (3): 222–234. https://doi.org/10.1016/j.gee.2016.10.003.
65. Hao, L., Li, P., and Chung, T.S. (2014). PIM-1 as an organic filler to enhance the gas separation performance of Ultem polyetherimide. *Journal of Membrane Science* 453: 614–623. https://doi.org/10.1016/j.memsci.2013.11.045.
66. Miltner, M., Makaruk, A., and Harasek, M. (2016). Selected methods of advanced biogas upgrading. *Chemical Engineering Transactions* 52: 463–468. https://doi.org/10.3303/CET1652078.

7

Digestate from Agricultural Biogas Plant – Properties and Management

Wojciech Czekała
Department of Biosystems Engineering, Poznań University of Life Sciences, ul. Wojska Polskiego 50, 60-627, Poznań, Poland

7.1 Introduction

Processing biomass into biofuels is one of the directions of renewable energy production [1]. Biogas plants are installations in which biogas can be produced from many different substrates. The main component of biogas is methane [2, 3]. Biogas is produced in agricultural biogas plants, biogas plants processing municipal waste, on landfills and on sewage treatment plants. Regardless of the type of installation, biogas is produced as a result of the anaerobic digestion process [4]. As a result of this process, part of the organic matter in the substrates will be transformed into biogas, which is a renewable source of energy. The second product is digestate, otherwise known as digestate pulp [5].

Despite the fact that digestate is perceived as a byproduct of the anaerobic digestion process, its properties and the amount require lawful management. The same like with biogas. However, it should be emphasized that the mass of digestate produced is usually 70–90% of the mass of substrates used for biogas production. For a biogas plant with a capacity of 1 MW, it will be tens of thousands of Mg per year. In addition to the amount of digestate, its properties are an extremely important issue [6]. Properties of digestate and its fractions indicate the preferred direction of their use. The properties are affected by the type of substrates used for biogas production and their fragmentation and properties, both

Biogas Plants: Waste Management, Energy Production and Carbon Footprint Reduction,
First Edition. Edited by Wojciech Czekała

physical and chemical. The digestate consists primarily of water, organic compounds not decomposed in the fermentation process, mineral compounds, and biomass of organisms. The aim of the paper was to discuss issues related to the properties and management of digestate from agricultural biogas plants.

7.2 Digestate from Agricultural Biogas Plant – Production, Properties, and Processing

7.2.1 Production

Anaerobic digestion is a natural phenomenon that occurs in nature [7]. Examples include wetlands or the decomposition of biodegradable waste stored in landfill quarters. The decomposition of organic matter in anaerobic conditions will result mainly in methane and carbon dioxide. Methane is the source of energy and the main product of the process. Due to the specific course of anaerobic digestion, regardless of whether it is carried out in uncontrolled or controlled conditions (biogas plants), only a part of the organic matter contained in the substrates will be decomposed. This is due to the fact that some of it is not degradable under anaerobic conditions. The number and variety of substrates that can be used for biogas production is very large. They differ from each other, e.g. chemical composition [8]. In practice, this means that substrates will be available that will decompose in more than 90%, such as beet pulp. In contrast, there will also be raw materials such as energy crops, where the degree of decomposition is much lower (https://www.sciencedirect.com/science/article/pii/S092666902300184X). Irrespective of the raw materials from which biogas is produced, the produced digestate should be managed in accordance with applicable legal regulations.

Every day, a new portion of substrates is directed to the biogas plant. For this reason, a certain amount of digestate must also be removed from the fermenters each day. The mass of digestate produced depends mainly on the mass of raw materials introduced into the fermentation chambers and their properties. The mass of digestate, depending on several factors, is usually 70–90% of the mass of the charge [9]. For a biogas plant with a capacity of 1 MW, it will be several dozen thousand Mg of digestate to be used annually. In addition to the mass of substrates used for biogas production, the type of technology used affects the amount of digestate. Depending on such factors as the applied pretreatment or the temperature of the process, the amount and properties of the digestate will vary [10].

The most important function of the biogas plant (Figure 7.1) is the production of electricity and heat using cogeneration. An alternative solution that is gaining in importance every year is the production of biomethane [11]. However, regardless of the direction of use of the biogas produced as a result of the anaerobic digestion process, the second product, which is digestate, should also be managed.

7.2.2 Properties

Substrates used in biogas plants and biogas production technology are the main factors affecting the properties of digestate. The biomass (including waste) used to produce agricultural biogas consists of sugars, proteins, and fats. The decomposition time of individual compounds varies. They are also characterized by different biogas efficiency [12].

Figure 7.1 *Small agriculture biogas plant in Switzerland.*

An example may be the difference resulting from the processing of raw materials of plant and animal origin. If their composition is significantly diversified, the digestate created on their basis will be characterized by different properties. Another aspect will be that some of the substrates will be easily degraded, e.g. beet pulp. Corn silage or straw decompose much slower in anaerobic conditions. In addition, some of them will not decompose, which will result in them remaining in the digestate [13]. The digestate includes organic matter, mineral substances, water, and microorganisms not decomposed in the process of anaerobic digestion.

Total solids: Total solids is one of the basic and most important properties of digestate. The value characterizing this parameter indicates how much the digestate is hydrated. In agricultural biogas plants, the total solids of digestate is usually in the range of 3–8%, but there are cases when the total solids is below 3%, e.g. when liquid substrates are used that decompose very well, e.g. distillery stillage. In contrast, there may be a situation when the value exceeds 10%. This possibility occurs with anaerobic decomposition of substrates such as maize silage or straw.

Volatile solids: This parameter indicates what part of the digestate is organic matter and what part is mineral matter. Most often, these values are given in relation to total solids. As mentioned earlier, for many substrates, only part of the organic matter is degraded under anaerobic conditions. Research results indicate the effect of using digestate on soil organic matter [14]. Therefore, some organic matter will remain in the digestate.

For this reason, it is recommended that the solid fraction be additionally subjected to the composting process. This action will allow you to obtain fertilizers in solid and liquid form.

pH: Typically, the pH of the digestate and its fractions is alkaline. It was mentioned, among others, in the studies of Cao et al. [15], where the pH of digestate was 8.46. The value of the pH is primarily influenced by the type of raw materials used and the time and conditions of anaerobic digestion. A high pH value is often seen as an advantage from a fertilizer point of view. This is due to the fact that much of the soil is acidic.

Elements content: Digestate is a source of numerous elements, both those necessary for plants and those that may prove to be a threat. Among all the elements, the content of nitrogen, phosphorus, and potassium in the digestate is most often emphasized [16]. These elements have a significant impact on the growth and development of plants. In addition, the role of fermentation in the context of micronutrients should not be overlooked. Due to the fact that digestate is made of many different substrates, it is the source of all micronutrients. In addition to macronutrients and micronutrients, there are also heavy metals in digestate [17]. Due to the fact that agricultural biogas plants mainly use products of agricultural origin and those from agri-food processing, there are usually no problems with exceeding the permissible standards for heavy metals [18].

7.2.3 Processing

The digestate can be used without processing for fertilization purposes. However, there may be a situation when its processing will be necessary or recommended [19]. For example, such a situation may occur when there is no sufficient number of areas for the distribution of the substance in question, which are located in the vicinity of the biogas plant. Then, a rational action will be, for example, dehydration of digestate. As a result of this process, water will be separated, which can be used in the biogas plant. The densified solid part can be used or transported over long distances. Another, and probably the most common, reason for processing digestate is to produce other products that can be sold for a higher price than raw digestate.

The primary way of processing digestate is its separation [20]. Separation consists in subjecting the raw digestate to processing, as a result of which at least two different fractions will be obtained. However, there may be more of these fractions [2], but two are most often produced. The liquid and solid fraction will differ significantly in properties, including total solids, mineral content, and organic matter. Water is a valuable substance that is contained, among others, in the digestate. An alternative way to process the digestate is to dehydrate it. This process involves the use of technologies that allow the separation of some water from the raw digestate [21]. As a result of this process, nutrients and organic matter are concentrated in a smaller volume. This is especially important for logistical reasons, including, e.g. digestate transport. To reduce the costs associated with this process, heat from cogeneration can be used [22], which in the case of agricultural biogas plants is often not utilized. However, it should be emphasized that the quantity is less than the demand. Another option is to thicken the digestate using an evaporator, where the process takes place under reduced pressure.

The processing of digestate is associated with numerous benefits. The most important of them is to increase the number of directions of use of individual fractions or products based on them. Many obtained products, such as compost or solid biofuel, will not be biodegradable, unlike digestate. Another reason may be to obtain fertilizers with a higher total solids than the raw digestate, which will reduce transport costs [23]. However, it should be remembered that the processing of digestate, like any raw material, is associated with incurring additional costs, mainly related to expenses for additional equipment and the consumption of electricity or heat in the event of water evaporation. For this reason, environmental, financial, and technical aspects should be taken into account before choosing the direction of using the digestate.

Digestate management is an element directly related to biogas production. Both the digestate itself and the fraction resulting from its processing can be used in many ways. It should also be mentioned that digestate and products based on it can be a source of further income for biogas plant owners. That is why it is so important to take rational actions aimed at obtaining positive effects both for the environment and for the owner of the biogas plant.

7.3 Digestate from Agricultural Biogas Plant – Management

7.3.1 Raw Digestate Fertilization

Fertilization is one of the ways to increase yields and improve their quality. Providing nutrients will enable plants to grow and develop properly, which will also have a direct impact on the yield. Fertilizers can be classified into mineral, natural, and organic. The advantages of mineral fertilizers include quick effect, precise composition, and ease of storage and transport. The disadvantages include the possibility of a negative impact on the environment during their production or improper use. Another issue is their high price and problems with availability, which may be related to, among others, with the increase in the prices of raw materials necessary for their production. Therefore, natural and organic fertilizers are gaining an importance. Natural fertilizers include mainly liquid manure and manure, and their use in production is a popular solution used all over the world [24]. Organic fertilizers include compost, biochar, or digestate. Digestate, both in raw and processed form, can be successfully used in agriculture.

The use of digestate as a fertilizer is a relatively new issue in the context of scientific work [9]. However, this issue is gaining importance every year, which is confirmed by the systematically increasing number of scientific papers. It is the fertilizing use of digestate or its solid fraction that is the main way of its management. This effect takes place due to the abundance of digestate in nutrients, pH above 7, and the content of organic matter. Tambone and Adani [25] in a laboratory incubated soil evaluated the properties of digestate as a nitrogen fertilizer in comparison with sewage sludge, compost, and one of typical mineral nitrogen fertilizer – urea. Based on the obtained results, the authors concluded that the mineralization of nitrogen in digestate is very similar in terms of both quantity and quality to that of urea. This is a confirmation that digestate could replace typical mineral fertilizers. In turn, Walsh et al. [26] indicated that replacing inorganic fertilizers with liquid digestate can maintain or improve yields from grassland systems. Another aspect will be to concurrently reduce the potential for losses of nutrients to the environment. In this case, the use of digestate may reduce agricultural dependence on inorganic fertilizer and positively affect the energy and economic costs associated with their use.

However, it should be mentioned that before the application of digestate, it is necessary to check the content of selected pathogens and heavy metals. In digestate from agricultural biogas plants, there are rather no problems with achieving the required levels. This is due to the fact that selectively collected waste or agricultural products are used to produce biogas. Cucina et al. [27] conducted research in Colombia in psychrophilic conditions. The results of the research showed that all analyzed digestates were characterized by physicochemical properties, nutrients, and concentrations of heavy metals appropriate for their reuse as biofertilizer. The authors mentioned that heavy metals were under the detection limit of the analytical method (Pb, Hg, Ni, Mo, Cd, and Chromium VI) or present at low concentration (Cu, Zn, As, and Se) in all the digestate. Porterfield et al. [28] indicate that another extremely important aspect in the context of the use of digestete is microplastic. The authors indicate that the land application of contaminated organic amendments is one of multiple potential pathways by which microplastics may enter agricultural soils. It is worth mentioning, however, that this problem primarily concerns digestate resulting from anaerobic digestion of biowaste from the municipal sector.

Chemical composition, pH, and availability are some of the advantages of using digestate in fertilization. A sudden increase in the prices of artificial fertilizers and problems with their availability may turn out to be another argument in favor of the use of organic fertilizers, including digestate. When producing biogas as a renewable source, the possibility of using biodegradable waste should also be mentioned, which is perceived as an additional benefit.

7.3.2 Liquid Fraction Management

As mentioned earlier, the easiest way to process digestate is separation process. As a result of this process, at least two fractions are produced, including the liquid fraction. The fraction in question is characterized primarily by a reduced content of total solids in relation to raw digestate. As pointed out by Wei et al. [29], total solids for digestate liquid fraction was only 1.97%. Another feature will be the reduced content of organic matter [30]. Organic matter in the form of mostly undecomposed substrate elements will remain in the solid fraction.

There are two basic directions of liquid fraction management. The first is fertilization [31]. The liquid fraction has definitely more similar properties to the raw digestate than the solid fraction (Figure 7.2). The low content of total solids and the abundance of elements necessary for plant development should be mentioned here. Especially, the latter feature is important when using digestate and liquid fraction as fertilizer. Due to similar properties, the method of application of this fraction will also be similar to the raw digestate.

The second possibility of using the liquid fraction will be to use it to hydrate a new batch of substrates. In wet technology, the total solids of the batch should be below 15%. When using waste with a higher content of total solids, such as manure or maize silage, it may be necessary to add substrates in liquid form, such as the discussed digestate fraction or slurry. By using digestate liquid fraction with the listed substrates, there will be no need to use water to reduce the total solids of the feedstock, which will achieve not only economic profit but also environmental benefits. This solution is additionally suggested for other reasons.

Figure 7.2 Liquid fraction of digestate.

Chen et al. [32] conducted research to evaluate the improving effects of digestate recirculation on the performance, energy recovery, and microbial community for swine manure and rice straw anaerobic digestion. In the digestate, recirculation increased total methane production and organic matter removal. Both digestate, its liquid fraction, and slurry are substrates containing microorganisms necessary in the process of methane fermentation. Thanks to this, these substrates can be used, for example, to start a new biogas plant. Another alternative solution is to use digestate liquid fraction in microalgae cultivation. Zielinska et al. [33] conducted research aim was membrane filtration of the liquid fraction of digestate to produce permeate that will be an effective medium for the cultivation of *Chlorella vulgaris*. The authors indicated that membrane filtration followed by the use of permeates for microalgae growth may be considered a way for the valorization of digestate liquid fraction.

7.3.3 Solid Fraction Management

The solid fraction is the second fraction formed in the separation process. The main characteristic of the solid fraction is the higher total solids content (Figure 7.3). Most often, it is 25–35%. The second feature that characterizes the solid fraction is the high content of organic matter. Depending on the substrates used for anaerobic digestion and the technology of decomposition, the content of organic matter in the solid fraction can

Figure 7.3 Solid fraction of digestate.

be very diverse. There are cases when this value is over 90%. The mass of the solid fraction produced is definitely lower than that of the liquid fraction. However, the number of directions and possibilities of its development is much greater. The solid fraction can also be subjected to the pelleting and briquetting process. This is confirmed by both own research [34] and research conducted by other authors [35]. Pellets and briquettes will have a reduced water content and a more concentrated content of nutrients and organic matter. This is due to the technology of their production and the increase in density compared to the digestate solid fraction. Another advantage will be the possibility of their easier and longer storage in relation to unprocessed raw material. Cathcart et al. [36] conducted research on biosafety aspects of the digestate pellets production. The aim of this research was to determine the effect that each step in the pellet production process has on bacteria numbers. The studies included anaerobic digestion, mechanical separation, solid drying, and pelletization. Based on the research, it was found, among others, that the pelletization reduced *Enterobacteriaceae* numbers to below detectable levels.

The simplest way to use the solid fraction as fertilizer is to use it without processing. The product in question is loose, both in a fresh and dry state, so there is no problem with its application. In addition to the supply of organic matter, another advantage will be the supply of nutrients. It should also be emphasized that the solid fraction of digestate can improve the soil structure and have a positive effect on water conditions. Caspersen et al. [37] found that digestate solid fraction obtained after solid–liquid separation of digestate is interesting as a potential fertilizer. According to the authors, another application is as well as a peat substitute in horticultural growing substrates. This raw material can be formed both alone and with the addition of other raw materials. This will allow you to prepare the right fertilizer, dedicated to specific crops. The granulation process itself will be similar, regardless of the fertilization or energy use.

The second possibility of using the solid fraction of digestate for fertilization is the composting process. Due to the low content of total solids in the raw digestate, only the solid fraction or dehydrated digestate will be useful for composting. In both cases, a minimum total solids content of 20–25% is recommended. As part of own research, it was found that digestate solid fraction, even without additives, can be subject to the composting process [38]. However, it is recommended to use it as a co-substrate. The most rational solution will be to add nitrogen-rich substrates, such as sewage sludge, to the mix. Digestate solid fraction will also play a structural role in the compost heap. The topic of sustainable management and recycling of anaerobic digestate solid fraction by composting was discussed in detail in the review paper by Czekała et al. [39].

Another possibility of using digestate solid fraction for fertilization will be the production and use of biochar. Biochar is a solid product of pyrolysis, i.e. one of the types of thermal waste processing. Due to its specific properties, it is used as a component of fertilizers, as fuel, or in many aspects related to environmental protection. Ayaz et al. [40] examined the effect of pig manure digestate-derived biochar as a soil amendment with N fertilizer on soil and plant heavy metal levels and nutrient availability under various moisture regimes. The authors showed that biochar applications significantly decreased heavy metals in the spring wheat plants. The decrease compared to non-biochar (control) treatments was significant, reaching, among others, 90% for Cr. As a conclusion, it was indicated that the use of pig manure digestate-derived biochar with N fertilizer under normal moisture conditions was able to reduce heavy metal availability to plants and thus could be used in contaminated soils to maintain better crop growth and development [40]. An alternative direction of digestate solid fraction management is to use them as bedding [41], both in loose form and in the form of granules.

7.3.4 Energy Management of the Solid Fraction

Despite the fact that as a result of anaerobic decomposition, organic matter decomposes, a certain part of it remains that can be used. This results, among others, from the fact that digestate is produced every day and is an element related to the production of biogas. This is confirmed by a research by Ülgüdür and Demirer [42], who analyzed residual biogas potential of digestates obtained from five full-scale farm-based digesters. Based on the obtained results, the authors indicated that biogas yields obtained by further anaerobic digestion of the digestates were found to be comparable to the raw feedstocks such as cattle, dairy cattle, and horse manure. Recently, the interest in digestate and products resulting from its use has increased significantly. Despite the fact that fertilization is by far the most popular direction [43], the role of the solid fraction in the production of solid biofuels, which are energy sources, should not be overlooked.

The digestate solid fraction consists primarily of substrate components not decomposed in the anaerobic process. Therefore, there is a possibility of its further use, e.g. for the production of pellets and briquettes. The energy value of pellets and briquettes made from digestate solid fraction is similar or slightly lower than for typical biomass used for the production of solid biofuels, e.g. sawdust or straw. In own research [44], the lower heating value for digestate solid fraction was as much as 19.394 MJ kg^{-1}. This result was very high and may indicate small decomposition of biomass subjected to the anaerobic digestion process. An extremely important aspect of the production of solid biofuels from digestate is the cost of the process. Cathcart et al. [45] conducted research on determining the financial viability of digestate fuel pellet production. The research covered and compared two mechanical separation technologies: screw press and decanting centrifuge. Pellet production cost was, respectively, £122 t^{-1} using the screw press and £95 t^{-1} using the decanting centrifuge.

7.4 Conclusion

The production of biogas as one of the renewable energy sources is becoming more and more important every year. In addition to the production of biogas, biomethane, heat, or electricity, more and more attention is targeted to digestate, which was not so obvious before. Due to the positive properties of the product in question, it can be used in agriculture to improve soil properties and increase yield. Numerous scientific studies, including the author's own research, indicate that digestate can be successfully used as a fertilizer. The fertilization of raw digestate is the most commonly used method of its management around the world. However, this is not the only possibility. As a result of physical, biological, and thermal transformations, further products can be produced, the essential or only component of which may be individual fractions of post-ferrite. Despite the fact that the most popular way to use digestate is to use it as fertilizer, new directions are being developed. Fertilization methods include composting or production of specialized fertilizers. Energy methods include production of pellets, briquettes, and biochar. With the increase in the number of biogas plants in the world, it increases the amount of digestate produced. Therefore, there is room for the use of other alternative methods of digestate management.

References

1. Hakeem, I.G., Sharma, A., Sharma, T. et al. (2022). Techno-economic analysis of biochemical conversion of biomass to biofuels and platform chemicals. *Biofuels, Bioproducts & Biorefining* https://doi.org/10.1002/bbb.2463.
2. Czekała, W. (2022). Biogas as a Sustainable and Renewable Energy Source. In: *Clean Fuels for Mobility, Energy, Environment, and Sustainability* (ed. G. Di Blasio, A.K. Agarwal, G. Belgiorno, and P.C. Shukla), 201–214. Singapore: Springer https://doi.org/10.1007/978-981-16-8747-1_10.
3. Keerthana Devi, M., Manikandan, S., Oviyapriya, M. et al. (2022). Recent advances in biogas production using agro-industrial waste: a comprehensive review outlook of techno-economic analysis. *Bioresource Technology* 363: 127871. https://doi.org/10.1016/j.biortech.2022.127871.
4. Ferreira, L.O., Astals, S., and Passos, F. (2022). Anaerobic co-digestion of food waste and microalgae in an integrated treatment plant. *Journal of Chemical Technology and Biotechnology* 97 (6): 1545–1554. https://doi.org/10.1002/jctb.6900.
5. Okaze, T. and Tada, C. (2022). Development and performance evaluation of a micro anaerobic digestion system for household use. *Japan Architectural Review* 5 (4): 644–648. https://doi.org/10.1002/2475-8876.12289.

6. Tan, F., Zhu, Q., Guo, X., and He, L. (2020). Effects of digestate on biomass of a selected energy crop and soil properties. *Journal of the Science of Food and Agriculture* 101 (3): 927–936.
7. Kozlowski, K., Dach, J., Lewicki, A. et al. (2018). Laboratory simulation of an agricultural biogas plant start-up. *Chemical Engineering and Technology* 41 (4): 711–716. https://doi.org/10.1002/ceat.201700390.
8. Calbry-Muzyka, A., Madi, H., Rüsch-Pfund, F. et al. (2022). Biogas composition from agricultural sources and organic fraction of municipal solid waste. *Renewable Energy* 181: 1000–1007. https://doi.org/10.1016/j.renene.2021.09.100.
9. Czekała, W., Jasiński, T., Grzelak, M. et al. (2022). Biogas plant operation: digestate as the valuable product. *Energies* 15: 8275. https://doi.org/10.3390/en15218275.
10. Wang, W., Chang, J.-S., and Lee, D.-J. (2023). Anaerobic digestate valorization beyond agricultural application: current status and prospects. *Bioresource Technology* 373: 128742. https://doi.org/10.1016/j.biortech.2023.128742.
11. Nguyen, L.N., Kumar, J., Vu, M.T. et al. (2021). Biomethane production from anaerobic codigestion at wastewater treatment plants: a critical review on development and innovations in biogas upgrading techniques. *Science of the Total Environment* 765: 142753. https://doi.org/10.1016/j.scitotenv.2020.142753.
12. Czekała, W., Łukomska, A., Pulka, J. et al. (2023). Waste-to-energy: biogas potential of waste from coffee production and consumption. *Energy* 276: 127604. https://doi.org/10.1016/j.energy.2023.127604.
13. Dale, B.E., Bozzetto, S., Couturier, C. et al. (2020). The potential for expanding sustainable biogas production and some possible impacts in specific countries. *Biofuels, Bioproducts and Biorefining* 14 (6): 1335–1347. https://doi.org/10.1002/bbb.2134.
14. Rosace, M.C., Veronesi, F., Briggs, S. et al. (2020). Legacy effects override soil properties for CO_2 and N_2O but not CH_4 emissions following digestate application to soil. *GCB Bioenergy* 12 (6): 445–457. https://doi.org/10.1111/gcbb.12688.
15. Cao, W., Wang, M., Liu, M. et al. (2018). The chemical and dynamic distribution characteristics of iron, cobalt and nickel in three different anaerobic digestates: effect of pH and trace elements dosage. *Bioresource Technology* 269: 363–374. https://doi.org/10.1016/j.biortech.2018.08.094.
16. Tavera, C.G., Raab, T., and Trujillo, L.H. (2023). Valorization of biogas digestate as organic fertilizer for closing the loop on the economic viability to develop biogas projects in Colombia. *Cleaner and Circular Bioeconomy* 4: 100035. https://doi.org/10.1016/j.clcb.2022.100035.
17. Golovko, O., Ahrens, L., Schelin, J. et al. (2022). Organic micropollutants, heavy metals and pathogens in anaerobic digestate based on food waste. *Journal of Environmental Management* 313: 114997. https://doi.org/10.1016/j.jenvman.2022.114997.
18. Czekała, W. (2018). Agricultural biogas plants as a chance for the development of the agri-food sector. *Journal of Ecological Engineering* 19 (2): 179–183. https://doi.org/10.12911/22998993/83563.
19. Kovačić, Đ., Lončarić, Z., Jović, J. et al. (2022). Digestate management and processing practices: a review. *Applied Sciences* 12: 9216. https://doi.org/10.3390/app12189216.
20. Feiz, R., Carraro, G., Brienza, C. et al. (2022). Systems analysis of digestate primary processing techniques. *Waste Management* 150: 352–363. https://doi.org/10.1016/j.wasman.2022.07.013.
21. Salamat, R., Scaar, H., Weigler, F. et al. (2022). Drying of biogas digestate: a review with a focus on available drying techniques, drying kinetics, and gaseous emission behawior. *Drying Technology* 40 (1): https://doi.org/10.1080/07373937.2020.1781879.
22. Weinand, J.M., McKenna, R., Karner, K. et al. (2019). Assessing the potential contribution of excess heat from biogas plants toward decarbonizing residential heating. *Journal of Cleaner Production* 238: 117756. https://doi.org/10.1016/j.jclepro.2019.117756.
23. Waggoner, A.L., Bottomley, P.J., Taylor, A.E., and Myrold, D.D. (2021). Soil nitrification response to dairy digestate and inorganic ammonium sources depends on soil pH and nitrifier abundances. *Soil Science Society of America Journal* 85 (6): 1990–2006. https://doi.org/10.1002/saj2.20325.

24. Borek, K. and Romaniuk, W. (2020). Biogas installations for harvesting energy and utilization of natural fertilisers. *Agricultural Engineering* 24: 1–14. https://doi.org/10.1515/agriceng-2020-0001.
25. Tambone, F. and Adani, F. (2017). Nitrogen mineralization from digestate in comparison to sewage sludge, compost and urea in a laboratory incubated soil experiment. *Journal of Plant Nutrition and Soil Science* 180 (3): 355–365. https://doi.org/10.1002/jpln.201600241.
26. Walsh, J.J., Jones, D.L., Edwards-Jones, G., and Williams, A.P. (2012). Replacing inorganic fertilizer with anaerobic digestate may maintain agricultural productivity at less environmental cost. *Journal of Plant Nutrition and Soil Science* 175 (6): 840–845. https://doi.org/10.1002/jpln.201200214.
27. Cucina, M., Castro, L., Escalante, H. et al. (2021). Benefits and risks of agricultural reuse of digestates from plastic tubular digesters in Colombia. *Waste Management* 135: 220–228. https://doi.org/10.1016/j.wasman.2021.09.003.
28. Porterfield, K.K., Hobson, S.A., Neher, D.A. et al. (2023). Microplastics in composts, digestates, and food wastes: a review. *Journal of Environmental Quality* 52 (2): 225–240. https://doi.org/10.1002/jeq2.20450.
29. Wei, Y., Lan, Y., Li, X. et al. (2022). Effect of wheat straw pretreated with liquid fraction of digestate from different substrates on anaerobic digestion performance and microbial community characteristics. *Science of the Total Environment* 818: 151764. https://doi.org/10.1016/j.scitotenv.2021.151764.
30. Chuda, A. and Ziemiński, K. (2021). Digestate mechanical separation in industrial conditions: efficiency profiles and fertilising potential. *Waste Management* 128: 167–178. https://doi.org/10.1016/j.wasman.2021.04.049.
31. Reuland, G., Sigurnjak, I., Dekker, H. et al. (2021). The potential of digestate and the liquid fraction of digestate as chemical fertiliser substitutes under the RENURE criteria. *Agronomy* 11 (7): 1374. https://doi.org/10.3390/agronomy11071374.
32. Chen, H., Zhang, W., Wu, J. et al. (2021). Improving two-stage thermophilic-mesophilic anaerobic co-digestion of swine manure and rice straw by digestate recirculation. *Chemosphere* 274: 129787. https://doi.org/10.1016/j.chemosphere.2021.129787.
33. Zielińska, M., Rusanowska, P., Zieliński, M. et al. (2022). Liquid fraction of digestate pretreated with membrane filtration for cultivation of *Chlorella vulgaris*. *Waste Management* 146: 1–10. https://doi.org/10.1016/j.wasman.2022.04.043.
34. Czekała, W., Bartnikowska, S., Dach, J. et al. (2018). The energy value and economic efficiency of solid biofuels produced from digestate and sawdust. *Energy* 159: 1118–1122. https://doi.org/10.1016/j.energy.2018.06.090.
35. Ogwang, I., Kasedde, H., Nabuuma, B. et al. (2021). Characterization of biogas digestate for solid biofuel production in Uganda. *Scientific African* 12: e00735. https://doi.org/10.1016/j.sciaf.2021.e00735.
36. Cathcart, A., Smyth, B.M., Forbes, C. et al. (2022). Effect of anaerobic digestate fuel pellet production on *Enterobacteriaceae* and *Salmonella* persistence. *GCB Bioenergy* 14 (9): 1055–1064. https://doi.org/10.1111/gcbb.12986.
37. Caspersen, S., Oskarsson, C., and Asp, H. (2023). Nutrient challenges with solid-phase anaerobic digestate as a peat substitute – storage decreased ammonium toxicity but increased phosphorus availability. *Waste Management* 165: 128–139. https://doi.org/10.1016/j.wasman.2023.04.032.
38. Czekała, W., Dach, J., Dong, R. et al. (2017). Composting potential of the solid fraction of digested pulp produced by a biogas plant. *Biosystems Engineering* 160: 25–29. https://doi.org/10.1016/j.biosystemseng.2017.05.003.
39. Czekała, W., Nowak, M., and Piechota, G. (2023). Sustainable management and recycling of anaerobic digestate solid fraction by composting: a review. *Bioresource Technology* 375: 128813. https://doi.org/10.1016/j.biortech.2023.128813.
40. Ayaz, M., Stulpinaite, U., Feiziene, D. et al. (2021). Pig manure digestate-derived biochar for soil management and crop cultivation in heavy metals contaminated soil. *Soil Use and Management* 38 (2): 1307–1321. https://doi.org/10.1111/sum.12773.

41. Setoguchi, A., Oishi, K., Kimura, Y. et al. (2022). Carbon footprint assessment of a whole dairy farming system with a biogas plant and the use of solid fraction of digestate as a recycled bedding material. *RCR Advances* 15: 200115. https://doi.org/10.1016/j.rcradv.2022.200115.
42. Ülgüdür, N. and Demirer, G.N. (2019). Anaerobic treatability and residual biogas potential of the effluent stream of anaerobic digestion processes. *Water Environment Research* 91 (3): 259–268. https://doi.org/10.1002/wer.1048.
43. Weckerle, T., Ewald, H., Guth, P. et al. (2022). Biogas digestate as a sustainable phytosterol source for biotechnological cascade valorization. *Microbial Biotechnology* 16 (2): 337–349. https://doi.org/10.1111/1751-7915.14174.
44. Czekała, W. (2021). Solid fraction of digestate from biogas plant as a material for pellets production. *Energies* 14: 5034. https://doi.org/10.3390/en14165034.
45. Cathcart, A., Smyth, B.M., Lyons, G. et al. (2021). An economic analysis of anaerobic digestate fuel pellet production: can digestate fuel pellets add value to existing operations? *Cleaner Engineering and Technology* 3: 100098. https://doi.org/10.1016/j.clet.2021.100098.

8

Environmental Aspects of Biogas Production

Yelizaveta Chernysh[1,2], Viktoriia Chubur[1,2], and Hynek Roubík[1]

[1] *Faculty of Tropical AgriSciences, Department of Sustainable Technologies, Czech University of Life Sciences Prague, Suchdol, Czechia*

[2] *Department of Ecology and Environmental Protection Technologies, Faculty of Technical Systems and Energy Efficient Technologies, Sumy State University, Sumy, Ukraine*

8.1 Introduction

The production of biohydrogen and biomethane from waste is currently a major focus in the world. Biogas and biomethane can potentially reduce CO_2 emissions through renewable energy production by replacing fossil fuels and limit the emission of greenhouse gas methane from organic waste generated by livestock farms and municipalities. In addition, it also reduces emissions by storing carbon in soils, producing green fertilizers, and enabling carbon reuse. A negative carbon footprint can be achieved with a combined pathway to mitigate greenhouse gas emissions through biogas and biomethane production. Carbon dioxide emitted by biogas energy consumption corresponds to the amount of carbon dioxide that plants remove from the atmosphere during the growing season. The amount of biogas production is relatively unlimited and is constantly available, so these technologies fit perfectly into the long-term perspective of reducing greenhouse gas emissions. Agricultural facilities constantly generate and accumulate organic waste of plant and animal origin [1].

This trend is supported by the growing demand for green energy [2] and the significant optimization of anaerobic digestion (AD) technologies in recent years. Furthermore, EU countries pay great attention to this direction, and a powerful project cluster funded

Biogas Plants: Waste Management, Energy Production and Carbon Footprint Reduction, First Edition. Edited by Wojciech Czekała

by international grant systems is being formed, and involving partner organizations from different countries [3] is being formed. The long-term development of the bioeconomy in the countries of the European Union is determined by the Bioeconomy Development Strategy until 2030, The Bioeconomy to 2030: Designing a Policy Agenda [4]. The EU bioeconomy strategy seeks to shift to circularity and sustainability from substitution.

In general, most types of waste can be considered renewable resources. However, at this point, recycling all types of waste is not cost-effective, although their reuse can reduce the technogenic impact on the environment. Biogas production is an important technology for the sustainable use of organic waste as a resource for the production of renewable energy. Bioenergy units use locally available feedstock, contributing to the conservation of nonrenewable energy resources. Fossil fuels can be replaced directly with biogas and nondirectly reducing the demand for fossil fuels in fertilizer production processes. Biogas technologies allow the utilization of organic waste, providing a high bioenergy yield and a positive environmental effect.

According to the data analysis in Figure 8.1, research activity related to the environmental impact of plant biogas production has shown a progressive growth of research since 2003. For the last 20 years, interest in this topic has increased significantly. According to the Scopus database, the previous five years' tendency to 80–100 publications per year has been noticed. The Web of Science database has shown noticeably increased interest since 2015, compared to the Scopus database. The number of Web of Science publications has grown

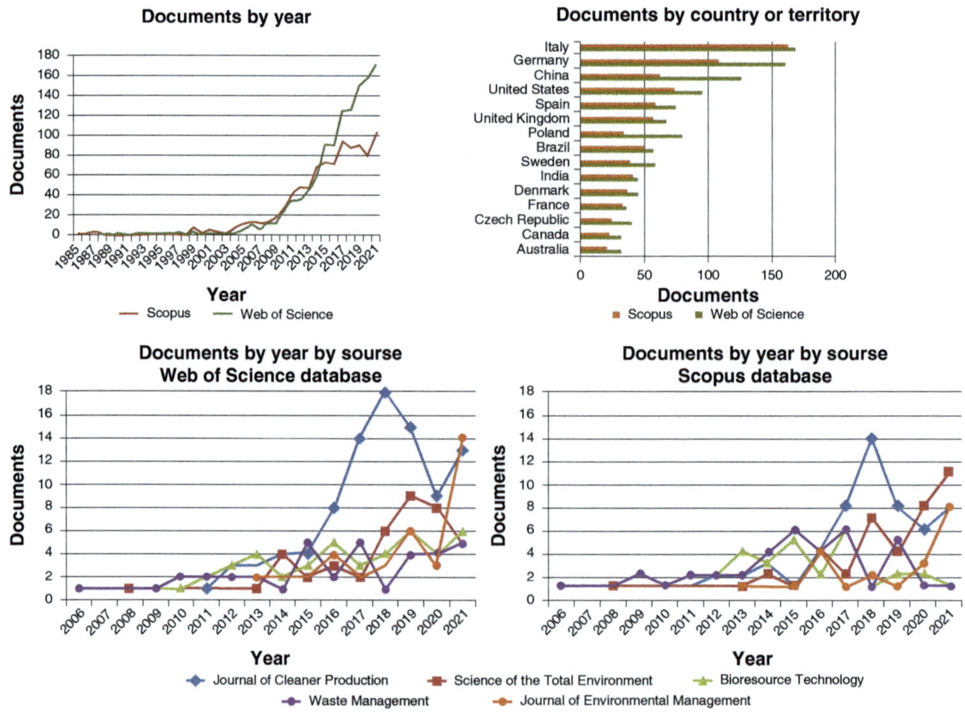

Figure 8.1 Analysis of the Scopus and Web of Science database search with the key words "environmental impact," "plants," and "biogas". Source: Scopus – Elsevier [5]/with permission of Elsevier.

exponentially, and the publication activity since 2017 has become120–170 publications per year. The published journals have a strong focus on environmental protection and waste management technologies. Researchers from Italy, Germany, and China have been highly published in these journals, and the United States, Spain, United Kingdom, Brazil, and Sweden are leading the course.

Significant theoretical developments provide the fundamental basis for the intensification of research in the last decade and for determining the directions and opportunities for the formation of a bioeconomy in the world.

8.2 Impact of Farms and Livestock Complexes on the Environment

The share of greenhouse gas emissions from agriculture (Figure 8.2) is currently approximately 13%, a substantial percentage of the total. Agricultural emissions are expected to increase along with the growing demand for food, fuel, fiber, and other materials supplied by agriculture. At the same time, new technologies and agricultural practises are interconnected with the responsibility to reduce greenhouse gas emissions from the agricultural sector [7].

Methane emission sources are divided into branch groups. The first group is livestock, which produces large amounts of methane during digestion, precisely during the conversion of carbohydrates by microorganisms to simple molecules for the absorption of the bloodstream by ruminants. The second branch is the cultivation of rice. Anaerobic decomposition in flooded rice paddies releases methane. During the growing season, methane is released into the atmosphere by diffusive transport through the rice plant. At the same time, rice grown in the mountains does not emit significant amounts of CH_4. Furthermore,

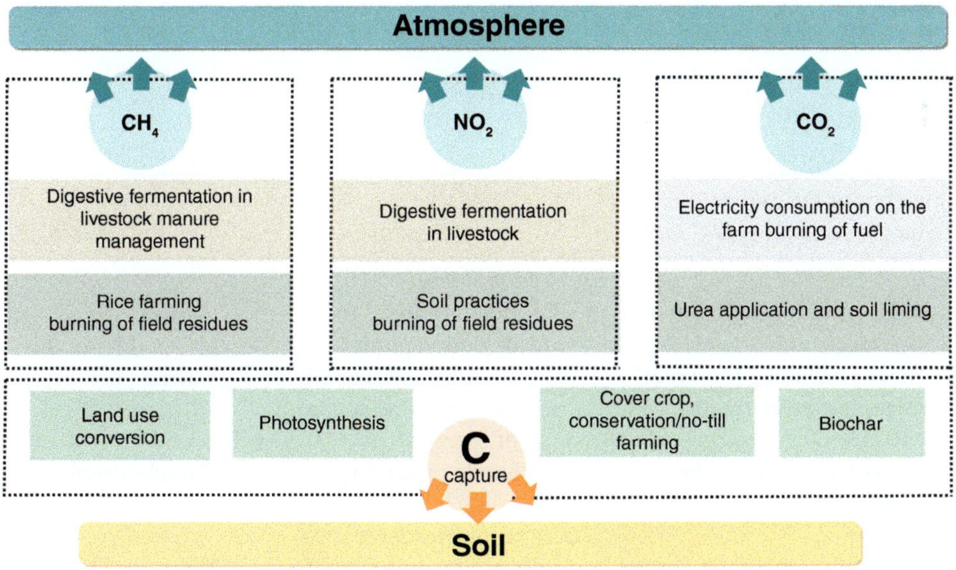

Figure 8.2 Greenhouse gas emissions in the agricultural sector. Source: Adapted from Aggeek | Actual Knowledge [6].

the recommended burning of savannas occurs on average every one to four years. Burning of savannas produces instantaneous emissions of CO_2, as well as CO, CH_4, N_2O, and nitrogen oxides. Burning agricultural residues in the field also causes significant methane emissions [8].

The flow of carbon and nitrogen in agroecosystems can be managed in several ways. More efficient N uptake by crops to mitigate N_2O emissions can be promoted, livestock and livestock diets or nutrients, and water management in rice cultivation can be controlled to improve carbon sequestration and mitigate CH_4 emissions [9].

Carbon sequestration by biological systems is generally considered a method of carbon conservation. Other technologies convert atmospheric CO_2 into other chemicals, such as methanol and similar organic substrates. However, carbon sequestration in the agricultural system is linked to crop performance and is considered one of the best ways to store carbon in the biological system. It is defined as binding carbon in a solid and stable form through direct or indirect fixation of atmospheric carbon dioxide. CO_2 sequestration for carbon binding is a scientific and technical approach to mitigate the effects of atmospheric CO_2 release [10, 11].

Biogas technologies are used to reduce greenhouse gas emissions. In addition to the fact that agricultural waste AD plants offer the possibility to avoid emissions from animal waste storage, they also provide the possibility to obtain renewable electricity with negative carbon emissions from biogas. At the same time, they also have a number of environmental aspects to consider in their implementation. It is necessary to consider methane emissions in the context of the balance of greenhouse gases and the overall sustainability of bioenergy systems, as there is a risk of methane emissions from biogas systems and the application of the necessary safety measures to prevent leakage [12].

8.3 The Environmental Benefits of Biogas Production

The environmental benefits of biogas production and utilization are global and local.

Specifics of the use of biogas technologies	Benefit	References
	Global level	
Reducing the consumption of fossil fuels	Extending fossil fuels depletion period	O'Shea [13]
Biomass (including bio-organic waste) is constantly renewed	Obtaining a practically inexhaustible source of energy	Kucher [14]
No carbon footprint in the atmosphere	Climate protection	Daniel-Gromke [15]
	Local territories level	
Reducing the amount of waste and the volume of its accumulation	Reducing pollution of the environment	Tanigawa [16]

Specifics of the use of biogas technologies	Benefit	References
Reduction of local air pollution due to lower amount of harmful emissions in comparison with fossil fuels	Environmental safety of the area located in close proximity to agro-industrial enterprises	WBA [17], Trypolska [18]
Shortening the time for disposal of animal wastes	Reduction of organic substances in waste and wastewater	Fernando-Foncillas [19], Saleh [20])
Integration of biogas production into the nutrient cycle	Ecological closure of production	Koppelmäki [21]
Reduction in the areas allocated for waste storage and disposal, and, in rural areas	Rational use of agricultural land	De Clercq [22]
Application of digested sludge	Improved soil fertility or rehabilitation of disturbed land	Hamdi [23]
Reducing unpleasant odors, improving the epidemiological situation as a result of the death of pathogenic microflora contained in waste, etc.	Solving a number of hygiene and sanitation tasks	Pietrangeli [24]

For example, manure, one of the most large-tonnage types of organic waste, belongs to the category of unstable organic contaminants and, according to the World Health Organization, is a factor of transmission of more than 100 species of various pathogens of animal and human diseases. [25] Furthermore, manure is characterized by a high content of environmentally hazardous substances: ammonia, hydrogen sulphide, mercaptan, phenol, heavy metal salts, etc., and the level of chemical pollution of the environment is 10 times more dangerous compared to municipal solid waste. [26] There is so much manure in livestock complexes that it is often not even used as fertilizer but accumulated on farms [27].

In the form of a flowchart, the figure compares the greenhouse gas emissions from conventional manure storage and AD according to Pucker [28] and Valli [29]. More detailed direct ecological effects and benefits of biogas complexes are shown in Figure 8.3.

In general, the following main elements of biogas technology with a focus on the environmental effect can be identified:

i. *First of all, biogas plants are an effective way to solve the problems of utilizing agricultural waste, including byproducts of animal origin (manure).*

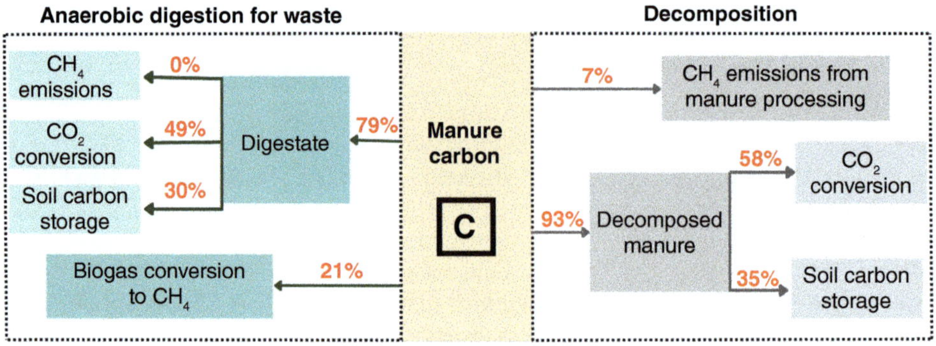

Figure 8.3 Comparison diagram of the ecological effect of anaerobic digestion and conventional manure decomposition. Source: Based on Pucker [28] and Valli [29].

The raw material for the biogas production could be either plant or animal origin waste. However, the most significant environmental effect is related to the ability of biogas plants to deal with the problem of manure and dung disposal. The conversion of organic waste to biogas occurs as a result of a complex set of biochemical transformations (fermentation of biomass). In the production of such facilities, farmers receive an environmentally friendly liquid or solid biofertilizer, which is free of unpleasant odors, helminth eggs, weed seeds, and nitrates. Biogas plants improve the environmental situation by recycling waste and improving sanitary and hygienic conditions. The constant availability of organic matter allows continuous production of biogas.

ii. *Solving the problem of storage and transportation of raw materials*

The implementation of biogas complexes allows the processing of animal wastes and the nonexploitation of anaerobic rates. For example, manure removed from livestock buildings can be stored in anaerobic ponds for six months in the case of dairy farms or 12 months in the case of pig farms [30]. Furthermore, the bottom of anaerobic ponds must be lined with material that prevents manure from entering groundwater [31].

In addition to being a major source of airborne pollutants and can be soil and groundwater pollutants, stakes and storage facilities also occupy large areas. Biogas projects can either reduce waste storage areas or eliminate such manure and manure storage by supplying them immediately not to open ponds and storage facilities but to biogas plants, and the accumulated methane will be burnt in a co-generator or flare.

iii. *Fermentation residue from biogas production in biogas plants as quality*

Organic waste from livestock farms and the processing industry itself is already a fertilizer. However, the efficiency coefficient of such fertilizers is only 10 to 15% of possible. When this waste is processed via a biogas plant, its properties are significantly improved [32]. The fermentation residue from biogas production is a quality fertilizer which can be sold or used instead of a mineral fertilizer. For example, an average biogas installation, located on cattle farms and processing around 37 000 tons/year of manure, gives approximately 35 000 tons of valuable biofertilizer per year. One ton of such fertilizers contains on average 3.5 kg of total nitrogen N [33].

iv. *Advantages of environmentally friendly biofertilizer over other organic fertilizers*

Biofertilizer is several times better than other organic fertilizers (manure, dung, and peat). Unprocessed organic fertilizers are more harmful to the soil, polluting it and groundwater. At the same time, biofertilizer is a clean and environmentally friendly fertilizer. Pig manure, cattle manure, and peat manure usually contain many weed seeds. A ton of fresh manure contains up to 10 000 seeds of various weeds. This leads to a yield loss of 5–7 quintals of cereal crops per hectare [34].

- The absence of pathogenic microflora but the presence of active microflora contributes to intensive plant growth. Meanwhile, because of its form, the biofertilizer begins to work effectively immediately when applied. In contrast to mineral fertilizers, biohumus can be applied in any amount. Its use does not cause mineralization of the soil, as it is an environmentally friendly product.

v. *Storage and transportation of fertilizers*

As we aforementioned, a great advantage of biofertilizers is the absence of an adaptation period. Manure and other organic matter require long preparation (6–12 months) before application to the soil. The useful substances contained in them are partially lost, and the rest begin to act in the soil only two to four years after their application [35]. Furthermore, it should be noted that, depending on the method and duration of storage, organic waste loses between 25% and 50% of organic matter and nutrients (primarily nitrogen N). Even greater losses are observed during freezing with subsequent thawing of up to 70% [36]. Therefore, biofertilizers not only have an advantage in quality but also significantly save another resource, time, and, as an indirect effect, a saving of storage area. It should be noted that the aforementioned properties of biofertilizer help reduce cases, environmental crimes, when manure or manure is removed from the fields without prior preparation. As a result, not only irreparable damage to the environment, especially the soil, but also the local population suffers from unpleasant smells [37].

vi. *Reduction of greenhouse gas emissions*

Agribusiness is one of the largest sources of methane and other greenhouse gases in the atmospheric air, so biogas plants can be seen as an implementation of emission reduction under Joint Implementation projects (Kyoto Protocol, Paris Agreement). Emission reductions will be achieved by replacing energy produced from nonrenewable sources with energy produced from alternative sources [38]. On many pig and cattle farms, manure is stored in anaerobic ponds, resulting in methane emissions directly into the atmosphere. The reduction in methane emissions in biogas complexes will be achieved by capturing biogas and then combusting it in a cogeneration plant [39].

Furthermore, there will be a reduction in the emissions of another greenhouse gas, CO_2, because the production of electricity and heat from renewable sources (biogas) will lead to the replacement of the equivalent amount of energy obtained by burning fossil fuels in power plants that supply power to the grid [40].

8.4 Environmental Safety of the Integrated Model of Bioprocesses of Hydrogen Production and Methane Generation in the Stages of Anaerobic Fermentation of Waste

Considering that the biogas plant is a potentially dangerous object for workers, it is necessary to constantly monitor the parameters of the digester to ensure the technogenic and environmental safety of engineering structures. Biogas mixed with air in a proportion of 5–15% in the presence of a flash source with a temperature of 600 °C or higher can lead to an explosion. A list of safety requirements must be followed, such as the prohibition of having a flame near the plant and a safe distance of 10 m for welding works. After the biogas digester is empty for repair, the reactor should be ventilated to avoid the risk of explosion of a mixture of biogas and air. The pressure of the gas supplied through the gas pipeline to the place of consumption should not exceed 0.15 MPa, and in front of the gas equipment should not exceed 0.013 MPa [41]. The reactor must be equipped with shutters and hydraulic gates, which, if necessary, could disconnect the reactor from the main discharge of excess pressure in the gas system in case it exceeds the norm. Electrical equipment must be grounded. The long-term inhalation of biogas in large quantities can cause poisoning due to the highly toxic hydrogen sulphide content in biogas. Therefore, all rooms with biogas appliances must be ventilated regularly. Gas pipes must be regularly checked for leakages and protected against damage. Gas leaks must be detected with a soap emulsion or special devices. The use of an open flame to detect gas leaks is prohibited [42].

The high priority development of the biogas energy sector over other sectors of bioenergy in the world has been due to the facts of environmental safety in biogas use, the high heating value with simple production technology, and the significant amount of waste that must be utilized. The problems of accumulation of different types of waste require their processing in a short time, achieving the lowest economic and energy costs with minimal ecological impact [43]. The environmental impact factors of biogas plants are shown in Figure 8.4.

Biotechnology productions, including anaerobic fermentation, use various raw materials and chemicals and produce many types of products in different forms. The most reliable way to ensure biosafety in biotechnology production facilities is to organize production according to aseptic rules. However, in anaerobic decontamination of liquid and solid organic waste of various origins (municipal, industrial effluent, municipal solid waste, and agricultural waste), it is not necessary to use sterilization methods.

In the case of anaerobic methane fermentation, nearly organic matter of any origin (except for lignin) can be converted into methane and carbon dioxide. The methane produced during open fermentation of surplus activated sludge and inadequately treated wastewater is captured in a gas holder, reducing the release of the gas into the atmosphere, and then is used as a sustainable fuel. The carbon dioxide released is collected and used as "dry ice" in the food industry [44].

Practically, all raw materials are stored before AD but could be fed directly into the anaerobic bioreactor. The anaerobic bioreactor can include more than one gastight vessel. The digestate may be stored in a gastight container, an open tank, or other storage facilities.

In addition to sanitary decontamination of the waste, AD technology preserves the nutrient content of the waste. Therefore, after methane fermentation (approximately 40% of the initial quantity) solid residues are used as fertilizer to cultivate crops. Fermented biomass

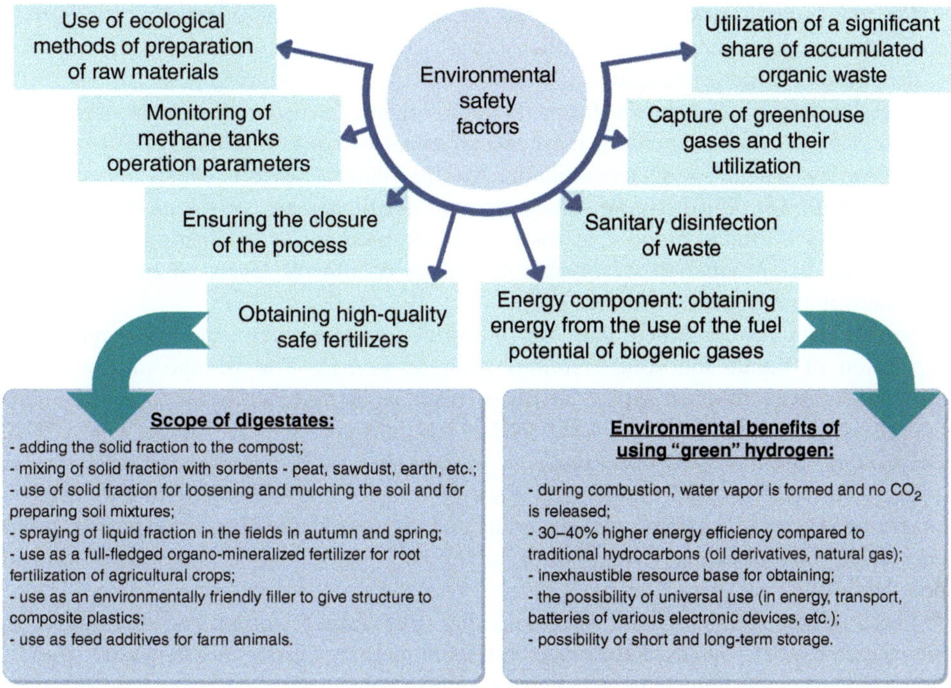

Figure 8.4 *Factors of ecological safety of anaerobic fermentation processes (author's development).*

has a mineralization of 60% and minerals are converted into a plant-available form. The application of such biofertilizers allows the reduction of the share of mineral fertilizers used in agriculture. The biofertilizers obtained by most criteria are several times better than other organic fertilizers such as pus, manure, or peat. There is an absence of weed seeds and pathogenic microflora and the presence of active microflora (contains about 1014 microflora colonies per gram) that intensifies plant growth. As a positive aspect is the absence of the adaptation period and therefore digested fertilizers in their form are effective immediately after application to the soil and fairly resistant to leaching nutrients from the soil (washed out less than 15%, so fertilizers do not lose their effectiveness for three to five years longer). Generally, the ecological impact on the soil is positive. The receipt of biofertilizer is pollution-free. Due to the AD of organic waste in biogas plants, it saves the total amount of nitrogen and increases the soluble nitrogen content of NH_4 by 10–15% [45].

To ensure the technogenic and environmental safety of the biogas plant engineering facilities, continuous monitoring of the reactor operating parameters is necessary [46]. It is important to purify technological water after separating digested sediment and reuse it in the technological process to ensure the completeness of the biogas production process and reduce the consumption of process water used for moisture stabilization. Since such water has a high concentration of nitrogen compounds, Kozlovets [47] suggested a two-stage purification in an anaerobic process using nitrification and denitrification stages, where nitrogen compounds are converted to molecular nitrogen and released into the atmosphere [47].

The main sources of environmental risk are the presence of helminth eggs, *Escherichia coli* bacteria, and other pathogenic microflora in organic waste. Therefore, it is necessary to use preventive measures against contamination.

Digestates are a bioproduct of fermentation as organic fertilizers that can be used instead of inorganic fertilizers, increase the concentrations of some nutrients, self-decompose, and have fewer chances of accumulating toxicity, with a lower negative impact on the environment [48]. Furthermore, since organic fertilizers are produced from organic sources, only a limited amount of fossil fuels are used in the production process. This means that the production of organic fertilizers has lower levels of greenhouse gas emissions than inorganic fertilizers [49].

Storage can be combined with or followed digestate treatment, for example, the separation of liquid and solid fractions prior to storage. The storage of raw materials and digestate in open tanks and their mechanical treatment can potentially cause emissions of nitrogen (N), ammonia (NH_3), and nitrogen oxide (NO), as well as N_2O, CH_4, odors, and dust. AD takes place in gastight vessels, so uncontrolled emissions are unlikely to occur, except for the processes of transferring substances to and from the anaerobic reactor and storing raw materials and digestate. However, uncontrolled biogas emissions from emergency ventilation valves and poorly sealed water separators are possible [50].

There is a potential for NH_3 emissions during feedstock storage and digestate at the biogas plant site. Crops used for biogas production (energy crops) are generally stored as silage [51]. Because the pH of the silage is low for conservation purposes, the NH_3 emissions from the storage of energy crops before AD are negligible.

Since the anaerobic bioreactor is completely closed, there should be no NH_3 emissions. However, cases of overpressure can occur during the operation of the biogas plant. In these cases, the pressure valves can release some biogas (approximately 1% of the gas produced). For greenhouse gas calculations, these losses are relevant because about 60% of the gas volume is methane. The concentration of NH_3 in biogas is much lower (0.1–1%), depending on the substrates that undergo fermentation [52].

The development of energy efficient and environmentally friendly technologies for the production of hydrogen and methane is a current urgent problem [53]. One of the ways to achieve these requirements is the production of biotechnological processes, as shown in Figure 8.5.

The combination of stages of anaerobic fermentation is an important aspect. At the acetogenic stage, two groups of acetogenic microbes are involved. The first one forms acetate with the release of hydrogen (acetogens that form hydrogen from soluble products in the preliminary stage of acid formation). The second group of acetogenic bacteria leads to the formation of acetic acid by hydrogen use to reduce carbon dioxide (acetogenes using hydrogen).

In the methanogenic stage, methane bacteria produce methane in two ways: by breaking down acetate and by reducing carbon dioxide with hydrogen. The first pathway has 72% methane and the second has 28% [54, 55].

Trophic systems during AD are described using metabolic products of some bacteria groups by other bacteria groups. Therefore, it is necessary to organize the successful passage of each stage in a separate unit or a separate zone of the unit, with different conditions.

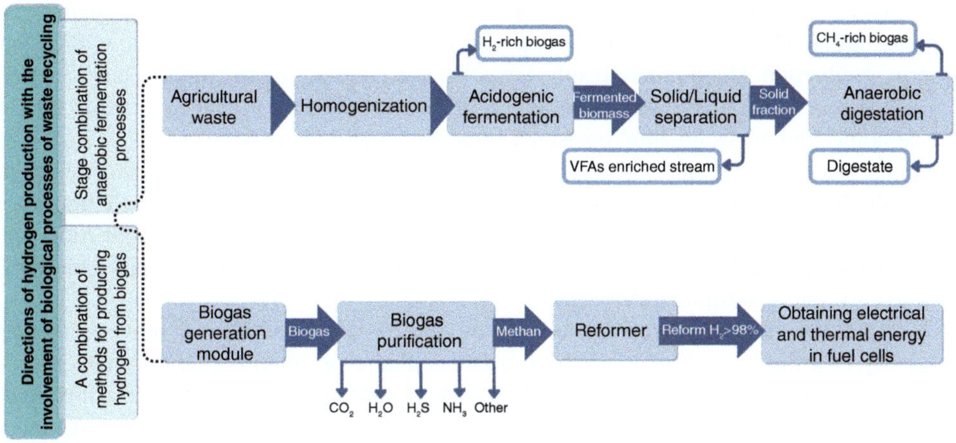

Figure 8.5 Model of the combination of bioprocesses for hydrogen and methane production (author's development).

The scientific basis for the concept of phase distribution technology is based on the different requirements of acid and methane-producing microorganisms under medium conditions and the difference in the physiological characteristics of these microorganisms.

The combination of these methods with the production of hydrogen from biogas is also quite promising. According to the technological scheme of the process (Figure 8.3), this combination contains the first stage of biogas production from organic waste, its subsequent accumulation and purification to methane to produce high-purity hydrogen in a fuel processing unit; experiments in this direction are carried out very actively [56].

8.5 Life Cycle Assessment for Biogas Production

The interest in evaluating the environmental efficiency of biogas production as a clean and renewable energy source using the life cycle assessment (LCA) methodology is strong. Comprehensive research on biogas plant LCA can further expand biogas potential as a sustainable renewable energy resource. The studies by Aziz et al. [57] and Hijazi et al. [58] reviewed an analysis of LCA studies of biogas production around the world based on the methods described in ISO 14040 and 14 044.

In general, LCA studies of biogas systems have different purposes and are conducted by different functional evaluation units, making it difficult to compare the life cycles of production from different feedstocks. The scanned biogas production life cycle impact assessment studies had unconnected functional assessment units.

The limits of the possibility of comparing different LCA studies are representative impact indicators, and incorrect interpretations due to the high uncertainty of the approach to the use of performance coefficients were identified by Aziz et al. [57]. The economic effect of biogas production was confirmed using animal and agricultural waste, solid waste, and wastewater for energy production. Organic waste recycling is defined as an effective ecological investment for the benefit of society and the environment. Compared to European studies, the study by Hijazi et al. [58] summarizes the LCA implementation process in four stages: definition of goals and scope, life cycle inventory, impact assessment, and interpretation. The research cycle begins with feedstock production and ends with biogas energy production. Quantitative results have become challenging to compare environmental performance due to significant variables, different allocation methods, and differences in economic and environmental pressures between geographic regions.

Studies by D'Imporzano et al. [59], Sun et al. [60], Wang et al. [61], Ioannou-Ttofa et al. [62], Singh et al. [63], and Cahyani et al. [64] mainly evaluated the environmental impact of the biogas system, taking into account pollutant reduction and clean energy production, and focused on mid-level impacts such as climate change, negative greenhouse gas emissions, ecotoxicity, and maximum nutrient use.

Comparison of energy crop intensification scenarios with additional process impacts study by D'Imporzano et al. [59] showed that a dual crop intensification system causes increased impact categories (nutrient management and agricultural emissions, increased overall costs). On the contrary, crop replacement significantly reduces the impacts of energy crops. Crop replacement significantly reduces the impact of energy crops more due to high biomass productivity per unit area, low agronomic maintenance, fuel and chemical costs for production, and reduced nitrogen and phosphorus emissions to the environment due to lower demand, allowing better utilization of nutrients.

An approach to comparing results by unifying the uncertainty of complex processes is demonstrated in study by Sun et al. [60] biofuel production systems based on microalgae. Summarizing the analysis of biogas production processes, it is determined that more than 70% of the energy consumption is in energy demand in fertilizer production and bioenergy conversion, representing 50% of the total amount of emissions. The pretreatment anaerobic fermentation system is determined to be the most industrially feasible and environmentally safe among algae bioenergy conversion systems.

Based on the analysis of prospective livestock waste treatment in China [61], large-scale domestic biogas plants showed good environmental sustainability in terms of reducing pollutant emissions and producing clean energy. Large-scale plants had better ecological performance than domestic plants in terms of energy use and environmental impact. The complexity of the biogas production system and the use of byproducts contributed to the improved performance of large plants in terms of productivity period, CH_4 production rate, and biogas efficiency. In comparison, large-scale biogas plants have a higher global warming and photochemical oxidation potential due to increased CO_2 and volatile organic compounds emissions. On the contrary, domestic plants show a greater potential for acidification, eutrophication, and human toxicity, mainly due to increased SO_2 and NH_3 emissions. The higher environmental impact potential is related to biogas combustion and digestate treatment during application in different-scale biogas plants.

The research by Ioannou-Ttofa et al. [62] focuses on the environmental sustainability of domestic biogas plants that operate on animal feedstock in Egypt to make recommendations from an environmental perspective. According to the study, environmental sustainability is optimized by combining two best-case scenarios: minimum percentage of biogas leakage and absence of intentional emissions, reducing the overall environmental footprint of the system by 60% compared to the basic scenario. At the regional level, the total environmental impact of a biogas plant in Egyptian households has been shown to be twice that of a large biogas plant operating in Europe. The environmental impact of household biogas plants is mainly due to the use of animal feedstock and systemic emissions into the atmosphere. In general, the low level of impact on the environmental sustainability of the system is due to the emissions to the water and the emissions to the soil from digestate leaks. In addition to the fact that having a small digester is of little importance in the environmental sustainability of the system, the system is sensitive to biogas leaks and intentional emissions, which are common with large digesters. Therefore, it is necessary to properly size the digester according to local needs. The environmental profile of the system can be improved by burning excess biogas instead of releasing it and using additional digestate as a biofertiliser, which will significantly reduce the environmental impact.

LCA of impacts of urban wastewater-based biogas plants was conducted by Singh et al. [63] showing that the construction of the biogas plant has a negligible effect on the entire life cycle, and the digestate obtained is a good option to replace chemical fertilizers. Emissions from feedstock supplies had an almost negligible contribution to the total impact of the life cycle; therefore, CH_4 and CO_2 from the plant contributed to a significant reduction in climate change, giving negative greenhouse gas emissions. Biogas production and distribution in agriculture contributed to negative CFC-11 emissions (−3059E-08 kg of CFC-11 eq m^{-3}), which is a beneficial effect that can be attributed to the phasing out of the use of electricity and fertilizers. In general, the biogas plant based on sewage sludge has a favorable impact on the environment. A small industrial biogas digester to process industrial tapioca wastewater with an anaerobic pond system [64] could potentially reduce greenhouse gas production by up to 296 kt CO_2-equivalent over one operating period.

The assessment of the life cycle of biogas technologies is strategically important for the development of renewable energy and waste management. At the same time, several studies [57–65] modify the methodological approach to LCA with different measurement values for parametric variables. It should be noted that it leads to the complexity of comparative analysis of data behind studies from different regions of the world. It remains important to form a unified approach to the LCA of biogas technologies with the development of a unified basis with an assessment of the impact on climate change using technologies for the adaptation and optimization of AD of various types of waste. Based on the number of studies analyzed, the LCAs of biogas production on different origin substrates, the cycle is graphically represented in Figure 8.6, and additional characteristics are given in Table 8.1.

Therefore, research is required to better understand clean biogas practises and environmentally friendly pretreatment processes to maximize biogas production. Environmentally friendly technologies are essential to stimulate the production and use of biogas to reduce the environmental impact and increase the efficiency of biogas utilization. A large-scale biogas plant shows greater environmental efficiency compared to a domestic plant and has excellent potential to become a major type of biogas production in the future.

8.6 Environmental Issue of Biogas Market in Ukraine – Case Study

According to the Ukrainian data from the State Statistics Service, in 2020, agricultural waste accounts for 1.15% of the structure of the total amount of waste generated by the types of economic activity, presented in Figure 8.7 [66].

Despite many policies and strategies on environmental protection in Ukraine, waste management still needs to be fully reflected in these, even if waste is harmful to human health and the environment. The major part of the economic activity of agricultural

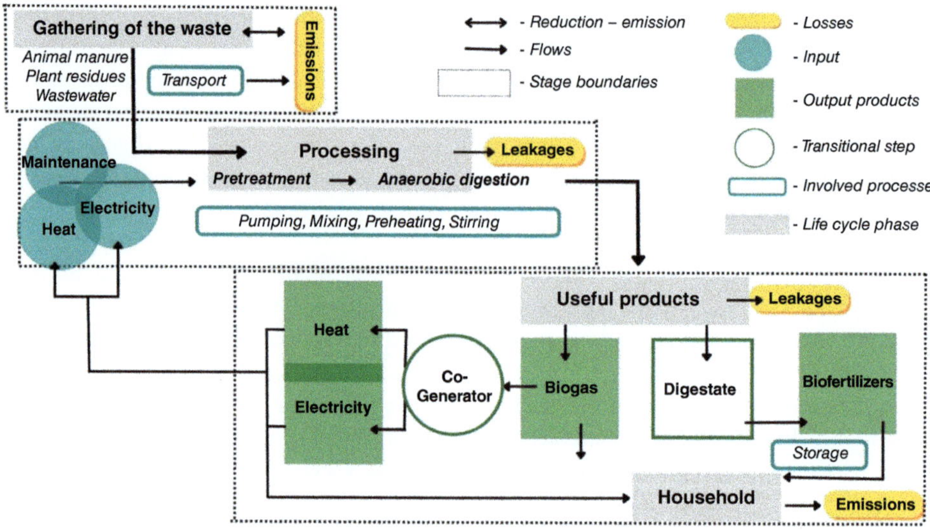

Figure 8.6 Model of the life cycle phases of biogas production (author's development).

companies is waste management of plant and animal origin, mixed food waste, animal feces, urine, manure, etc.

During the analysis of agricultural waste, animal wastes, especially poultry wastes, are the most dangerous for the environment because large complexes are concentrated in a limited area, leading to environmental disasters. Due to increased production and the lack of effective cleaning systems, waste from these companies ends up in water, soil, and air. Some researchers have written about this problem in their publications, so Goncharuk I.V. in a paper [67] studied in detail the emission structure of agriculture industrial complexes in Ukraine, where the fact that the livestock industry entails 18% of greenhouse gas emissions; in particular, methane emissions from this sector are about 16% of annual global emissions, nitrogen oxide – 17%, and many other environmentally hazardous substances and compounds (Figure 8.8).

Today, much attention is paid to the search for sustainable methods of recycling waste. However, there is not enough research on the general aspects of using waste as a recycling source. More and more waste is being used as a raw material for other industries, for example, animal waste from agriculture is a source of biogas and different types of biofuels. The use of animal waste for bioenergy purposes allows one to strengthen or provide the energy independence of the industry and reduce the anthropogenic impact on the environment.

Waste management is the main type of animal waste treatment; this method eliminates twice as much waste as the production of animal waste/mixed food products. Incineration is the most common method of handling plant waste, which means that 434 200 tons of energy were burned in 2020. The processing of the three main types of waste generated in rural farms must be practiced more in Ukraine (Table 8.2) [66].

Koletnik [68] in his research proved the importance of the development of the agro-industrial complex of Ukraine in solving the state's energy and environmental

Table 8.1 *Characteristics of the life cycle phases of biogas production.*

System boundaries	Emission	Sources of influence	Environmental benefit
Gathering of the waste *Animal manure* *Plant residues*	Airborne emissions: CO_2, CH_4, CO NH_4, NOx, and SO_2	Livestock feed production, transportation, manure management, plant collection, and transportation	Reduce airborne emissions and global warming potential
Proccessing *Pretreatment* *Anaerobic digestion*	Airborne emissions: CO_2, CH_4, CO NH_4, NOx, and H_2S Waterborne emissions: NH_4, K, and P Emissions to soil: NH_4, K, and P	Leakages from the digester unit or intentional releases to the atmosphere, in cases of biogas production exceeding its consumption, and biogas pressure is created in the digester	Ways to reduce environmental burdens: the production of high value-added coproducts and the reduction of pollutant emissions from manure management
Useful products *Biogas* *Digestate* *Cogeneration*	Airborne emissions: CO_2, CH_4, CO NH_4, NOx, and H_2S Waterborne emissions: NH_4, K, and P Emissions to soil: NH_4, K, and P	Leakages from nooks and cracks in the pipping	Manure treated under anaerobic digestion can supply electricity consumption and produce a biofertilizer with N content. Efficiency can be achieved by using digestate as a fertilizer, avoiding the impact of minerals fertilizers. The environmental efficiency of the biogas cogeneration option is higher.

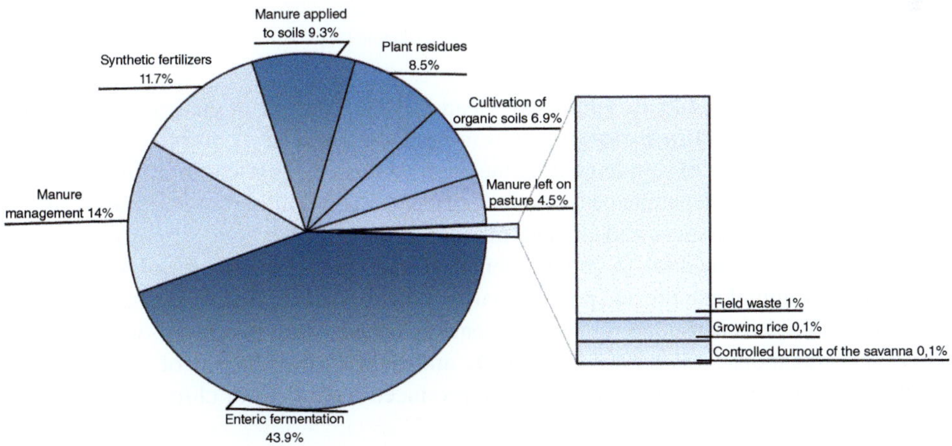

Figure 8.7 *The amount of waste generation in Ukraine by type of economic activity in 2019. Source: Adapted from Official site of the State Statistics Service of Ukraine [66].*

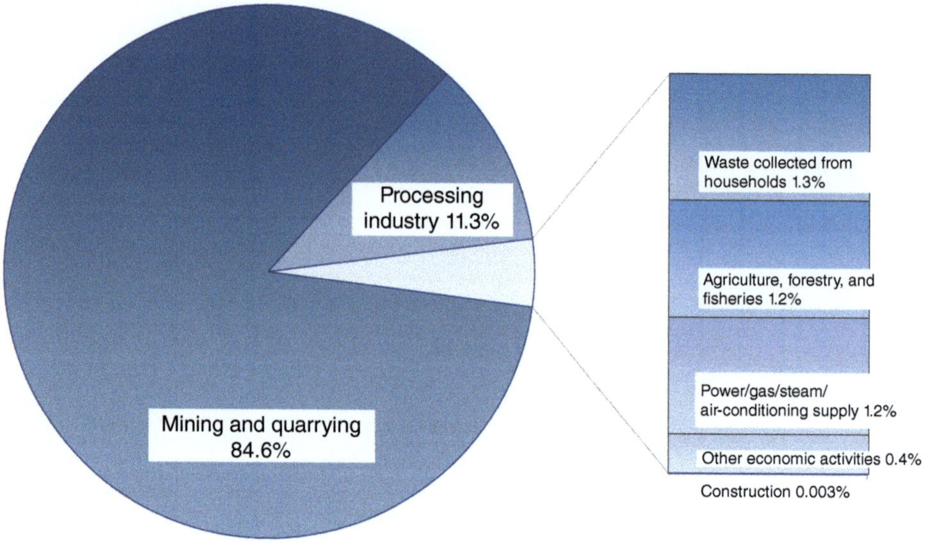

Figure 8.8 Structure of emissions of the agro-industrial complex of Ukraine, average indicator 1990–2017, CO_2 equivalent. Source: Adapted from Honcharuk [67].

Table 8.2 Generation and management of waste of the I–IV hazard classes by categories of materials in Ukraine's agriculture in 2019, 1000 tons.

Type	Formed	Recycled	Burned	Removed to special places or facilities
Animal waste and mixed food waste	405.4	203.4	4.5	1.9
Plant waste	6101.8	1502.5	480.2	15.5
Animal excrement, urine, and manure	3314.7	2324.6	–	0

Source: Adapted from Official site of the State Statistics Service of Ukraine [66].

security problems. In general, scientists considered the concept of energy independence by using agricultural waste as raw materials to produce biofuels.

An additional source of supplementing the energy balance of agricultural enterprises and ensuring the energy independence of the industry can be agricultural waste, mainly animal waste, such as cattle manure and chicken manure.

Ukraine has the potential to produce energy from renewable sources by processing animal waste (animal manure and poultry manure) for the production of biogas. Biogas produced from biomass is used as an environmentally friendly fuel because it does not cause additional greenhouse gas emissions of CO_2 and reduces the amount of organic waste. Unlike wind and solar energy, biogas can be produced regardless of climatic and weather conditions [69].

Biogas, as a product of biotransformation with high methane content, can be used to produce electricity, which is sold at a "green tariff," heat for use or sale, hot water, and also as fertilizer for organic waste biogas plants (digestate) [70].

Goncharuk I.V. and Tomashuk I.V. [71], while studying the prospects of using agricultural waste for energy independence, explained that every year large livestock and poultry farms could obtain environmentally safe biofertilizers by AD and significantly improve wastewater quality. The conversion of a certain amount of manure produced by one cow per year can be about 500 m^3 of biogas. From 1 ton of fresh cattle manure, it is possible to obtain 30–50 m^3 of biogas, from pigs 50–80 m^3, and straw and grass 30–60 m^3. Biotechnology includes the complex processing and biological utilization of waste. The use of AD of manure allows 36 kg of nitrogen from 37 kg of manure to be returned to the soil as fertilizer and 12–15 kg under normal fermentation conditions.

In summary, the utilization of agricultural waste, particularly livestock waste through conversion to biogas, is an important aspect of environmental safety and energy independence, which means the use of renewable raw material resources and the elimination of fossil energy sources. However, the economic benefits of using biogas in each case will depend on the type of waste available for processing, investment opportunities, the availability of the local energy market, and government initiatives.

Today agricultural waste is, on the one hand, a source of environmental pollution. On the other hand, it is the type of waste that, in an environmentally safe way, transforms into energy and fuel, thus ensuring energy independence. After analyzing a number of European and national normative documents concerning the regulation of waste generation and agricultural waste, it was found that there is no clear definition of the category "agricultural waste" both in normative legal acts and in the opinion or theory of scientists, so these authors proposed a personal approach to understanding the essence of the category of labeled waste [72, 73].

Ukraine has unused potential for the development of renewable energy sources, the production of biogas from animal waste, which means the manure of animals and birds. One of the main points of biogas production from animal waste is the opportunity to solve environmental problems of waste management and obtain economic benefits. Dung and manure are byproducts of animal husbandry, and their formation in large amounts is the main source of nitrate pollution of soils, surface, and groundwater.

Animal waste also releases greenhouse gases into the air, significantly impacting the environment and exacerbating global environmental problems such as global warming and climate change.

The Ukrainian biogas market is in its development stage but has good growth potential [74]:

1. The level of domestic waste processing in Ukraine in 2020 does not exceed 6–7%, this opens up a perspective for the share growth of waste processing into biogas.
2. Ukrainian agriculture (the main source of raw material for biogas production) is developing rapidly. The beginning of agricultural land sales is expected to lead to an intensification of investments in agriculture and a subsequent increase in the share of industrial waste recycling of farming enterprises. Currently, no more than 5% of the residue in the fields is used for processing and energy production.
3. The land market will allow an inventory to be made and allocate the territories that are not suitable for the cultivation of cereals but suitable for the cultivation of particular energy crops for the production of biogas. This land is estimated to be 4 million hectares

and the potential volume of natural gas substitution is 20 billion cubic meters by growing bioenergy crops.

4. The intended legislative changes, according to which "green" auctions will begin to work in Ukraine for the guaranteed purchase of electricity from renewable energy sourcess (RES) by the State for 20 years, including biogas power plants. With the payback period of five to eight years of biogas projects, such state support is substantial and allows one to recoup the investments. Furthermore, until 2030 the "green" tariff for biomass and biogas power plants remains unchanged, and from 2020 it becomes higher than for wind and ground-based solar power plants.
5. Unlike other types of renewable energy, biogas power plants do not depend on solar weather, overcast days, nights, wind gusts, etc., but provide an uninterrupted alternative energy source if there is access to a stable supply of raw materials. This fact is important not only for an investor in terms of predictability of cash flow and biogas production but also for the stability of the energy system as a whole in the network connection of the biogas power plant and for the work in the market balancing capacity.
6. Biogas can be used not only for power generation. Heat is also generated during biogas production. From 1 cubic meter of biogas, about 2–2.5 kWh of electricity and up to 2.5–3 kWh of heat energy can be produced by cooling engines after burning biogas for electricity production. Furthermore, when biogas is purified from CO_2 (carbon dioxide), it becomes natural gas (biomethane), which can be used to fuel cars. Moreover, after producing biogas from biomass, it becomes a biofertilizer that, in its composition, in many cases is not lower than chemical fertilizers and is environmentally friendly [75].

8.7 Conclusion

An important side of biogas production is the use of renewable energy sources, which are waste at the same time. The use of organic waste creates an ecological effect on transportation, storage, and use. The ecological effect of biogas production is the environmentally safe processing of organic waste and byproducts through AD. The agro-industrial industry is one of the largest sources of supply of methane and other greenhouse gases into the atmospheric air, and biogas production plants can be considered as emission reduction implementations. Reducing greenhouse gas emissions is achieved by replacing the energy produced from nonrenewable sources with energy produced from alternative sources.

Production of electricity from the combustion of biogas in cogeneration units is a solution to the issues of energy independence of the enterprise. A significant aspect is that fermentation residues from biogas production in biogas plants are a quality fertilizer that can be sold and used instead of mineral fertilizer.

References

1. European Biogas Association. European Biogas Association. https://www.europeanbiogas.eu (accessed September 2021)
2. Eurostat | European Commission. Renewable energy statistics – Statistics Explained. https://ec.europa.eu/eurostat/statistics-explained/index.php?title=Renewable_energy_statistics (accessed September 2021)

3. European Environment Agency. Greenhouse gas emissions from agriculture in Europe. https://www.eea.europa.eu/ims/greenhouse-gas-emissions-from-agriculture (accessed October 2021)
4. OECD. (2009). The Bioeconomy to 2030: designing a policy agenda – OECD. https://www.oecd.org/futures/long-termtechnologicalsocietalchallenges/thebioeconomyto2030designingapolicyagenda.htm
5. Scopus – Elsevier. Expertly curated abstract & citation database. https://www.scopus.com/search/form.uri?display=basic&zone=header&origin=#basic (accessed June 2021)
6. Aggeek | Actual Knowledge. (2021). Accurate farming technologies that reduce greenhouse gas emissions. https://aggeek.net/ru-blog/tehnologii-tochnogo-zemledeliya-umenshayuschie-vybrosy-parnikovyh-gazov
7. US EPA. Global Greenhouse Gas Emissions Data | US EPA. https://www.epa.gov/ghgemissions/global-greenhouse-gas-emissions-data (accessed August 2021)
8. Leahy, S., Clark, H., and Reisinger, A. (2020). Challenges and prospects for agricultural greenhouse gas mitigation pathways consistent with the Paris agreement. *Front Sustain Food System* 22 (4): 69. https://doi.org/10.3389/fsufs.2020.00069.
9. Radicetti, E., Osipitan, O., Langeroodi, A. et al. (2019). CO_2 flux and C balance due to the replacement of bare soil with agro-ecological service crops in Mediterranean environment. *Agriculture* 9 (4): 71. https://doi.org/10.3390/agriculture9040071.
10. Carrera, G.V.S.M., Branco, L.C., and da Ponte, M.N. (2017). Bio-inspired systems for carbon dioxide capture, sequestration and utilization. In: *Recent Advances in Carbon Capture and Storage* (ed. Y. Yun). InTech http://www.intechopen.com/books/recent-advances-in-carbon-capture-and-storage/bio-inspired-systems-for-carbon-dioxide-capture-sequestration-and-utilization.
11. Gayathri, R., Mahboob, S., Govindarajan, M. et al. (2021). A review on biological carbon sequestration: a sustainable solution for a cleaner air environment, less pollution and lower health risks. *Journal of King Saud University – Science* 33 (2): 101282. https://doi.org/10.1016/j.jksus.2020.101282.
12. Bioenergy | International Collaboration in Bioenergy. (2017). https://www.ieabioenergy.com/wp-content/uploads/2018/01/Two-page-summary----Methane-emissions-from-biogas-plants-R1.pdf
13. O'Shea, R., Lin, R., Wall, D.M. et al. (2020). Using biogas to reduce natural gas consumption and greenhouse gas emissions at a large distillery. *Applied Energy.* 279: 115812. https://doi.org/10.1016/j.apenergy.2020.115812.
14. Kucher, O., Hutsol, T., Glowacki, S. et al. (2022). Energy potential of biogas production in Ukraine. *Energies.* 15 (5): 1710. https://doi.org/10.3390/en15051710.
15. Daniel-Gromke, J., Liebetrau, J., Denysenko, V., and Krebs, C. (2015). Digestion of bio-waste – GHG emissions and mitigation potential. *Energ Sustain Soc.* 5 (1): 3. https://doi.org/10.1186/s13705-014-0032-6.
16. Environmental and Energy Study Institute | Ideas. Insights. Sustainable Solutions. (2017). Biogas: Converting Waste to Energy | EESI. https://www.eesi.org/papers/view/fact-sheet-biogasconverting-waste-to-energy Fact Sheet | Biogas: Converting Waste to Energy. Environmental and Energy Study Institute (EESI) https://www.eesi.org/papers/view/fact-sheet-biogasconverting-waste-to-energy
17. World Biogas Association | Making Biogas Happen. https://www.worldbiogasassociation.org/wp-content/uploads/2018/07/WBA-Urban-Air-Quality-Biogas-factsheet1.pdf
18. Trypolska, G., Kyryziuk, S., Krupin, V. et al. (2021). Economic feasibility of agricultural biogas production by farms in Ukraine. *Energies* 15 (1): 87. https://doi.org/10.3390/en15010087.
19. Fernando-Foncillas, C., Estevez, M.M., Uellendahl, H., and Varrone, C. (2021). Co-management of sewage sludge and other organic wastes: a scandinavian case study. *Energies* 14 (12): 3411. https://doi.org/10.3390/en14123411.
20. Saleh, H.M. and Hassan, A.I. (2021). The potential of sustainable biogas production from animal waste. In: *Advanced Technology for the Conversion of Waste into Fuels and Chemicals*, 115–134. Elsevier https://linkinghub.elsevier.com/retrieve/pii/B9780128231395000034.

21. Koppelmäki, K., Parviainen, T., Virkkunen, E. et al. (2019). Ecological intensification by integrating biogas production into nutrient cycling: modeling the case of agroecological symbiosis. *Agricultural Systems.* 170: 39–48. https://doi.org/10.1016/j.agsy.2018.12.007.
22. De Clercq, D., Wen, Z., and Song, Q. (2019). Innovation hotspots in food waste treatment, biogas, and anaerobic digestion technology: a natural language processing approach. *Science of The Total Environment.* 673: 402–413. https://doi.org/10.1016/j.scitotenv.2019.04.051.
23. Hamdi, H., Hechmi, S., Khelil, M.N. et al. (2019). Repetitive land application of urban sewage sludge: effect of amendment rates and soil texture on fertility and degradation parameters. *CATENA* 172: 11–20. https://doi.org/10.1016/j.catena.2018.08.015.
24. Pietrangeli, B. and Lauri, R. (2018. http://www.intechopen.com/books/advances-in-biofuels-and-bioenergy/biogas-production-plants-a-methodological-approach-for-occupational-health-and-safety-improvement). Biogas production plants: a methodological approach for occupational health and safety improvement. In: *Advances in Biofuels and Bioenergy* (ed. M. Nageswara-Rao and J.R. Soneji). InTech. doi: 10.5772/intechopen.72819.
25. (2019). *Safety and Practice for Organic Food.* Elsevier. Science Direct. Composted Manure https://www.sciencedirect.com/topics/agricultural-and-biological-sciences/composted-manure.
26. Working Group Compost – Consulting & Development (2004). Heavy metals and organic compounds from wastes used as organic fertilisers. (Austria). Ref. Nr. TEND/AML/2001/07/20, pp. 73–74. http://ec.europa.eu/environment/waste/compost/pdf/hm_
27. Zupančič, M., Možic, V., Može, M. et al. (2022). Current status and review of waste-to-biogas conversion for selected European countries and worldwide. *Sustainability* 14 (3): 1823. https://doi.org/10.3390/su14031823.
28. Pucker, J., Jungmeier, G., Siegl, S., and Pötsch, E.M. (2013). Anaerobic digestion of agricultural and other substrates – implications for greenhouse gas emissions. *Animal.* 7: 283–291. https://doi.org/10.1017/S1751731113000840.
29. Valli, L., Rossi, L., Fabbri, C. et al. (2017). Greenhouse gas emissions of electricity and biomethane produced using the BiogasdonerightTM system: four case studies from Italy. *Biofuels, Bioprod Bioref.* 11 (5): 847–860. https://doi.org/10.1002/bbb.1789.
30. Guidelines for sustainable manure management in Asian livestock production systems. Vienna (Austria): IAEA; 2008. 125. https://www-pub.iaea.org/MTCD/publications/PDF/TE_1582_web.pdf
31. Chastain, J.P. and Henry, S. (2015). *Swine Training Manual Table of Contents.* Clemson University. Management of lagoons and storage structures for swine manure; pp. 4–31. https://www.clemson.edu/extension/camm/manuals/swine/sch4_03.pdf.
32. AgSTAR (2018). *Market Opportunities for Biogas Recovery Systems at U.S. Livestock Facilities.* Washington, D.C.: U.S. Department of Agriculture & U.S. Department of Energy. 42. https://www.epa.gov/sites/default/files/2018-06/documents/epa430r18006agstarmarketreport2018.pdf
33. Penn State Extension | The Pennsylvania State University. Biogas from Manure. https://extension.psu.edu/biogas-from-manure (accessed October 2021)
34. American Chemical Society. (2020). Cow Power: A Climate Change Solution From Manure - American Chemical Society. https://www.acs.org/content/acs/en/education/resources/highschool/chemmatters/past-issues/2019-2020/apri-2020/cow-power.html
35. RHS - Inspiring everyone to grow / RHS Gardening. Organic matter: how to use in the garden / RHS Gardening. www.rhs.org.uk/soil-composts-mulches/organic-matter-how-to-use-in-garden (accessed August 2021)
36. (2021). *Soft Computing Techniques in Solid Waste and Wastewater Management.* Elsevier. Science Direct. Organic Waste https://www.sciencedirect.com/topics/earth-and-planetary-sciences/organic-waste.
37. Jones DD, Koelsch RK, Mukhtar S, Sheffield R, Worley JW. Closure of earthern manure structures (including basins, holding ponds and lagoons). *Conference Presentations and White Papers: Biological Systems Engineering.* 2006;4. https://digitalcommons.unl.edu/biosysengpres/4

38. Zhu T, Curtis J, Clancy M. Promoting agricultural biogas and biomethane production: lessons from cross-country studies. *Renewable and Sustainable Energy Reviews* 2019 114:109332. https://doi.org/10.1016/j.rser.2019.109332
39. Kim, D., Kim, K.T., and Park, Y.K. (2020). A comparative study on the reduction effect in greenhouse gas emissions between the combined heat and power plant and boiler. *Sustainability* 12 (12): 5144. https://doi.org/10.3390/su12125144.
40. Gasum. Renewable biogas for efficient emission cuts I Gasum. https://www.gasum.com/en/our-operations/biogas-production/biogas-emissions (accessed August 2021)
41. Semenenko, I.V. (1996). Design of biogas plants. Sumy: PF "McDen", IPP "Mriya-1" Ltd., p. 347.
42. Lyubin, M.V., Tsurkan, O.V., and Tokarchuk, D.M. (2012). Fundamentals of launching and operation of biogas plants for farms. *Collection of Scientific Works of Vinnytsia National Agrarian University*. 10(58):69. http://repository.vsau.vin.ua/repository/getfile.php/6608.pdf.
43. Semenenko, I.V. and Zinchenko, M.G. (2012). Equipment and processes of methane digestion of organic waste : monograph. Kharkiv: Pidruchnik NTU "KhPI", p. 272. http://library.kpi.kharkov.ua/files/new_postupleniya/semenenko_oborud.pdf
44. Janke, L., McCabe, B.K., Harris, P. et al. (2019). Ensiling fermentation reveals pre-treatment effects for anaerobic digestion of sugarcane biomass: an assessment of ensiling additives on methane potential. *Bioresource Technology*. 279: 398–403. https://doi.org/10.1016/j.biortech.2019.01.143.
45. Karpenko, V.I., Kozlov, V.V., Golodok, L.P., and Gorlinsky, O.V. (2012). Waste utilization with the production of biofuels and fertilizers. *Problems of Ecological Biotechnology* 2: 97–123. http://nbuv.gov.ua/UJRN/peb_2012_2_10.
46. Opolinsky, I. (2019). Improving the technology of organic waste utilization by anaerobic fermentation with preliminary destruction of the substrate [dissertation for degree of Candidate of Technical Sciences]. Kyiv: KPI, 194 p.
47. Kozlovets, O.A. (2017). Biotechnology of biogas production in coenzyme of bird droppings [dissertation for degree of Candidate of Technical Sciences]. Kyiv: KPI, 189 p.
48. Barłóg, P., Hlisnikovský, L., and Kunzová, E. (2020). Effect of digestate on soil organic carbon and plant-available nutrient content compared to cattle slurry and mineral fertilization. *Agronomy* 10 (3): 379. https://doi.org/10.3390/agronomy10030379.
49. Koszel, M. and Lorencowicz, E. (2015). Agricultural use of biogas Digestate as a replacement fertilizers. *Agriculture and Agricultural Science Procedia* 7: 119–124. https://doi.org/10.1016/j.aaspro.2015.12.004.
50. Baral, K.R., Jégo, G., Amon, B. et al. (2018). Greenhouse gas emissions during storage of manure and digestates: key role of methane for prediction and mitigation. *Agricultural Systems*. 166: 26–35. https://doi.org/10.1016/j.agsy.2018.07.009.
51. Tilvikiene, V., Venslauskas, K., Povilaitis, V. et al. (2020). The effect of digestate and mineral fertilisation of cocksfoot grass on greenhouse gas emissions in a cocksfoot-based biogas production system. *Energ Sustain Soc.* 10 (1): 13. https://doi.org/10.1186/s13705-020-00245-6.
52. Czubaszek, R. and Wysocka-Czubaszek, A. (2018). Emissions of carbon dioxide and methane from fields fertilized with digestate from an agricultural biogas plant. *International Agrophysics*. 32 (1): 29–37. https://doi.org/10.1515/intag-2016-0087.
53. Łukajtis, R., Hołowacz, I., Kucharska, K. et al. (2018). Hydrogen production from biomass using dark fermentation. *Renewable and Sustainable Energy Reviews*. 91: 665–694. https://doi.org/10.1016/j.rser.2018.04.043.
54. Ding, C., Yang, K.L., and He, J. (2016). Biological and fermentative production of hydrogen. In: *Handbook of Biofuels Production*, 303–333. Elsevier https://linkinghub.elsevier.com/retrieve/pii/B9780081004555000114.
55. Markov, S.A., Protasov, E.S., Bybin, V.A., and Stom, D.I. (2013). Hydrogen production by microorganisms and microbial fuel cells using wastewater and waste products. *International Scientific Journal*

for Alternative Energy and Ecology 1/2 (118): 108–116. https://www.researchgate.net/publication/281626594.
56. Chusov, A.N., Molodtsov, D.V., Fedorov, M.P., and Maslikov, V.I. (2011). Experimental complex for the production of hydrogen from organic waste for use in fuel cells. *Energy, Electrical Engineering* 4 (135): 35–41.
57. Aziz, N.I.H.A., Hanafiah, M.M., and Gheewala, S.H. (2019). A review on life cycle assessment of biogas production: challenges and future perspectives in Malaysia. *Biomass and Bioenergy.* 122: 361–374. https://doi.org/10.1016/j.biombioe.2019.01.047.
58. Hijazi, O., Munro, S., Zerhusen, B., and Effenberger, M. (2016). Review of life cycle assessment for biogas production in Europe. *Renewable and Sustainable Energy Reviews.* 54: 1291–1300. https://doi.org/10.1016/j.rser.2015.10.013.
59. D'Imporzano, G., Pilu, R., Corno, L., and Adani, F. (2018). *Arundo donax* L. can substitute traditional energy crops for more efficient, environmentally-friendly production of biogas: a life cycle assessment approach. *Bioresource Technology.* 267: 249–256. https://doi.org/10.1016/j.biortech.2018.07.053.
60. Sun, C.H., Fu, Q., Liao, Q. et al. (2019). Life-cycle assessment of biofuel production from microalgae via various bioenergy conversion systems. *Energy* 171: 1033–1045. https://doi.org/10.1016/j.energy.2019.01.074.
61. Wang, Y., Wu, X., Tong, X. et al. (2018). Life cycle assessment of large-scale and household biogas plants in Northwest China. *Journal of Cleaner Production.* 192: 221–235. https://doi.org/10.1016/j.jclepro.2018.04.264.
62. Ioannou-Ttofa, L., Foteinis, S., Seifelnasr Moustafa, A. et al. (2021). Life cycle assessment of household biogas production in Egypt: influence of digester volume, biogas leakages, and digestate valorization as biofertilizer. *Journal of Cleaner Production.* 286: 125468. https://doi.org/10.1016/j.jclepro.2020.125468.
63. Singh, A.D., Upadhyay, A., Shrivastava, S., and Vivekanand, V. (2020). Life-cycle assessment of sewage sludge-based large-scale biogas plant. *Bioresource Technology.* 309: 123373. https://doi.org/10.1016/j.biortech.2020.123373.
64. Cahyani, D., Haryanto, A., Putra, G.A. et al. (2019). Life cycle assessment of biogas digester in small scale tapioca industry. *IOP Conf Ser: Earth Environ Sci.* 258: 012017. https://doi.org/10.1088/1755-1315/258/1/012017.
65. Rivas-García, P., Botello-Álvarez, J.E., Abel Seabra, J.E. et al. (2015). Environmental implications of anaerobic digestion for manure management in dairy farms in Mexico: a life cycle perspective. *Environmental Technology.* 36 (17): 2198–2209. https://doi.org/10.1080/09593330.2015.1024758.
66. Official site of the State Statistics Service of Ukraine. www.ukrstat.gov.ua (accessed September 2021)
67. Honcharuk, I. (2020). Use of wastes of the livestock industry as a possibility for increasing the efficiency of AIC and replenishing the energy balance. *Visegrad Journal on Bioeconomy and Sustainable Development.* 9 (1): 9–14.
68. Kaletnik, G.M. (2009). The role of the agro-industrial complex of Ukraine in solving the problems of energy and environmental security of the state. *Agrosvit* 22: 2–5. http://nbuv.gov.ua/UJRN/agrosvit_2009_22_2.
69. Andreichenko, A.V. (2017). Typology of waste in agriculture: domestic and European experience. *Economic Space* 124: 67–76. http://nbuv.gov.ua/UJRN/ecpros_2017_124_8.
70. Honcharuk, I. (2020). Biogas production in the agricultural sector — the way to increase energy independence and soil fertility. *Agrosvit.* 15: 18.
71. Honcharuk, I.V. and Tomashuk, I.V. (2019). Economic efficiency of energy autonomy of agro-industrial complex due to the use of biofuels. *Economy, Finances, Management: Current Issues of Science and Practice* 2: 7–19.
72. (2002). *Shomin AA Biogas na selskom podvorie: kniga / AA Shomin*, 68. Balakleya: Information and Publishing Company "Balakliyshchina" http://nbuv.gov.ua/UJRN/efmapnp_2019_2_3.

73. Home Guides | SF Gate. Inorganic Fertilizer Vs. Organic Fertilizer. https://homeguides.sfgate.com/inorganic-fertilizer-vs-organic-fertilizer-39528.html (accessed September 2021)
74. Green energy technologies market. (2019). Investing in alternative energy: how biogas is earned in Ukraine. Доступно на: https://getmarket.com.ua/ru/news/investicii-v-al-ternativnuyu-energetiku-kak-v-ukraine-zarabatyvayut-na-biogaze
75. Geletukha, G.G., Kucheruk, P.P., and Matveev, Y.B. (2014). Prospects for the production and use of biomethane in Ukraine. *Analytical note of UAB.* 11: 42. https://uabio.org/wp-content/uploads/2020/04/position-paper-uabio-11-ua.pdf.

9

Hybrid Environmental and Economic Assessment of Biogas Plants in Integrated Organic Waste Management Strategies

Amal Elfeky[1], Kazi Fattah[2], and Mohamed Abdallah[1]

[1]Department of Civil and Environmental Engineering, University of Sharjah, Sharjah, United Arab Emirates

[2]Department of Civil, Environmental, and Architectural Engineering, University of Kansas, Kansas, United States of America

9.1 Introduction

The rapid economic development, urbanization, and population growth have resulted in a significant amount of generated of municipal solid waste (MSW). In 2016, over 2 billion tons of MSW were generated worldwide, and the number is expected to increase to 3.4 billion tons by 2050 [1]. Although landfills are known to pose significant environmental and health hazards, they continue to be the most widely used MSW management facilities handling 75% of the total waste generated globally. This constitutes the third largest cause of anthropogenic methane in the world, a greenhouse gas (GHG) which has a 28-fold greater global warming impact than CO_2 over 100 years [2]. It is expected that existing open dumps will account for 10% of global GHG emissions by 2025 [3]. More recently, the recovery of materials and/or energy from waste via waste-to-energy (WTE) systems has gradually replaced conventional disposal practices in the waste management sector to achieve sustainability [4]. Waste conversion provides a sustainable solution to

Biogas Plants: Waste Management, Energy Production and Carbon Footprint Reduction, First Edition. Edited by Wojciech Czekała

reduce the environmental impacts of landfills and supply clean energy that diversifies the energy mix along with fossil fuels [5].

The sustainable hierarchy of waste management includes the reduction of generated waste quantities, waste reuse or recycling, waste treatment using cost-effective technologies, and safe disposal in landfills. WTE technologies can be classified under two categories: biochemical and thermochemical conversion processes. Biochemical systems, e.g. anaerobic digestion (AD) and fermentation, involve the use of bacteria or other microorganisms to break down biomass into methane-rich biogas [6]. In contrast, thermochemical systems utilize heat and chemical reactions in the production of energy products from biomass; examples of those systems are incineration and pyrolysis. This chapter compares AD to incineration as well-established representatives of biochemical and thermochemical processes, respectively. To strategically prioritize AD plants, it is important to assess their environmental and financial impacts, particularly in developing countries. There are different methods to thoroughly assess the environmental impacts of WTE facilities, one of which is the life cycle assessment (LCA) [7, 8]. For the financial aspects, a life cycle costing (LCC) is typically conducted to assess the potential economic impacts of those projects [9].

Throughout the literature, a limited number of studies have assessed AD plants as a part of an integrated solid waste management (ISWM) strategy. A comprehensive analysis of ISWMs is essential to cover potential modifications imposed by AD plants, including changes to the collection and segregation programs, and handling the byproducts of those facilities. To date, no previous work has been published in the literature that covers a comparative eco-efficiency study of those technologies in the Middle East and North Africa (MENA) region. Despite the present plans to shift toward renewable energy sources and foster WTE projects, there is a lack of knowledge and very limited full-scale applications of those technologies in the region. The main objective of this chapter is to conduct a thorough life cycle analysis of WTE strategies toward sustainable management of organic wastes. The specific objectives included:

(1) Propose ISWM strategies, incorporating AD plants, and conduct relevant material and energy balances.
(2) Assess the life cycle environmental footprint of the selected WTE systems in terms of multiple LCA impact categories using the systemic ISO guidelines.
(3) Conduct LCC analysis to evaluate the potential economic impacts of the examined ISWM strategies via various financial indicators.
(4) Combine the LCA and LCC findings through an eco-efficiency assessment to assess the examined WTE-based strategies for the management of organic wastes in the United Arab Emirates (UAE).

9.2 Methodology

9.2.1 Overview

A systematic methodology is followed in the present chapter to conduct LCA of multiple waste management systems as shown in Figure 9.1. At first, the proposed waste management strategies incorporating WTE systems for the treatment of the organic fraction of MSW are defined. Next, a life cycle model of the examined strategies is developed including

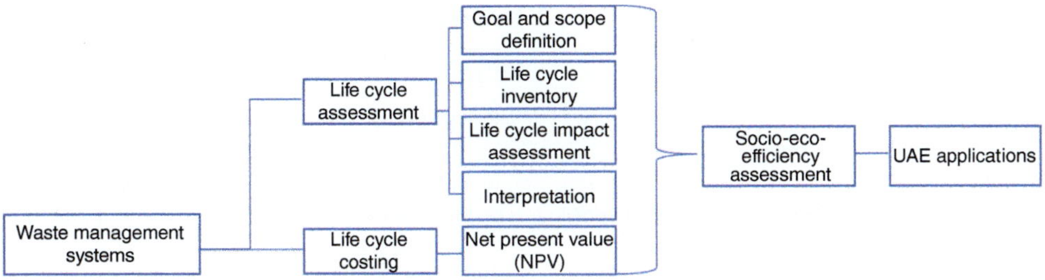

Figure 9.1 *Proposed methodology framework of the present chapter.*

various processes, material/energy streams, and project phases. The current and projected environmental footprints are computed through six environmental impact categories, while economic aspects are computed through a set of financial indicators. It should be noted that the proposed methodology is applied on the UAE as an example of a developing country shifting toward WTE-based waste management systems.

9.2.2 Waste Management Scenarios

In this chapter, the assessed WTE-based waste management strategies constitute incineration and/or AD compared to the conventional practices, i.e. landfilling. The following assumptions were considered in the proposed strategies:

1. One bin collection system of organic waste (120 L wheeled bin) is used to segregate the required feedstocks at source to increase the material recovery efficiency.
2. As the segregation efficiency at source is rarely perfect, i.e. the organic waste stream would include other materials that are falsely classified, the analysis assumes a 100% segregation efficiency at source for simplicity.
3. Although the proposed strategies are based on material and energy recovery, the landfill remains an essential element for the disposal of waste processing byproducts or unprocessed wastes.
4. It is assumed that 90% of the waste processed at an incinerator is recovered as heat and gas, while the remaining 10% is converted into ash that is disposed at landfill. In contrast, in the AD plant, 55% of the processed waste is converted to digestate that can be marketed, incinerated, or landfilled with an estimated moisture content of 18%. It is also assumed that 5.5% of the waste input to the AD plant is rejected and disposed in the landfill.
5. Incinerators produce heat to generate electricity through steam turbines, while AD plants produce biogas that is combusted to produce electricity.

The proposed scenarios for the treatment of organic waste are presented in Figure 9.2 and discussed as follows:

1. BAU: the current waste management system is set as the business-as-usual (BAU) strategy, i.e. conventional landfilling, compared to the proposed alternative ISWM strategies.
2. LFGR: similar to the BAU scenario, organic waste is disposed of in a landfill, but with recovering the landfill gas (LFG) to assess the impact of its utilization on the eco-efficiency performance of the traditional strategy.

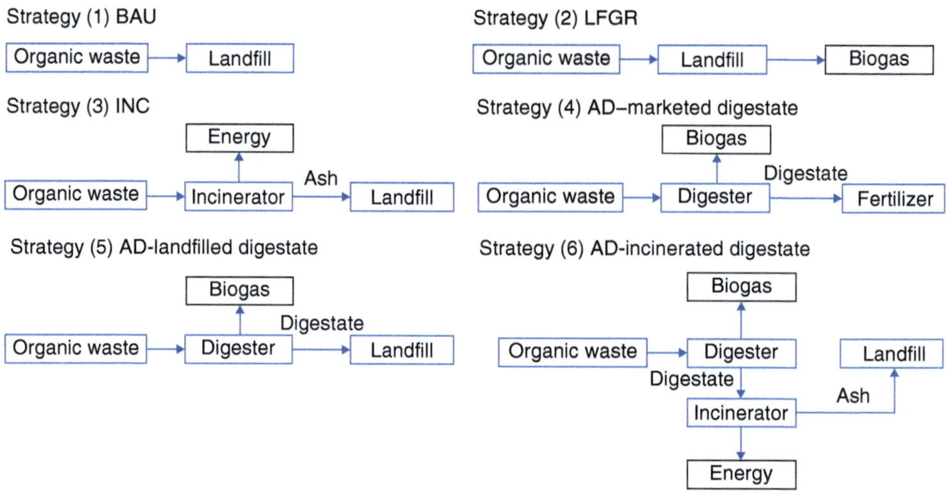

Figure 9.2 Schematic flow diagram of the proposed BAU and WTE-based waste management strategies.

3. INC: organic waste is processed at an incineration facility. The recovered heat is used to generate electricity, while the produced ash is disposed of in the landfill.
4. AD-marketed digestate: organic waste is processed at an anaerobic digester and the recovered digestate is marketed as a fertilizer.
5. AD-landfilled digestate: organic waste is processed at an anaerobic digester and the recovered digestate is disposed of in a landfill.
6. AD-incinerated digestate: organic waste is processed at an anaerobic digester and the recovered digestate is processed at an incineration facility. The recovered heat from the incinerator is utilized to generate electricity, while the produced ash is disposed of in the landfill.

9.2.3 Life Cycle Assessment

A LCA assesses all aspects from raw material extractions, production, use, and disposal to define the hotspots and focus on the decision-making process related to the environmental impact assessment of the overall process [10]. According to ISO 14040 guidelines, LCA has four mandatory steps required to assess any product/system: (i) goal and scope definition, (ii) life cycle inventory (LCI), (iii) life cycle impact assessment, and (iv) interpretation [11].

9.2.3.1 Goal and Scope Definition

This chapter aims to determine the environmental implications of selected scenarios for managing the organic MSW fraction compared to existing practices. Management of one ton of organic waste generated was selected as the functional unit to which all inputs and outputs of the study are referred to. As shown in Figure 9.3, the proposed LCA system boundary comprises the waste treatment and disposal processes. It should be noted that the collection process was excluded from the present study as it is common among all

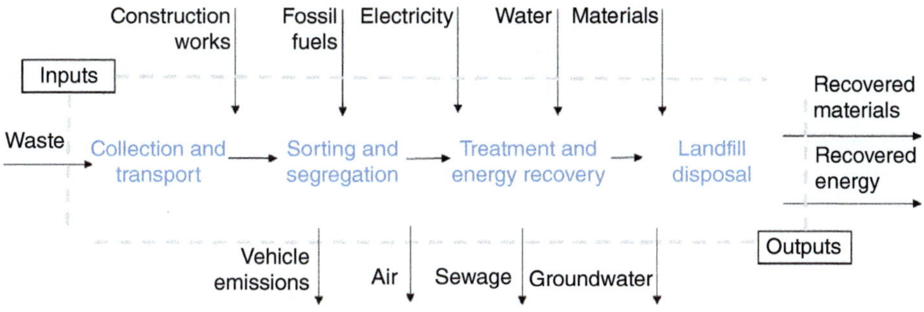

Figure 9.3 General schematic of the LCA system boundary for the examined strategies.

examined strategies and generally has minimal impact relative to other processes [12]. Moreover, since a one-bin collection system of organic waste is assumed in this study, there were not any sorting/segregation facilities. The analysis takes into consideration the recovered byproducts (digestate) through the displacement of equivalent masses of materials that would have been produced from virgin resources. Likewise, the recovered energy from waste transformation is counted as substitution of equivalent amounts of energy that would have been produced from conventional fossil fuels.

9.2.3.2 *Inventory Analysis*

The specific waste characteristics of the study area, such as composition and energy content, as well as the local market and operation parameters, particularly those related to energy consumption and transportation, are used in the LCI. As reliable local data are limited, the Ecoinvent database was utilized to quantify the energy, resource use, and emissions of the processes and materials involved. The main LCI inputs of incineration plants included MSW materials, electricity, fuels, water, and activated carbon (for air pollution control), whereas the outputs include flue gas, bottom ash, electricity generated, process water, and air pollution control residues. In the anaerobic digesters, the main LCI inputs included MSW materials, electricity, and water, whereas the biogas, digestate, exhaust gases, and residues are the main outputs. Moreover, in the landfill, the main inputs included MSW materials, construction and consumable materials, water, and fuels, whereas the main outputs are the LFG, flare/exhaust gases, and leachate. The transportation activities of waste materials, processed byproducts, and marketable products were also considered.

9.2.3.3 *Impact Assessment*

The environmental impacts are computed through the problem-oriented approach, known as the CML methodology [13]. The CML method has an equivalence unit and normalization factor for each impact assessment used for characterization. The assessment covers the environmental impact categories which are considered significant and depicting a wide range of environmental issues in the study area. The impact categories investigated were the global warming potential (GWP), abiotic depletion potential (ADP), acidification potential (ACP), freshwater aquatic ecotoxicity potential (FAETP), human toxicity potential (HTP), and eutrophication potential (ETP). The GWP represents the combined effect of GHGs which absorb heat radiation and consequently increase the atmospheric temperature. In the CML methodology, the reference time horizon for GWP is 100 years, and the values are reported in kg CO_2/ton equivalent. The LCA software utilized was the Waste and Resources Assessment Tool for the Environment (WRATE), which is specialized for waste management projects, and has an embedded database compiled from actual waste management facilities.

9.2.3.4 Interpretation

Analysis and interpretation of the outcomes regarding environmental impact can be found in the chapter discussion. In every scenario, the environmental hotspots are identified to propose possible improvements and/or mitigations.

9.2.4 Life Cycle Costing

The LCC is used to determine the total costs and revenues of the proposed waste management strategies throughout their lifetime. The different cost categories covered in the present study are as follows:

1. Capital cost (CAPEX): it includes the initial investment, planning, engineering and construction, source separation, mechanical handling costs as well as the capital cost of the equipment and construction of the facilities.
2. Operation and maintenance cost (OPEX): it covers the expenses incurred to ensure normal functioning of facilities during their lifespan. OPEX is classified into two categories, fixed and variable OPEX:
   (a) Fixed OPEX: it includes costs of labor, insurance, routine parts replacement, schedules maintenance, etc. Besides, the fixed OPEX is expressed as a percentage of CAPEX. According to the International Renewable Energy Agency (IRENA), the fixed part of OPEX ranges between 3% and 6% of the CAPEX. This research assumes that the fixed OPEX accounts for 4% of the total CAPEX cost [14].
   (b) Variable OPEX: it includes unscheduled maintenance, unplanned equipment replacement, ash disposal, and fuels costs. Moreover, the variable OPEX costs are related to the output of the WTE plant and, therefore, expressed as a per unit value of the WTE plant output, e.g. USD MWh^{-1} [14].

It should be noted that the tipping fees and energy recovery potential (ERP) are the main sources of revenues for the examined WTE-based management strategies.

The LCC of the strategies is examined based on the net present value (NPV). The NPV is the sum of all present values of costs and revenues over the assessment period and the key parameter in evaluating the economic viability of a system, where the negative NPV signifies an unprofitable system and is estimated as [15]:

$$NPV = \sum (CI_t - CO_t) \times (1 + i)^{-t} \qquad (9.1)$$

where the cash inflow for every year t is denoted by CI_t (USD), the cash outflow for every year t is denoted by CO_t (USD), and the percentage of discount rate is denoted by i (%).

9.2.5 Eco-Efficiency Analysis

The eco-efficiency assessment is developed to link the environmental and economic elements of systems and products [16]. In this chapter, an eco-efficiency assessment is employed to combine the environmental footprint and total cost of the ISWMs under consideration. The ISO guidelines 14045 state that the purpose of assessment is to achieve quantification of eco-efficiency on the basis of the system cost to complete environmental impact ratio [17]. The assessment represents a combination of the outcome of LCA calculations of carbon footprint and the LCC calculations of NPV to make a comparative assessment of how eco-efficient the proposed strategies are.

9.2.6 Case Study: The UAE

The environmental and energy challenges in the UAE can be related to having the third highest ecological footprint in the world and an energy mix consisting of 99% of fossil fuels [1]. The UAE is one of the top 10 oil producing countries with approximately 2.8 million barrels per day [18]. In preparation for the post-oil era, the UAE has planned to supply 7% of its energy from renewable sources by 2020, 25% by 2030, and 50% by 2050 [19]. However, the total GHG emissions increased from 180 to 200 million tCO_2e between 2014 and 2017, of which approximately 6% was generated from the waste sector. The UAE vision to reduce GHG emissions by 25% by 2030 [20] is rather challenging, considering that the current energy capacity must be at least doubled to meet the projected energy demand in 2030. WTE systems can be utilized to improve the environmental performance of the country and diversify the energy mix.

In 2018, the UAE waste generation rate was reported to be 1.76 kg $capita^{-1}$ d^{-1}, which translates to an annual total waste generation of around 6.6 million tons, most of which are disposed of in landfills [21]. Based on the annual population growth rate of 2.67%, waste generation is estimated to reach about 11 million tons per year by 2040. The waste profile of the UAE appears to be well suited for WTE conversion. The high organic waste fraction (39%) can allow energy production if the waste is managed in AD systems. The current BAU waste management strategy in the UAE constitutes a single-bin collection of commingled waste, followed by landfill disposal. There is a lack of local guidelines for WTE systems, and waste-related energy policies are not yet developed. However, there has been recent interest in waste conversion, and it was found techno-economically feasible to be implemented in the UAE [22].

As an example, for a developing country shifting to WTE, the following discussion applies the proposed framework, including the proposed WTE-based strategies and eco-efficiency assessment methodology, to local conditions in the UAE. The local waste characteristics along with the UAE-customized LCC input parameters and assumptions are presented in Table 9.1.

9.3 Results and Discussion

A comparative eco-efficiency assessment was performed for the six proposed waste management strategies to evaluate the material and energy streams, environmental burdens, financial feasibility, and social implications. Scenarios included three mono- (BAU,

Table 9.1 Input LCC parameters and assumptions of different waste management facilities.

Facility	Incinerator	Anaerobic digestor	Landfill	References
Population (capita)-2018	9,630,959			[23]
Population growth rate (%)	2.67			—
Waste generation (kg capita^{-1} d^{-1})	1.76			[24]
Total waste generation (ton y^{-1})	6,610,690			—
Fraction of organic waste (%)	39			[25]
ERP (kwh per ton of processed waste)	440.59	1,772.20	148.82	—
Discount rate (%)	10			
Digestate (%)	—	40	—	—
Ash (%)	10	—	—	—
Energy production efficiency (%)	30			[14]
LFG collection efficiency (%)	75			—
CAPEX (USD/kW)	4,000	3,000		[14]
CAPEX (USD ton d^{-1})	—	—	25,000	[26]
OPEX (%)	3.2	2.2	—	[14]
OPEX (USD ton^{-1})	—	—	10	[26]
Electricity tariff (USD kWh^{-1})	0.08			[22]
Landfilling cost (USD ton^{-1})	28			
Landfill tipping fee (USD ton^{-1})	28			
WTE plants tipping fee (USD ton^{-1})	14	14	—	
Digestate price (USD ton^{-1})	—	5	—	—
Transportation cost (USD km^{-1} t^{-1})	Household to Incinerator/AD		0.27	[27]
	Household/Incinerator/AD to Landfill		0.33	

AD, or INC), and one dual- (AD-incinerated digestate) management strategies, in all which waste processing byproducts were landfilled except for the AD-marketed digestate scenario. The individual impacts of the primary management processes (transportation, recycling, energy recovery, and disposal) were detailed to indicate the environmental hotspots of each strategy. Moreover, the costs and revenues of each strategy were computed. The discussion analyzes the potential energy production and corresponding revenues from the proposed strategies throughout the assessment period.

9.3.1 Material and Energy Recovery

One of the primary sources of GHG emissions in waste management is LFGs generated from the anaerobic biodegradation of organic wastes. Therefore, it was essential to determine the biodegradable fraction of landfilled waste in each scenario. Figure 9.4 shows the amounts (in million tons) of biodegradable and nonbiodegradable wastes disposed of in each scenario. The BAU strategy had the largest biodegradable landfilled waste stream of 6.6 Mt, whereas the lowest amount occurred in the INC scenario (zero due to complete incineration). Similarly, AD strategies achieved low biodegradable landfilled wastes (around 70–150 kt) when the digestate was incinerated or disposed in landfill. However, when digestate was marketed, the amount of biodegradable landfilled waste was decreased by 90%. Overall, for the same strategy, changing the destination of the waste processing byproducts resulted in a significantly different distribution of key polluting streams, leading to a different environmental performance, as emphasized throughout the discussion.

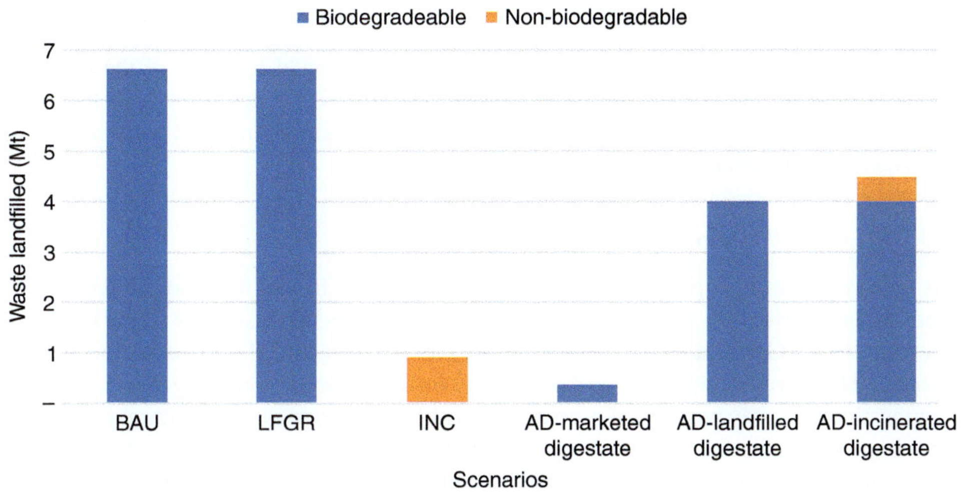

Figure 9.4 *Amounts of landfilled biodegradable and nonbiodegradable waste for the proposed waste management scenarios.*

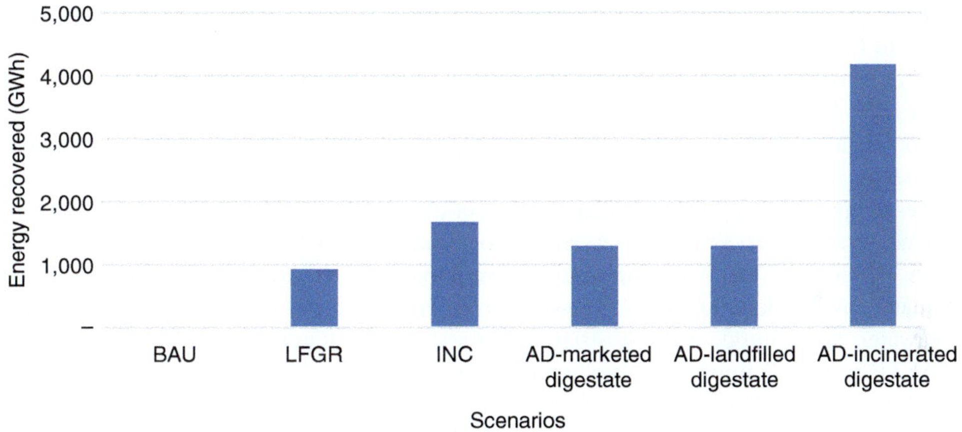

Figure 9.5 *Energy recovered in the examined waste management strategies.*

In terms of energy potential, Figure 9.5 shows the energy recovered in each of the examined strategies. The BAU scenario did not include any controlled WTE conversion process, i.e. no energy was recovered. Higher amounts of incinerated waste materials led to the recovery of more energy in the INC strategy. However, the highest energy recovery was achieved in AD-incinerated digestate (4172 GWh), in which all of the produced digestate was incinerated, followed by the INC scenario (1663 GWh). The difference between the INC and AD-marketed digestate and AD-landfilled digestate scenarios can be attributed to the higher energy yield of incinerated organic waste compared to being anaerobically digested. In contrast, processing the digestate in an incinerator increased the energy recovery in the AD-incinerated digestate scenario by 69%. Thus, marketing or disposal of digestate did not change the energy output.

9.3.2 Life Cycle Assessment

Six impact categories were used to describe the environmental footprint of the examined ISWM strategies: the GWP, ADP, ACP, FAETP, HTP, and ETP. Each strategy included several processes throughout its life cycle, including transportation, material recovery, energy recovery, and disposal.

9.3.2.1 Overall Impact Assessment

As clearly shown in Figure 9.6, the disposal of digestate in landfills had the worst GHG performance, primarily due to gas emissions from digesters and digester slurry storage as methane has a GWP index of 28 over 100 years. As a result, the incineration of waste produced the greatest benefit in terms of GWP, since the carbon dioxide generated through combustion is mainly biogenic (i.e. contained within the carbon cycle), and as a result, it was not counted toward the carbon footprint of marketing the digestate. Following AD-landfilled digestate, the BAU scenario was the second highest GWP contributor (986 kg CO_2-eq.). It should be noted that utilizing the LFG improved the overall GWP of the BAU strategy.

Acid gases typically emitted from incineration processes to some extent increased the ACP, which was not significantly offsetted by the energy recovery, and mainly related to the flue gas treatment process. All scenarios showed negative results for ACP ranging from 0.03 to 0.81 kg of SO_2 eq., primarily due to NH_3, NO_x, and SO_2 emissions. Emissions from electricity production or fertilization did not affect the results of WTE activities. Similarly, ETP was negatively impacted by all scenarios, especially those involving AD plants. Due to high ammonia emissions to aquatic and terrestrial ecosystems, AD-based strategies had greater potential for macronutrient enrichment. The avoided impacts from digestate production were not enough to offset the gross impacts. Gases such as NO_x produced during waste combustion were a major contributor to the overall impact of incineration. Moreover, FAETP measures the relative impact of toxic substances on the freshwater aquatic environment due to the emissions to environmental compartments air, freshwater, seawater, agricultural, and industrial soil. As shown in Figure 9.6, a positive favorable FAETP and HTP impact on the environment was depicted in AD-incinerated digestate scenario by −1.70 kg 1,4-DCB-eq., and −5.02 kg 1,4-DCB-eq., respectively, mainly due to Barite, Zinc, and Vanadium. However, the highest FAETP and HTP contributors were depicted when digestate was marketed (9.56 and 62.23 kg 1,4-DCB-eq., respectively). It should be noted that utilizing the LFG improved the overall FAETP and HTP of the BAU strategy.

In addition to the aforementioned categories, HTP is related to the negative effects of toxic substances such as volatile organic compounds, particulate matter, heavy metals, NO_x, SO_2, from the waste management process affecting the biological human system (excluding workplace exposures). All strategies contributed to negative HTP impacts on the environment, i.e. increased harmful effects of toxic substances, except for the AD-incinerated digestate scenario. Additionally, the INC scenario strategy presented slight increase in HTP impact due to Chromium, PAH, Arsenic, and Nickel in the air compartment and Barite, PAH, and Vanadium ions in the water compartment. These findings were in agreement with an LCA study that stated HTP emissions in incineration was 134 times greater than in landfill due to the air emissions of dioxins during incineration [28]. Moreover, ADP is defined

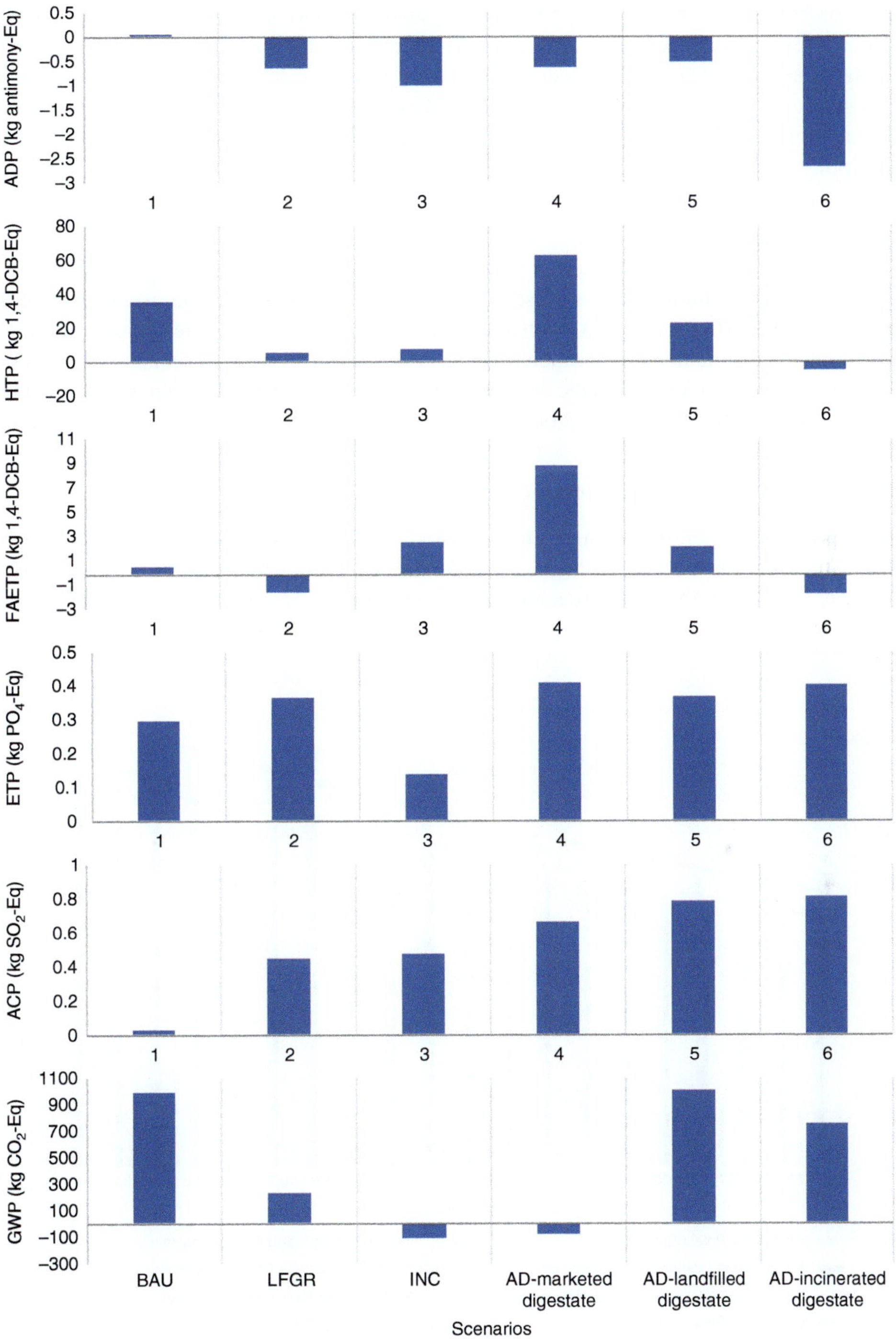

Figure 9.6 Environmental impact categories of the proposed waste management scenarios.

in terms of the annual rate of depletion of the stock of minerals and fossil fuels relative to ultimate reserves. Unlike HTP, all scenarios showed a reduction in the negative effects of resource depletion from ADP except for the BAU scenario. As a result of the replacement of fossil fuels, the scenarios with more landfilling yielded poor results. The scenario with the lowest reduced ADP impact was AD-incinerated digestate, with an impact of −5.74 kg Antimony-eq.

9.3.3 Life Cycle Costing

The LCC in this chapter was assessed via the NPV calculations. The assessment of all strategies was performed over a 20-year design period at a discount rate of 10%. To obtain a comprehensive financial assessment, the costs and revenues of each scenario were calculated. The discussion evaluates the potential energy production and corresponding revenues of the proposed strategies throughout the assessment period.

9.3.3.1 Cost and Revenue Streams

In the present chapter, the main revenue stream of WTE facilities was the ERP. Figure 9.7 shows the amount of energy that can be potentially produced from all proposed WTE strategies, as well as their sales revenues. For organic waste, the AD process generates more

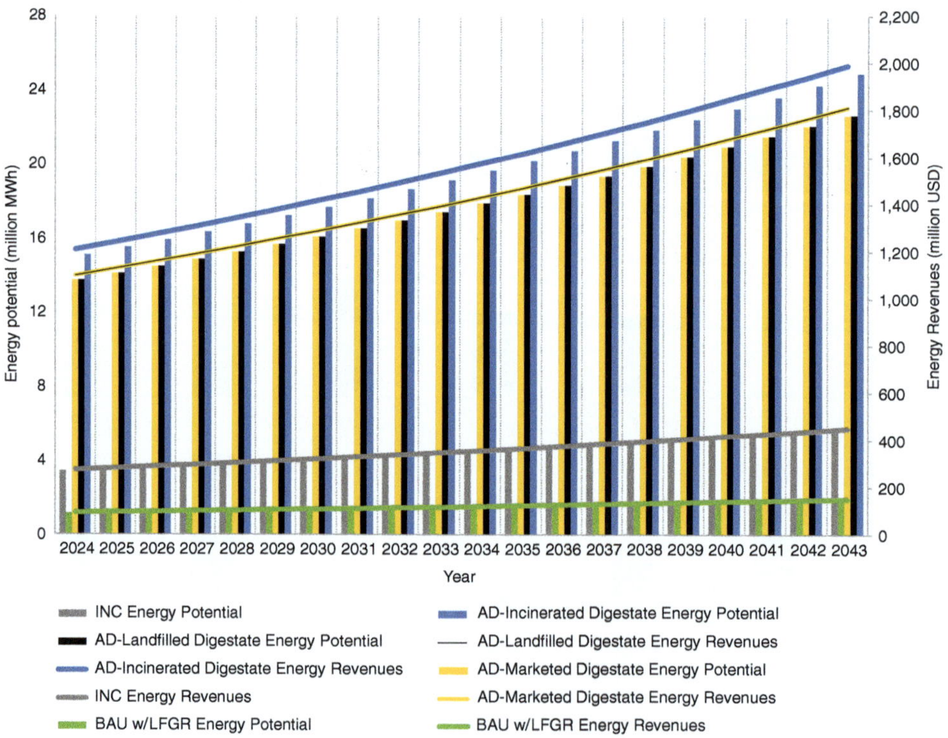

Figure 9.7 Potential energy production and revenues of the examined waste management strategies.

energy than incineration per ton of processed waste. The energy production in the INC scenario steadily increased throughout the assessment period mainly due to the slight increase in treated waste in incinerators compared to the significant increase of organic waste treated in anaerobic digesters. The latter significant increase in AD facilities was due to the corresponding increase of public participation in organic waste separation. In contrast, the LFGR scenario produced the least energy compared to all other scenarios. It should be noted that the BAU scenario has no energy utilization; therefore, it was excluded in the energy production comparison. The potential energy production was estimated to be around 1.15, 3.41, 13.72, 13.72, and 15.09 million MWh at the opening year and reaching around 1.90, 5.63, 22.64, 22.64, and 24.89 million MWh by the end of the assessment period for the LFGR, INC, AD-marketed digestate, AD-landfilled digestate, and AD-incinerated digestate strategies, respectively. Also, the total energy produced throughout the assessment period was 29, 88, 356, 142, and 392 TWh for the LFGR, INC, AD-marketed digestate, AD-landfilled digestate, and AD-incinerated digestate strategies, respectively. In terms of cash inflow, based on an electricity tariff of 0.08 USD kWh^{-1} in the UAE, the potential revenues generated from the energy production were calculated. In the opening year, LFGR, INC, AD-marketed digestate, AD-landfilled digestate, and AD-incinerated digestate strategies generated around 92, 273, 1,098, 1,098, and 1206 million USD and reached about 152, 450, 1,811, 1,811, and 1,991 million USD at the end of the study period, respectively.

9.3.3.2 Net Present Value

The main parameter utilized for the LCC analysis was the NPV, which is shown in Figure 9.8 shows for all evaluated strategies throughout the assessment period. The INC strategy was found to be financially infeasible with a negative NPV of −89.93 million USD, mainly due to the steep investments and operation costs. The INC scenario had greater investment

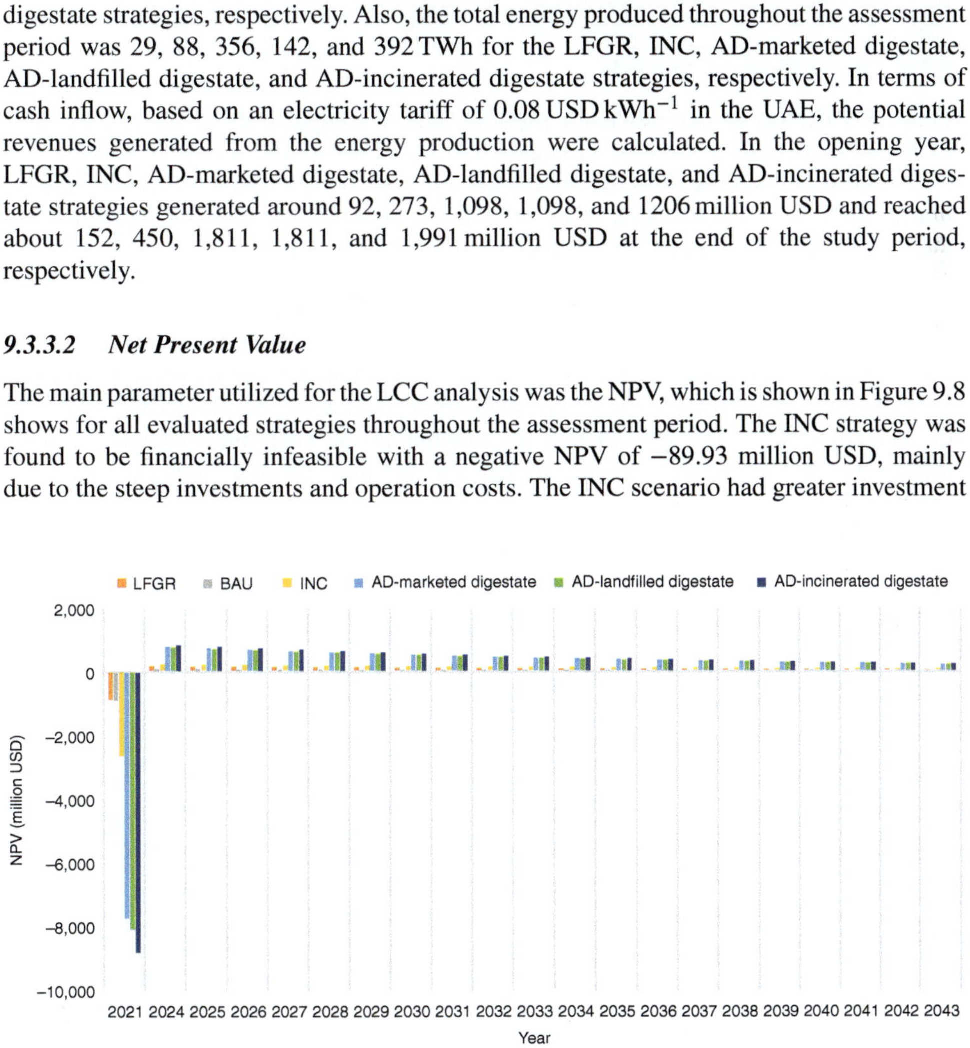

Figure 9.8 Annual net present values for the examined waste management strategies.

costs due to the high cost of building and equipment of the incinerators, and its higher operation costs were because of the more complex mechanical equipment, treatment units of fly ash, and potential consumption of the auxiliary coal. In contrast, the BAU strategy had the lowest CAPEX and OPEX costs due to the absence of power generating facilities. All other WTE strategies proved to be profitable with positive NPVs. Marketing digestate showed the highest positive NPV of 1348 million USD, which was mainly due to the digestate revenues, and the tipping fees obtained from the substantially larger quantity of processed waste. However, when digestate was landfilled and incinerated, the NPV was reduced by around 55% and 39%, respectively. Additionally, the larger present value of energy revenues had favored the strategies that included AD systems over incineration and BAU strategy.

9.3.4 Eco-Efficiency Analysis

Figure 9.9 depicts an eco-efficiency portfolio that is computed using the results of the LCA and LCC of the examined strategies. The figure shows four divided quartiles, with the bottom right quartile representing the highest system NPV and lowest environmental impact (highest eco-efficiency), while the top left quartile represents the strategies with the lowest NPV and highest environmental impacts (lowest eco-efficiency). The highest eco-efficient strategy was the AD-marketed digestate, followed by the LFGR. In contrast, the least sustainable scenarios were the BAU scenario followed by the AD-landfilled digestate. In contrast, the INC scenario had significantly low environmental impacts and NPV, i.e. semi-efficient. Oppositely, the AD-incinerated digestate scenario had high environmental footprint and NPV, i.e. semi-efficient as well. As shown in Figure 9.9, the eco-efficiency index showed that, by combining the environmental and economic aspects, the AD-marketed digestate strategy was the most favored among the examined strategies.

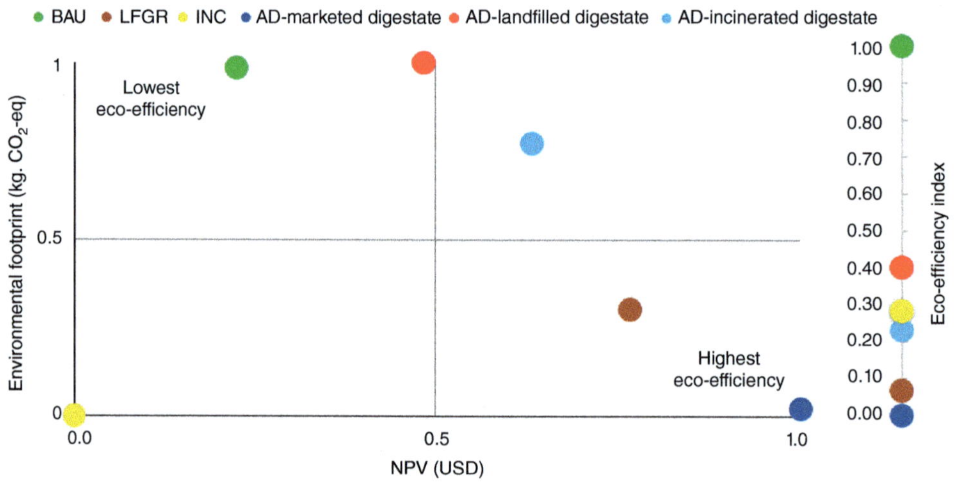

Figure 9.9 Eco-efficiency assessment for the examined waste management strategies.

9.4 Conclusion

The main objective of the present chapter was to evaluate the potential of implementing AD and incineration for the management of organic waste in the UAE as part of WTE-based ISWM strategies and compared to the BAU strategy. The environmental footprint along with social and economic impacts of WTE-based ISWM strategies were assessed via comprehensive LCA and LCC analyses to determine the most feasible management technique under local operating conditions. The LCI was compiled from multiple sources, including specific waste characteristics of the study area, such as composition and energy content, as well as the local market and operation parameters, literature, and international guidelines. WRATE software was used to build the LCA models and compute the environmental impact assessments, whereas the LCC was carried out through multiple financial indicators. The results of the LCA and LCC were used to compute the eco-efficiency of all proposed strategies. Findings showed that the AD-digestate-marketed scenario was the most eco-efficient due to the high NPV obtained compared to the low environmental impact.

References

1. Kaza, S., Yao, L., Bhada-Tata, P., and Van Woerden, F. (2018). *What a Waste 2.0: A Global Snapshot of Solid Waste Management to 2050. Urban Development*. Washington DC: World Bank https://doi.org/10.1596/978-1-4648-1329-0.
2. Maalouf, A. and El-Fadel, M. (2018). Aggregated and disaggregated data about default emission factors in emissions accounting methods from the waste sector. *Data in Brief* 21: 568–575. https://doi.org/10.1016/j.dib.2018.09.094.
3. Law, H.J. and Ross, D.E. (2019). International Solid Waste Association's 'closing dumpsites' initiative: status of progress. *Waste Management & Research* 37 (6): 565–568. https://doi.org/10.1177/0734242X19845755.
4. Abdallah, M., Arab, M., Shabib, A. et al. (2020). Characterization and sustainable management strategies of municipal solid waste in Egypt. *Clean Technologies and Environmental Policy* 22 (6): 1371–1383. https://doi.org/10.1007/s10098-020-01877-0.
5. Hadidi, L.A. and Omer, M.M. (2017). A financial feasibility model of gasification and anaerobic digestion waste-to-energy (WTE) plants in Saudi Arabia. *Waste Management* 59: 90–101. https://doi.org/10.1016/j.wasman.2016.09.030.
6. Hoornweg, D. and Bhada, P. (2012). What a waste. A global review of solid waste management. *Urban Development Series Knowledge Papers* 281 (19): 44. https://doi.org/10.1111/febs.13058.
7. Rebitzer, G., Ekvall, T., Frischknecht, R. et al. (2004). Life cycle assessment Part 1: Framework goal and scope definition inventory analysis and applications. *Environment International* no. 5: 701–720. https://doi.org/10.1016/j.envint.2003.11.005.
8. Elfeky, A., Abdallah, M., and Fattah, K. (2021). Environmental Assessment of Recyclable Waste Valorization in United Arab Emirates. *IOP Conference Series: Earth and Environmental Science* 897 (1): https://doi.org/10.1088/1755-1315/897/1/012024.
9. Shabib, A. and Abdallah, M. (2020). Life cycle analysis of waste power plants: systematic framework. *International Journal of Environmental Studies* 7233: https://doi.org/10.1080/00207233.2019.1708146.
10. Abeliotis, K. (2011). Life cycle assessment in municipal solid waste management. *Integrated Waste Management – Volume I* 2006: https://doi.org/10.5772/20421.
11. ISO (2006). Environmental Management - Life Cycle Assessment - Principles and Framework (ISO 14040:2006). *Environ. Manag. Syst. Requir.* 44 (0).

12. Abdallah, M. and Elfeky, A. (2020, 2021). Impact of waste processing byproducts on the carbon footprint of integrated waste-to-energy strategies. *Journal of Environmental Management* 280 (July): 111839. https://doi.org/10.1016/j.jenvman.2020.111839.
13. Tawatsin, A. (2014). Environmental assessment of waste to energy processes specifically incineration and anaerobic digestion using life cycle assessment.
14. IRENA, "IRENA: Renewable Power Generation Costs in 2014: An Overview," no. January, p. 92, 2015, https://www.irena.org/publications/2015/jan/renewable-power-generation-costs-in-2014 Costs in 2012.pdf
15. Zhao, X. gang, Jiang, G. wu, Li, A., and Wang, L. (2016). Economic analysis of waste-to-energy industry in China. *Waste Management* 48: 604–618. https://doi.org/10.1016/j.wasman.2015.10.014.
16. Abdeljaber, A., Zannerni, R., Masoud, W. et al. (2022). Eco-efficiency analysis of integrated waste management strategies based on gasification and mechanical biological treatment. *Sustainability* 14 (7): https://doi.org/10.3390/su14073899.
17. ISO14045 (2012). International Standard International Standard. *Int. Organ. Stand. Environ. Manag. Assess. Prod. Syst. Requir. Guid.*, p. 13.
18. APICORP Energy Research (2017). Solar Energy in the UAE: Impressive Progress. 03 (02).
19. Said, Z., Alshehhi, A.A., and Mehmood, A. (2018). Predictions of UAE's renewable energy mix in 2030. *Renewable Energy* 118: 779–789. https://doi.org/10.1016/j.renene.2017.11.075.
20. MOCCUAE (2017). *National climate change plan of the United Arab Emirates 2017-2050*. Abu Dhabi. www.moccae.gov.ae.
21. Abu Dhabi Statistics Centre (2015). *Waste Statistics 2015*. Abu Dhabi. https://www2.mst.dk/Udgiv/publications/2017/08/978-87-93614-20-8.pdf.
22. Abdallah, M., Shanableh, A., Shabib, A., and Adghim, M. (2018). Financial feasibility of waste to energy strategies in the United Arab Emirates. *Waste Management* 82: 207–219. https://doi.org/10.1016/j.wasman.2018.10.029.
23. Dubai Statistics Centre (2017). *Population Bulletin Emirate of Dubai 2017*. 1–12. https://www.dsc.gov.ae/en-us/Themes/Pages/Population-and-Vital-Statistics.aspx?Theme=42&year=2017 (accessed 9 November 2022)
24. Abu Dhabi Statistics Centre (2018). Waste statistics 2018. https://www2.mst.dk/Udgiv/publications/2017/08/978-87-93614-20-8.pdf.
25. Abu Dhabi Environment Agency (2016). *Waste and Environment Annual Report*. Abu Dhabi.
26. Abdallah, M., Hamdan, S., and Shabib, A. (2021). A multi-objective optimization model for strategic waste management master plans. *Journal of Cleaner Production* 284: 124714. https://doi.org/10.1016/j.jclepro.2020.124714.
27. Gregor, J., Šomplák, R., and Pavlas, M. (2017). Transportation cost as an integral part of supply chain optimisation in the field of waste management. *Chemical Engineering Transactions* 56: 1927–1932. https://doi.org/10.3303/CET1756322.
28. Leme, M.M.V., Rocha, M.H., Lora, E.E.S. et al. (2014). Techno-economic analysis and environmental impact assessment of energy recovery from Municipal Solid Waste (MSW) in Brazil. *Resources, Conservation and Recycling* 87: 8–20. https://doi.org/10.1016/j.resconrec.2014.03.003.

10

Reduction of the Carbon Footprint in Terms of Agricultural Biogas Plants

Agnieszka Wawrzyniak
Department of Biosystems Engineering, Poznań University of Life Sciences, Poznań, Poland

Acronyms

AD	anaerobic digestion
CF	carbon footprint
CHP	combined heat and power
FU	functional unit
GHG	greenhouse gas
GWP	global warming potential
IPCC	Intergovernmental Panel of Climate Change
ISO	International Organization of Standard
LCA	life cycle assessment
RES	renewable energy sources
WRI	World Resource Institute

Biogas Plants: Waste Management, Energy Production and Carbon Footprint Reduction,
First Edition. Edited by Wojciech Czekała

10.1 Introduction

10.1.1 Manure Management and Biomethane Potential in Poland and EU Countries

Biomass resources such as livestock and poultry manure, etc., can be used as organic fertilizers which have huge potentials in replacing chemical ones. For biomass nitrogen, the biggest contributor came from the manure and urine of livestock and poultry (54.5%) [1]. Poultry manure is rich in nitrogen but also contains significant quantities of phosphorous and potassium, and it can be applied as a fertilizer to improve soil properties and fertility [2]. Intensive livestock production is generated so much fertilizers that its current use in the fields is not possible [3]. The total amount of yearly swine manure production in 2018 in Poland reaches over 15 million Mg, and this is almost twice more than the slurry production (7.75 million) and yearly cattle manure production was 78 million Mg. The biogas potential of such amount manure and slurry was assumed on 5.04 billion m^3. Taking into consideration that a 60% share of methane in biogas, it gives the value of about 3.03 billion m^3 of methane [4]. In the Polish voivodeships, most manure is produced in mazowieckie, podlaskie, and wielkopolskie. Therefore, the potential of biogas production from swine and cattle manure and slurry in Poland 2018 are the biggest (in those regions) – Figures 10.1 and 10.2.

Production of poultry in 2019 Poland is estimated on average at 4 million ton per year – this included broiler chickens and turkeys, and egg-laying hens [5] because Poland is the leader in European Union in poultry production – 16.8% [6]. In comparison with the EU-28 by reviewing the farm manure production, it was estimated that approximately 1200 million tons wet manure is produced per year. The current situation of manure production in the EU countries is presented on the map in Figure 10.3 [6]. The theoretical biogas potential of manure was estimated at 26 billion m^3 biomethane in Europe (23 billion m^3 biomethane in the EU), and the realistic biogas potential, counting on collectible manure, was assessed at 18 billion m^3 biomethane in Europe (16 billion m^3 biomethane in the EU) [7].

The legal regulations forcing farmers to store manure and slurry in winter force investments related to the construction of tanks necessary for their storage [9]. The regulations allow farmers to management slurry, manure, or poultry manure to biogas plants without treating them as waste due to the requirements of the Regulation of the European Parliament and the Council (CE) No. 1069/2009 of 21 October 2009 [9] laying down health rules as regards animal byproducts and derived products not intended for human consumption and repealing Regulation (EC) No. 1774/2002 (Animal byproducts Regulation) and the Regulation of the European Parliament of the Council (CE) No. 142/2011 of 25 February 2011.

10.1.2 Substrates Used for Biogas Plants in Poland

Compared to other waste to energy solutions, anaerobic digestion (AD) is the most mature technology to upgrade manure's organic matter into renewable energy; however, the problems associated with high investment costs, operating parameters, manure collection, and digestate management have hindered its developments in rural areas in developing

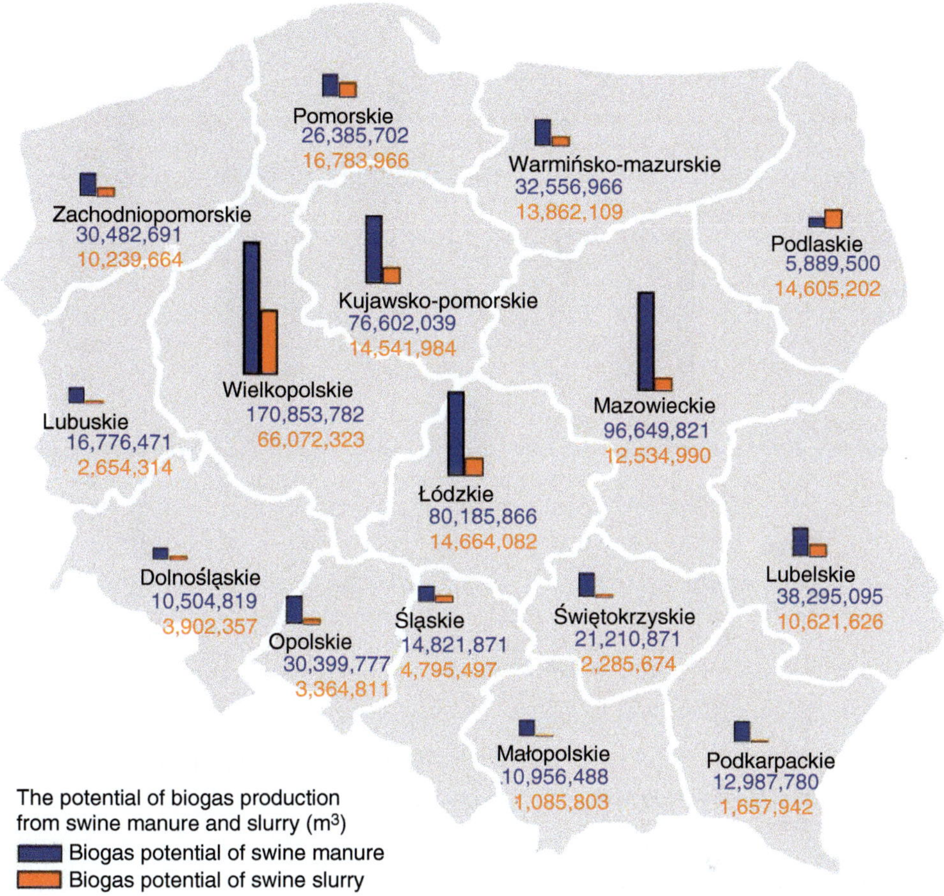

Figure 10.1 *The potential of biogas production from swine manure and slurry in Poland 2018. Source: Wawrzyniak et al. [4]/Journal of Ecological Engineering/Public Domain CC BY 4.0.*

countries [10]. The potential of manure for biogas production is not fully utilized due to the low and imbalance carbon to nitrogen (C/N) ratio in animal manures [11].

Within 15 years from the launch of the first biogas installation in Poland, the structure of the substrates used has diversified. The dominant substrates, such as slurry or corn silage in 2011 [12], began to be replaced with other products from the agri-food industry. According to the KOWR (The National Support Centre for Agricultural) data for 2019, the most frequently used substrates in biogas plants include distillers grains, fruit and vegetable residues, slurry, maize silage, and beet pulp [13].

In Figures 10.4 and 10.5, substrates used to biogas plant in 2011 (Figure 10.4) and in 2019 (Figure 10.5) are shown.

It is necessary to introduce more energetic substrate streams to improve the biogas installations efficiency [15]. In this variant, the slurry is mainly used as a diluent for substrates with a higher dry matter content, characterized by higher energy, while the manure is processed by the installation, making it possible to use its energy potential,

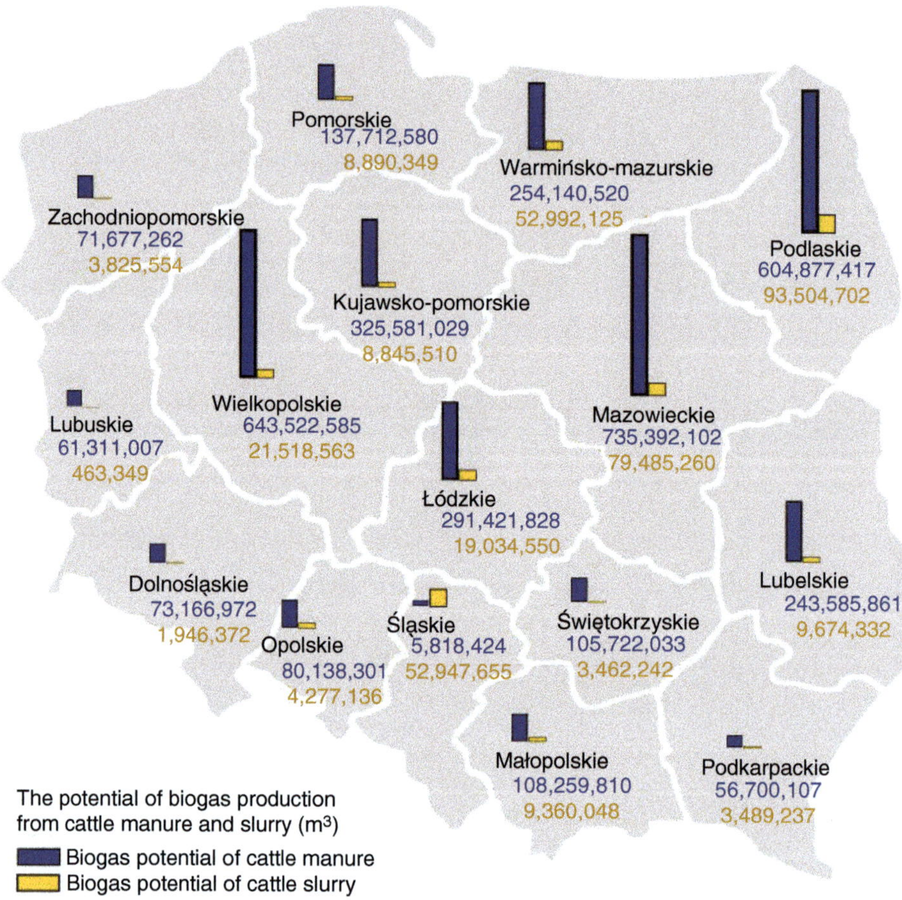

Figure 10.2 The potential of biogas production from cattle manure and slurry in Poland 2018. Source: Wawrzyniak et al. [4]/Journal of Ecological Engineering/Public Domain CC BY 4.0.

improving its fertilizing properties [16], and allows its lawful storage in the winter when it cannot be applied to the fields.

10.1.3 GHG Emissions from Agriculture and Biogas Plants as Tool for its Reduction

In terms of area, the majority of Polish farms are small. Fifty-four percent of farms are up to 5 ha and 76% cover the area of less than 10 ha, and only 2.4% of farms are over 50 ha [17]. Since 2013, there has been a downward trend in animal production. Animals are kept at 51% of farms. The % share of animal production is as follows: cattle were kept in 25% of farms, pigs in 12%, and poultry in 36% of farms.

According to the National Research Centre, in 2017, agriculture was a major producer of CH_4 and N_2O. About CH_4, emissions were 49.41 million tons of the equivalent amount of CO_2, while 25.9% of national emissions came from agriculture – mainly from

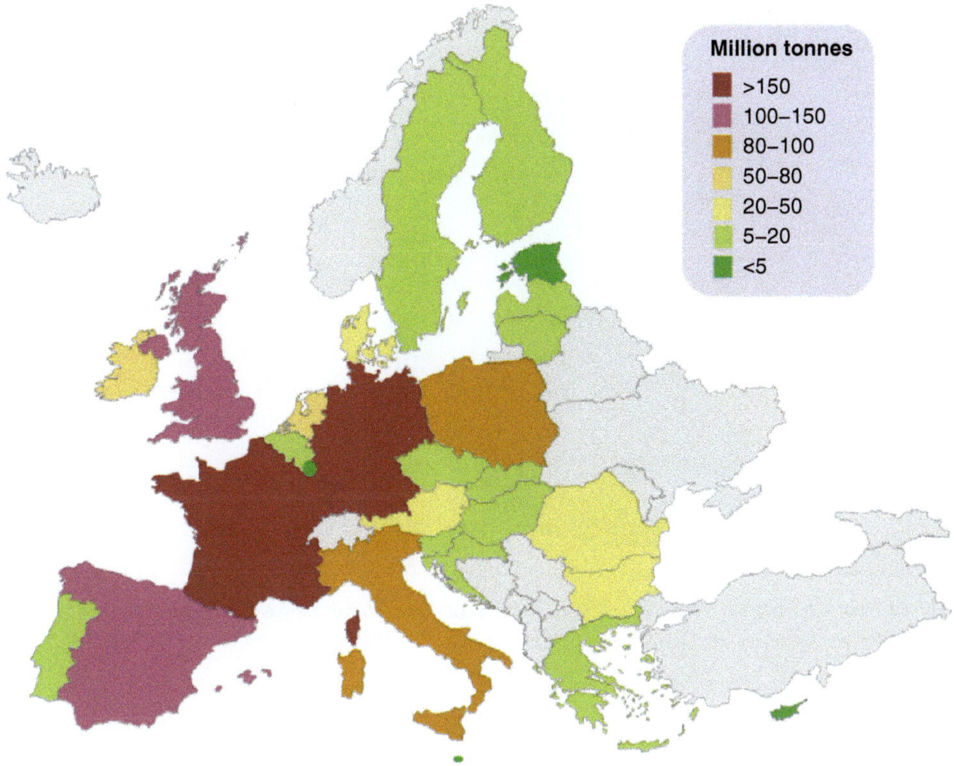

Figure 10.3 Map chart of farm manure produced in EU-28. Source: Scarlat et al. [8]/with permission of Elsevier.

enteric fermentation. In 2017, N_2O emissions amounted to 20.82 million tons of CO_2 equivalent [18]. The largest amount of CH_4 comes from manure produced by cattle (54%) and pigs (31%). N_2O emissions from manure management amounted to 9.4 kilotons (1 kiloton = 10^3 t) in 2019 and declined by 32% since 1988, which is mainly related to the decline in the population of livestock. In this category, direct emissions amount to approximately 49% and indirect emissions amount to 51% of N_2O emissions [19]. For comparison, Table 10.1 showed emissions from agriculture expressed in tons – the presented results come from the OECD report [20].

Compared to other countries in the EU, Polish agriculture occupies a leading position in terms of GHG emissions. Figure 10.6 presents GHG emissions by the EU member states in 2019, expressed in CO_2 equivalent [21]. Germany is the largest producer of GHG emissions per capita in the EU, in terms of the carbon footprint (CF) of the agricultural sector: 793,334.59 t CO_2. The United Kingdom, France, and Turkey have a slightly lower CF: 455,122.74 t CO_2 for the United Kingdom, 422,085.53 t CO_2 for Turkey 405,260.11 t CO_2. Poland is among the countries with a high CF from agriculture, which is 375,701.84 t [21].

Through a literature overview, it was noticed that emissions from agriculture can be reduced by biogas production through AD of manure, mainly together with other biowaste that can replace fossil energy and thus reduce CO_2 emissions and post-fermentation GHG

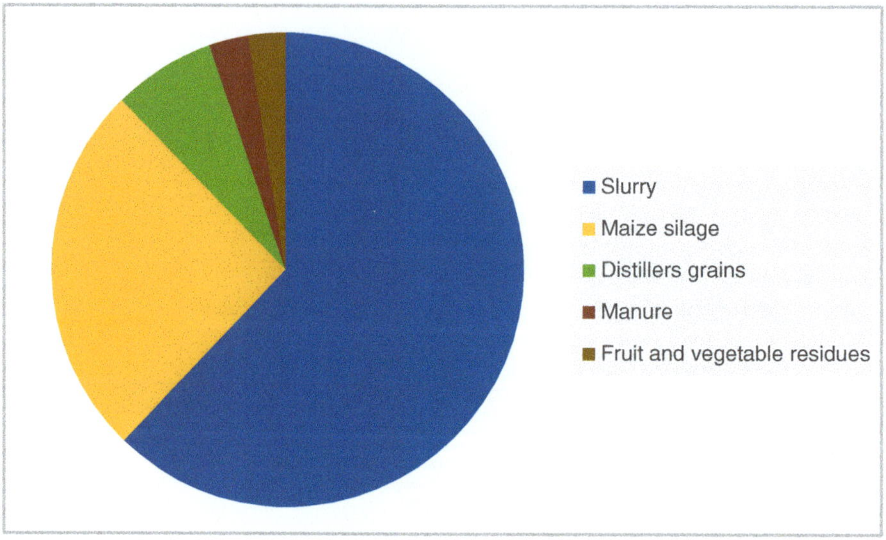

Figure 10.4 *Substrates used for the production of agricultural biogas in 2011. Source: Adapted from Krajowy Ośrodek Wsparcia Rolnictwa [13], https://www.gov.pl/attachment/b7c041c0-248c-4b4c-bc12-d22d291c5534.*

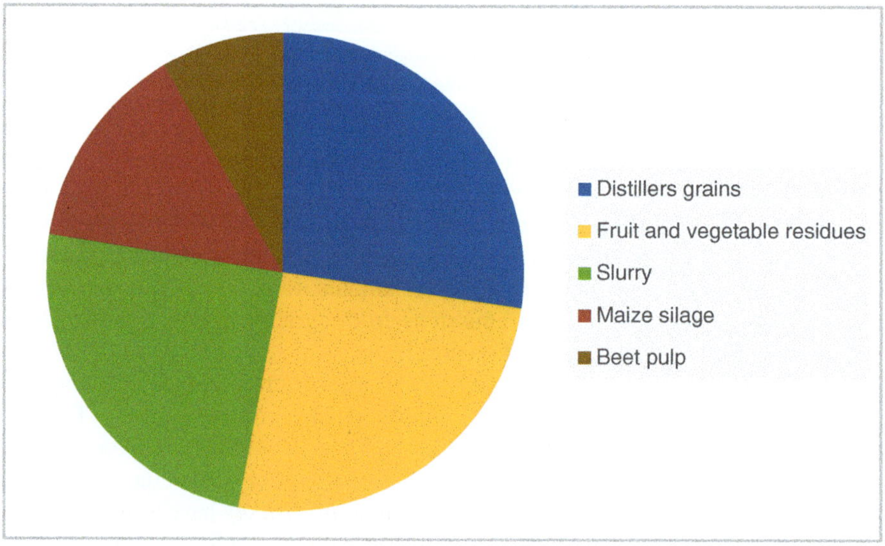

Figure 10.5 *Substrates used for the production of agricultural biogas in 2019. Source: Adapted from Krajowy Ośrodek Wsparcia Rolnictwa [14], https://www.gov.pl/attachment/b7c041c0-248c-4b4c-bc12-d22d291c5534.*

Table 10.1 Types of emissions.

Type of emission	Emissions (1000 t)
Non-methane gaseous organic compounds	94.7
NH_3	287.9
Dust from livestock manure	34.3
Dust from production plants harmful to the environment	5.247

Source: Adapted from OECD [20].

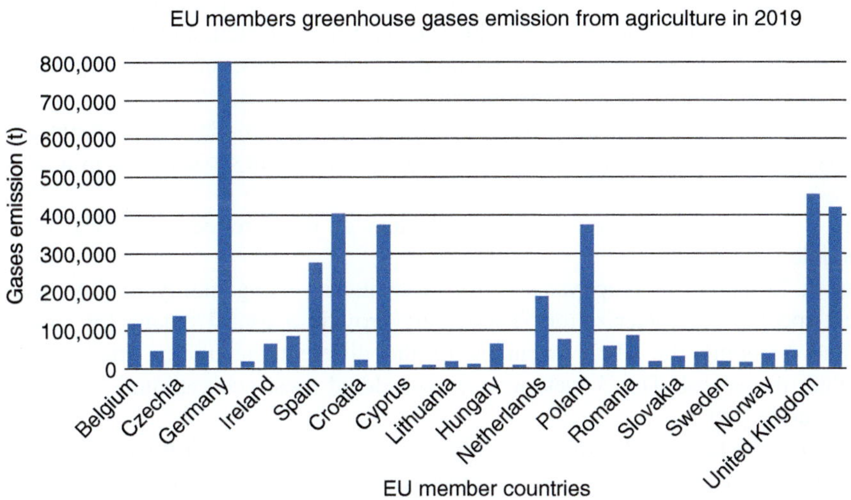

Figure 10.6 Greenhouse gases emission from agriculture in the EU countries in 2019. Source: Adapted from EUROSTAT [21], https://ec.europa.eu/eurostat/databrowser/view/ENV_AIR_GGE$DV_447/default/table?lang=en.

emissions [22]. In contrast, biogas production and use in electricity and heat production also led to GHG reductions compared to composting of feedstock, but the reductions are not as high as in transportation use biogas in most of the cases [23].

10.2 Methodology of CF

CF is an indicator based on Life Cycle Thinking. It is used as an usual indicator in an life cycle assessment (LCA) method. It is a measure of the total amount of GHG emissions of a system, considering all relevant sources, sinks, and storage within the spatial and temporal boundary of the system [24]. Calculating a CF involves a number of steps. It should be noticed that the term CF is frequently used in a marketing and communication context to promote to consumers products or services that help mitigate GHG emissions [25].

In contrast, the term GHG accounting often associated with precisely measuring emissions from specific practices or processes for the purpose of informing the business decisions of the emitter. The calculation of companies GHG emissions obligations under an emissions trading scheme is an example of such use [26].

Table 10.2 Global Warming Potential (GWP) values relative to CO_2.

GWP values for 100-yr time horizon	
Carbon dioxide CO_2	1
Methane CH_4	25
Nitrous oxide N_2O	298

Source: Adapted from World Business Council for Sustainable Development [29].

The methodology will often depend on the purpose of the enquiry and the availability of data and resources [27]. IPCC define three general tiers of methodologies based on their complexity and data requirements. The choice of tier depends, in part, on the significance of the emissions sources under consideration.

Tier 1: Simple, emission factor-based approach. Tier 1 emission factors are international defaults, although they will often have been based on studies conducted in a select few (mostly temperate) countries [28].

Tier 2: More region-specific emission factors or more refined empirical estimation methodologies [28].

Tier 3: Dynamic biogeophysical simulation models using multiyear time series and context-specific parameterization. These tiers provide a useful means for categorizing and understanding the likely accuracy of the different calculation methods that are available. In general, Tier 3 methods are considered most accurate and Tier 1 methods least accurate [28].

Choosing the tier decides of methodology to calculate the GHG fluxes. Next step is to use software with LCA to calculate the impact category: Global Warming Potential (GWP).

The GWP given by IPCC 2007 is stands for an index comparing the global warming effect of a given GHG to that of carbon dioxide [29]. The GWP is calculated from the 100-year global warming effect of 1 kg of a given gas compared to 1 kg of CO_2 – Table 10.2 base on IPCC 2007.

10.2.1 GHG Fluxes from Agriculture and Tools for its Calculations

GHG fluxes can be determined in different ways, ranging from the use of highly specialized, field-scale measurement equipment to global emission factors. There are four different types of calculation approaches can be used for nonmechanical sources:

- field measurements;
- emission factors;
- empirical models and process-based models.

Field measurement – direct techniques include controlled livestock chambers that measure the CH_4 emissions from enteric fermentation, flux chambers that measure the N_2O and CO_2 emissions from plots of land, and gas flux meters that measure the CH_4 emissions from certain livestock waste management systems (e.g. covered anaerobic lagoons). Indirect techniques include the measurement of carbon stocks before and after a change in management practices or land use. Indirect techniques are often much simpler and easier but may require additional planning ahead of time to capture the before'

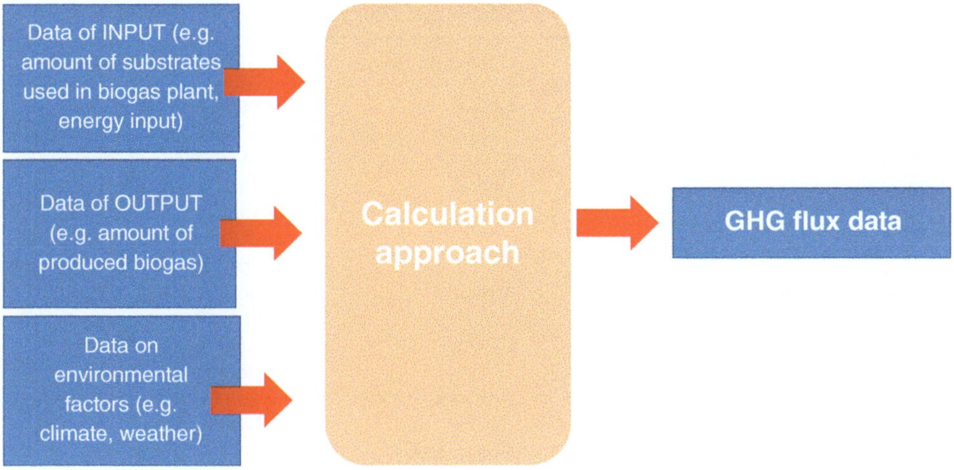

Figure 10.7 The general process for calculating GHG fluxes (own study).

scenario. While useful for research, both direct and indirect techniques are often far too costly for developing corporate inventories [30].

Emission factor, the simplest approach, involves the multiplication of management activity data by relevant emission factor, which is a coefficient describing the amount of GHG flux per GHG unit of activity [30]. Empirical models use field measurements to develop statistical relationships between GHG fluxes and agricultural management factors. In turn, process-based (or mechanistic) models mathematically link important biogeochemical processes that control the production, consumption, and emission of GHGs [30]. Figure 10.7 shows the general process for calculating GHG fluxes in biogas plant. To calculate GHG fluxes, there are a lot of publicly available tools – softwares and protocols based on emission factors of these approaches [30]. In Table 10.3, there are examples of tool dedicated to different geographic area.

10.2.2 System Boundaries for Biogas Plant and Data Collection

Before starting the data collection for CF analysis, the boundaries of the system and the Functional Unit (FU) should be determined. In this case, it is needed to identify all system inputs and outputs in relation to GHG emissions. The FU for an agricultural biogas plant is 1 ton of dry weight of biomass used in the AD process.

In Figure 10.8, AD process in agricultural biogas plant is presented. Figure shows GHG emissions for all stages in biogas plants in which emissions exist. The most important GHGs emission valued in CO_2 equivalents are CH_4, CO_2, and N_2O. To calculate emissions of those gages on each stages, it is necessary to use equation for Tier 2 methodology according to IPCC Directive dedicated to agriculture.

However, it should be emphasized that assessing the effect of AD as a potential mitigation measure require a whole-farm approach to estimate GHG emissions from the dairy or pig farm – the system boundaries for dairy farm is presented in Figure 10.8. The calculations must include evaluation of side effects in the form of increased NH_3 emissions, reduced,

Table 10.3 Examples of available tools for calculating agricultural GHG fluxes.

Tool	Geographic focus	Methodology
IPCC. 2006 Intergovernmental Panel on Climate Change Guidelines on National Inventories	Global	Three tiers of methods outlined. Tier 1 emission factors provided for wide range of sources
Cool farm tool [31]	Global	Combination of LCA emission factors, empirical models, Tier 1 and 2 methods and emission factors, and academic literature
Greenhouse in agriculture tools for dairy, sheep, beef, or grain farms	Australia	Emission factors from Australia's national inventory practices
Holos	Canada	Methodology is IPCC but customized to Canada
Carbon accounting for land managers (CALM)	United Kingdom	Emission factors from United Kingdom national inventory
Livestock analysis model	United States	Specific to cattle and buffalo

Source: Adapted from Greenhouse Gas Protocol Agricultural Guidance [30].

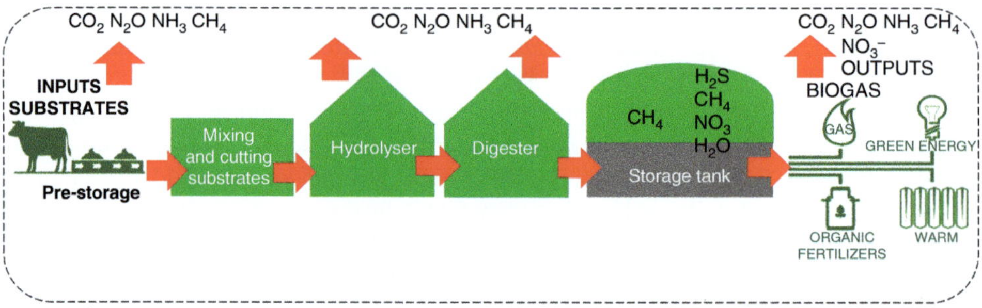

Figure 10.8 Anaerobic digestion process in agricultural biogas plant (own study).

leakage of CH_4 from the biogas plant. It is relevant to improve estimates of the potential of AD to reduce the negative GHG balance of livestock farming. In this case, the GHG gas balance included substituting CO_2 emission from power and heat production using fossil fuel, leakage of CH_4 from biogas production plants, CH_4 emissions during storage of animal manure and organic waste, N_2O emissions from stored and field applied manure, organic waste, and digestate, N_2O emissions related to NO_3 leaching and NH_3 emission, and N_2O emission from cultivating energy crops [32] – Figure 10.9.

10.3 Life Cycle CO_2 Footprints of Various Biogas Projects – Comparison with Literature Results

The template dedicated to collecting data from biogas plant was developed under the MilKey project. Table 10.4 presented the scope of data which should be collected to calculate CF of biogas plant.

Life cycle CO_2 footprint of renewable energy from biogas is typically higher than other renewable energy sources (RESs) [34]. In Table 10.5, values of CO_2 for different RES are presented.

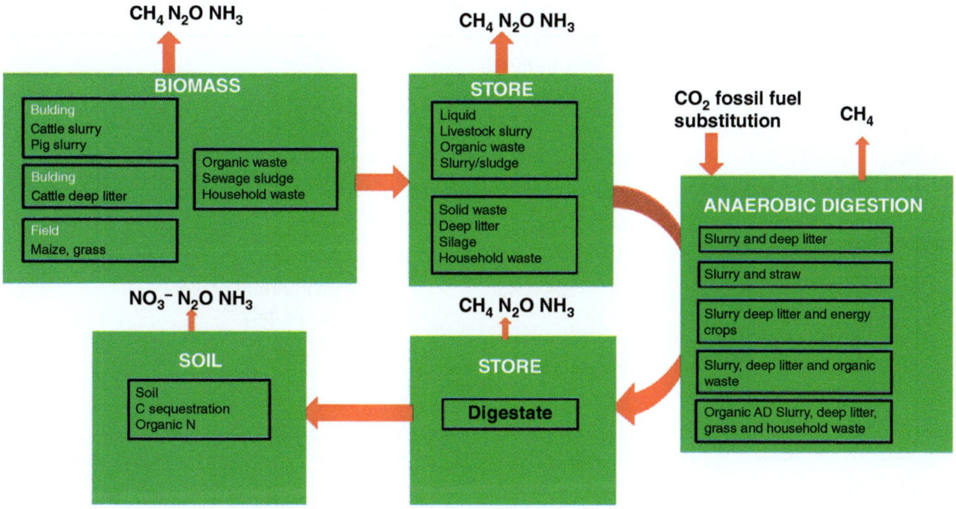

Figure 10.9 *Example of GHG fluxes emissions during AD in five manure management systems. Source: Moller et al. [22]/MDPI/Public Domain CC BY 4.0.*

Table 10.4 *Types of activity data that may be needed to calculate GHG fluxes.*

Source	Types of activity data needed
Biogas plant	*INPUT*
	Type of substrates, origin, distance from input production to digester, amount in tons
	STORAGE
	Capacity (m^3), length of storage (d)
	DIGESTER
	Biochemical methane potential (%), power of the digester (kW)
	OUTPUT
	Total biogas production ($m^3\ yr^{-1}$)
	Production of electricity ($MWh\ yr^{-1}$)
	Production of thermal energy ($MWh\ yr^{-1}$)
	Auto-consumption of the process (%)
	Electricity sales (local currency/yr)
	Digestate ($t\ yr^{-1}$)
	Spreading digestate without treatment (%)
	Exporting digestate with posttreatment (%)
	Exporting digestate without posttreatment (%)
	POST-STORAGE
	Capacity (m^3)
	Type of post-storage (d)
	Length of liquid output storage (d)
	Length of solid output storage (d)
	Length of storage (d)
	USE OF BIOGAS
	% of cogeneration
	% of injection in gas network (from biogas)

Source: Adapted from Baillet [33], https://www.researchgate.net/publication/279038112_The_carbon_footprint_of_a_biogas_power_plant.

Table 10.5 *CF (kgCO_2eq MW^{-1} h_{el}^{-1}) of the renewable energy sources (RES).*

Sort of RES	CO_2 footprint (kg CO_2eq MW^{-1} h_{el}^{-1})	Project characteristics	Literature
Wind	7–56	A group of wind farms	[34]
Hydro	10–200	A group of hydro plants	[34]
Solar PV	50–184	Average of a few solar farms	[34]
CSP	9–63	A group of solar farms	[36]
Solid biomass	10–260	A group of solid biomass plants.	[34]
Geothermal	380–1045	Four geothermal power plants in Italy	[37]

Source: Budzianowski and Postawa [35]/with permission of Elsevier.

Table 10.6 *CF (kg CO_2eq MW^{-1} h_{elp}^{-1}) of various biogas projects literature data.*

CO_2 footprint (kg CO_2eq MW^{-1} h_{el}^{-1})	Type of substrates in biogas plant	References
	Feedstock manure and maize	[34]
110	Feedstock: cattle slurry and silage corn; CHP; Brandenburg, Germany	[41]
222	CHP, 0.17 MW_{el}, feedstock: manure and agrowastes, digestate open storage	[41]
251	CHP, 0.17 MW_{el}, feedstock: maize and beet crops mixed with manure and cheese whey, digestate open storage	[41]
43	CHP, 0.17 MW_{el}, feedstock: manure and agrowastes, digestate covered storage	[41]
48	0.64 MW_{el}, Hungary	[42]
184 or 160 (if applying best available practices)	0.5 MW_{el}, feedstock maize silage, no credit for heat utilization, and tight gas cover	[43]
260	Energy crops (33%), dairy and swine manure (67%), 0.6 MW_{el}, CHP	[44]
413	Dairy and swine manure, 0.6 MW_{el}, CHP	[44]

Source: Budzianowski and Postawa [35]/with permission of Elsevier.

The typical life cycle CO_2 footprint of biogas reported in literature is between 50 and 450 kg CO_2 MW^{-1} h_{el}^{-1} (CO_2 footprint of biomethane) [35]. In Table 10.6, different values of CF for biogas were presented for various biogas projects base on literature data. Values vary depending on the type of substrate in the biogas plant and CHP.

LCA results are highly dependent on various inventorial and methodological assumptions regarding system boundaries, inclusion/exclusion of various agricultural practices, land use change effects, and plant species [38]. CO_2 footprint depends on regional parameters that are required for the LCA of biogas. These parameters change among regions, and always parameters for the relevant region need to be employed [39]. The most significant GHG emissions factor for the CF is the feedstock production. Agricultural wastes have meaningful potential to reduce CO_2 footprint (Table 10.6) [40].

One of literature case studies dedicated to liquid manure shows CO_2 footprint reduction by 720 kg CO_2eq MWh^{-1} (by displacing fossil fuels), while for energy crops like maize

Table 10.7 *Biogas plant as a practice that can reduce GHG emissions and improve farm performance.*

Description of AD	Potential GHG benefits	Potential environmental co-benefits	Potential agronomic/ business benefits	Potential trade-off or problems
Enclosed system in which organic material such as manure is broken down by microorganisms under anaerobic conditions	Reduced N_2O and CH_4 emissions from manure management Reduced scope 3 emissions from fertilizer manufacture	Reduced risk of accidental toxic leakages (pathogens killed) Reduced ammonia and VOC emissions	Processed solids can be used as bedding Reduced need for fertilizers (as nutrient availability in the digestate is increased) Electricity/heat generation	Digester technologies can be expensive

Source: Adapted from Greenhouse Gas Protocol Agricultural Guidance [30].

silage and grass silage admixed with manure it was only about 290 kg CO_2eq MWh^{-1}. Thus mono-digestion of manure reduced more CO_2 compared to co-digestion of liquid manure and energy crops [45]. Animal waste as input material can also substitute energy crops, which leads to less emissions of nitrous oxide and reduced consumption of fossil resources for fuel and mineral fertilizer [46].

Biogas plant is a project on farm that can reduce GHG emissions and improve farm performance, and the potential GHG benefits are N_2O and CH_4 reduction from manure management. The main benefits and problem are shown in Table 10.7.

10.4 Conclusions

Compared to other waste to energy solutions, AD is the most mature technology to upgrade manure's organic matter into renewable energy; however, the problems associated with high investment costs, operating parameters, manure collection, and digestate management have hindered its developments in rural areas in developing countries. CF is an indicator based on Life Cycle Thinking which is used as an usual indicator in an LCA method. It is a measure of the total amount of GHG emissions of a system, considering all relevant sources, sinks, and storage within the spatial and temporal boundary of the system. Before starting the data collection for CF analysis, the boundaries of the system and the FU should be determined. In this case, it is needed to identify all system inputs and outputs in relation to GHG emissions. The FU for an agricultural biogas plant is 1 ton of dry weight of biomass used in the AD process. The typical life cycle CO_2 footprint of biogas reported in literature is between 50 and 450 kg CO_2 MW^{-1} $h_{el}{}^{-1}$ (CO_2 footprint of biomethane). Biogas plant can reduce GHG emissions and improve farm performance, and the potential GHG benefits are N_2O and CH_4 reduction from manure management.

References

1. Cui, X., Guo, L., Li, C. et al. (2021). The total biomass nitrogen reservoir and its potential of replacing chemical fertilizers in China. *Renewable and Sustainable Energy Reviews* 135: 110215. https://doi.org/10.1016/j.rser.2020.110215.

2. Dróżdż, D., Wystalska, K., Malińska, A. et al. (2020). Management of poultry manure in Poland – current state and future perspectives. *Journal of Environmental Management* 264: 110327. https://doi.org/10.1016/j.jenvman.2020.110327.
3. Jugowar, J. L., Mielcarek, P., and Rzeźnik, W. (2019). Advisory Code of Good Agricultural Practice concerning the reduction of ammonia emissions edited by ITP in Falenty. Warsaw, 2019 Chapter IV. Low-emission natural fertilizer storage systems. https://www.gov.pl/web/rolnictwo/kodeks-dobrej-praktyki-rolniczej-w-zakresie-ograniczania-emisji-amoniaku (accessed 23 June 2022).
4. Wawrzyniak, A., Lewicki, A., Pochwatka, P. et al. (2021). Database system for estimating the biogas potential of cattle and swine feces in Poland. *Journal of Ecological Engineering*. ISSN 2299-8993. https://doi.org/10.12911/22998993/132426.
5. Tańczuk, M., Robert, J., Kolasa-Więcek, A., and Niemiec, P. (2019). Assessment of energy potential of chicken manure in Poland. *Energies* 12: 1244. https://doi.org/10.3390/en12071244.
6. Eurostat (2018). Poultry-annual data. https://ec.europa.eu/eurostat/databrowser/view/apro_ec_poula/default/table?lang=en (accessed 23 June 2022).
7. Achnas, S. and Euverink, G.J.W. (2020). Rambling facets of manure-based biogas production in Europe: a briefing. *Renewable and Sustainable Energy Reviews* 119: 109566. https://doi.org/10.1016/j.rser.2019.109566.
8. Scarlat, N., Fahl, F., Dallemand, J.F. et al. (2018). Spatial analysis of biogas potential from manure in Europe. *Renewable and Sustainable Energy Reviews* 94: 915–930. https://doi.org/10.1016/j.rser.2018.06.035.
9. Council of the European Union (1991). Council Directive (91/676/EEC) of 12 December 1991 concerning the protection of waters against pollution caused by nitrates from agricultural sources. https://eur-lex.europa.eu/legal-content/PL/TXT/PDF/?uri=CELEX:31991L0676&from=PL (accessed 23 June 2022).
10. Khoshnevisan B., Duan N., Tsapekos P., Kumar-Awasthi M., Liu Z., Mohammadi A., Angelidaki I, Tsang CW. D., Zhang Z., Pan J. Ma L., Anghbashlo M., Tabatabaei M., Liu H., A critical review on livestock manure biorefinery technologies: sustainability, challenges, and future perspectives *Renewable and Sustainable Energy Reviews* 135 2021, 110033. https://doi.org/10.1016/j.rser.2020.110033
11. Neshat, A., Mohammadi, M., Najafpour, G., and Lahijani, P. (2017). Anaerobic co-digestion of animal manures and lignocellulosic residues as a potent approach for sustainable biogas production. *Renewable and Sustainable Energy Reviews* 79: 308–322. https://doi.org/10.1016/j.rser.2017.05.137.
12. Magazyn Biomasa (2022). BIOGAS report in POLAND. Publisher Biomass Media Group Sp. z o. o. Kwiatowa 14/4 61–881, Poznań. https://magazynbiomasa.pl (accessed 23 June 2022).
13. Krajowy Ośrodek Wsparcia Rolnictwa (2011). National Support Center for Agriculture (2011). List of raw materials for the production of agricultural biogas in 2011. https://bip.kowr.gov.pl/uploads/pliki/oze/biogaz/surowce_w_2011.pdf (accessed 23 June 2022).
14. Krajowy Ośrodek Wsparcia Rolnictwa (2019). National Support Center for Agriculture (2019). List of raw materials for the production of agricultural biogas in 2019. https://bip.kowr.gov.pl/uploads/pliki/oze/biogaz/Surowce_w_2019_r.pdf (accessed 23 June 2022).
15. Nwokolo, N., Mukumba, P., Obileke, K., and Enebe, M. (2020). Waste to energy: a focus on the impact of substrate type in biogas production. *Processes* 8: 1224. https://doi.org/10.3390/pr8101224.
16. Czekała, W., Lewicki, A., Pochwatka, P. et al. (2020). Digestate management in Polish farms as an element of the nutrient cycle. *Journal of Cleaner Production* 242: 118454. https://doi.org/10.1016/j.jclepro.2019.118454.
17. Statistics Poland (2017). Characteristics of agricultural holdings in 2016. Central Statistical Office: Warsaw, Poland. https://stat.gov.pl/en/topics/agriculture-forestry (accessed 23 June 2022).
18. Poland's National Inventory Report 2019 (2019). Greenhouse gas inventory for 1988–2017. National Center for Balancing and Management of Emissions. https://www.kobize.pl/uploads/materialy/materialy_do_pobrania/krajowa_inwentaryzacja_emisji/NIR_POL_2019_23.05.2019.pdf (accessed 23 June 2022).

19. Witkowska-Dąbrowska, M. (2018). Changes in the amount of greenhouse gases and ammonia emissions to air from agricultural activities in Poland and the EU – analyzes using sustainable development indicators. *Problems of World Agriculture* 18: 303–314. https://doi.org/10.22630/PRS.2018.18.2.57.
20. OECD (2008). Report 2017. Impact of agriculture on the natural environment since 1990. https://www.oecd.org/poland/40806009.pdf (accessed 22 June 2022).
21. EUROSTAT (2020). Greenhouse gas emissions by source sector. Source: EEA. http://appsso.eurostat.ec.europa.eu/nui/show.do?dataset=env_air_gge&lang=en (accessed 23 June 2022).
22. Moller, H.B., Sorensen, P., Olesen, J.E. et al. (2022). Agriculture biogas production-climate and environmental impacts. *Sustainability* 14 (3): 1849. https://doi.org/10.3390/su14031849.
23. Uusitalo, V., Havukainen, J., Manninen, K. et al. (2014). CF of selected biomass to biogas production chains and GHG reduction potential in transportation use. *Renewable Energy* 66: 90–98. https://doi.org/10.1016/j.renene.2013.12.004.
24. ISO 14067:2018-10E Environmental Management (n.d.). Greenhouse gases. Carbon footprint of products. Quantification requirements and guidelines. English version.
25. Growcom, A. J. E. (2008). What is CF? An overview of definitions and methodologies. Horticulture Australia Ltd.
26. International Organisation for Standardization (2006). ISO 14064: greenhouse gas accounting and verification. International Organization for Standardization. Viewed 17 September 2008. http://store.payloadz.com/str-asp-i.105501n.ISO_140641_Green_House_Gases_Standard_eBooks_-end-detail.html (accessed 22 June 2022).
27. Wiedmann, T. and Minx, J. (2007). A definition of "CF", ISA Research and Consulting, Durham, United Kingdom. www.censa.org.uk/reports.html (accessed 22 June 2022).
28. World Resources Institute and World Business Council for Sustainable Development (2006). The Greenhouse gas Protocol for project accounting. https://ghgprotocol.org/sites/default/files/standards/ghg_project_accounting.pdf (accessed 22 June 2022).
29. World Business Council for Sustainable Development (2004). *The Greenhouse Gas Protocol. A Corporate Accounting and Reporting Standard*. World Business Council for Sustainable Development Revised edition. ISBN-156973-568-9. https://ghgprotocol.org/corporate-standard (accessed 22 June 2022).
30. Greenhouse Gas Protocol Agricultural Guidance (2006). Interpreting the Corporate Accounting and Reporting Standard for the agricultural sector. https://ghgprotocol.org/sites/default/files/standards/GHG%20Protocol%20Agricultural%20Guidance%20%28April%2026%29_0.pdf (accessed 22 June 2022).
31. IPCC Climate Change (2007). Synthesis report. Intergovernmental Panel on Climate Change, Geneva 2007. https://www.ipcc.ch/site/assets/uploads/2018/02/ar4_syr_full_report.pdf (accessed 23 June 2022).
32. Chadwick, D., Sommer, S., Thorman, R. et al. (2011). Manure management: implications for greenhouse gas emissions. *Animal Feed Science and Technology* 166–167: 514–531. https://doi.org/10.1016/j.anifeedsci.2011.04.036.
33. Baillet V. (n.d.). MilKey WP4 – models and emission factors for the life cycle inventory of the case studies.
34. IPCC (Intergovernmental Panel on Climate Change) Chapter 6: Energy systems. In: *Climate Change 2014: Mitigation of Climate Change. Working Group III Contribution to the IPCC 5th Assessment Report* (ed. O. Edenhofer, R. Pichs-Madruga, and Y. Sokona). Cambridge: IPCC. https://www.ipcc.ch/report/ar5/wg3 (accessed 23 June 2022).
35. Budzianowski, W.M. and Postawa, K. (2017). Renewable energy from biogas with reduced carbon dioxide footprint: implications of applying different plant configurations and operating pressures. *Renewable and Sustainable Energy Reviews* 68 (Part 2): 852–868. https://doi.org/10.1016/j.rser.2016.05.076.

36. Desideri, U., Zepparelli, F., Morettini, E., and Garroni, E. (2013). Comparative analysis of concentrating solar power and photovoltaic technologies: technical and environmental evaluations. *Applied Energy* 102: 765–784. https://doi.org/10.1016/j.apenergy.2012.08.033.
37. Bravi, M. and Basosi, R. (2014). Environmental impact of electricity from selected geothermal power plants in Italy. *Journal of Cleaner Production* 66: 301–308. https://doi.org/10.1016/j.jclepro.2013.11.015.
38. ISO 14040:2006 (2006). Environmental management – life cycle assessment – principles and framework. English version.
39. Dressler, D., Loewen, A., and Nelles, M. (2012). Life cycle assessment of the supply and use of bioenergy: impact of regional factors on biogas production. *The International Journal of Life Cycle Assessment* 17: 1104–1115. https://link.springer.com/article/10.1007/s11367-012-0424-9.
40. Gutierrez, M.C., Martin, M.A., Serrano, A., and Chica, A.F. (2015). Monitoring of pile composting process of OFMSW at full scale and evaluation of odour emission impact. *Journal of Environmental Management* 151: 531–539. https://doi.org/10.1016/j.jenvman.2014.12.034.
41. Whiting, A. and Azapagic, A. (2014). Life cycle environmental impacts of generating electricity and heat from biogas produced by anaerobic digestion. *Energy* https://doi.org/10.1016/j.energy.2014.03.103.
42. Szabó, G., Fazekas, I., Szabó, S. et al. (2014). The carbon footprint of a biogas power plant. *Environmental Engineering and Management Journal* 13 (11): 2867–2874. http://www.eemj.icpm.tuiasi.ro/pdfs/vol13/no11/Full/22_692_Szabo_14.pdf.
43. IFEU (Institute for Energy and Environmental Research Heidelberg GmbH) (2020). Basic data on GHG balances for biogas process chains and creation of new GHG balances. http://www.ifeu.de/oekobilanzen/pdf/THG_Bilanzen_Bio_Erdgas.pdf (accessed:22 June 2022).
44. Fuchsz, M. and Kohlheb, N. (2015). Comparison of the environmental effects of manure-and crop-based agricultural biogas plants using life cycle analysis. *Journal of Cleaner Production* 86: 60–66. https://doi.org/10.1016/j.jclepro.2014.08.058.
45. Lansche, J. and Muller, J. (2012). Life cycle assessment of energy generation of biogas fed combined heat and power plants: environmental impact of different agricultural substrates. *Engineering in Life Sciences* 12 (3): 313–320. https://doi.org/10.1002/elsc.201100061.
46. Bachmaier, J., Effenberger, M., and Gronauer, A. (2010). Greenhouse gas balance and resource demand of biogas plants in agriculture. *Engineering in Life Sciences* 10 (6): 560–569. https://doi.org/10.1002/elsc.201000073.

11

Financial Sustainability and Stakeholder Partnerships of Biogas Plants

To-Hung Tsui[1,2,3], Le Zhang[1,2,4], Jonathan T. E. Lee[1,2], Yanjun Dai[2,5], and Yen Wah Tong[1,2,6]

[1] *NUS Environmental Research Institute, National University of Singapore, Singapore*
[2] *Energy and Environmental Sustainability for Megacities (E2S2) Phase II, Campus for Research Excellence and Technological Enterprise (CREATE), Singapore*
[3] *Department of Engineering Science, University of Oxford, Oxford, UK*
[4] *Department of Resources and Environment, School of Agriculture and Biology, Shanghai Jiao Tong University, Shanghai, China*
[5] *School of Mechanical Engineering, Shanghai Jiao Tong University, Shanghai, China*
[6] *Department of Chemical and Biomolecular Engineering, National University of Singapore, Singapore*

11.1 Introduction

The global aspirations for a more sustainable future never get diminished, but the historical trends of developing biogas plants have already shown a couple of cycles over the last few decades. Particularly, biogas production had once been greeted as a promising alternative to reduce fossil fuel consumption in the 1980s, as there were remarkable advancements in anaerobic biotechnology [1]. However, a quick decline in global energy prices and more certainty of oil supply slowed down the development of biogas plants. One of many reasons was the consideration of their immediate financial returns. Meanwhile, there are a lot of developing and remote regions in which people apply household-scale biogas utilization for

Biogas Plants: Waste Management, Energy Production and Carbon Footprint Reduction,
First Edition. Edited by Wojciech Czekała

cooking and other purposes. As an infrastructure for energy, biogas plants can also provide nutrient-rich residues (commonly known as digestate) and appealing environmental benefits that directly link to the present advocacy of sustainable development goals by the United Nations.

Almost no stakeholders deny that biogas production can contribute to a future of less dependence on fossil fuels as well as a reduction of greenhouse gas emissions. It is also in alignment with the emerging context of a circular bioeconomy that strategizes economic growth of sustainability capacity. Despite these spectacular implications, the financial sustainability of biogas plants is weakly observed in many countries, when compared to countries like Denmark and Germany (known for their agricultural waste recycling) [2]. In extreme cases, biogas plants could go bankrupt before their expected service lifespan [3]. Primary reasons for those financial performances include the volatility of the energy market and also poor economic returns from other resource outputs.

From a technological aspect, biogas infrastructures have experienced rapid development in recent years [4]. However, at this moment, the economic profitability of biogas plants is still apparently low. The effective improvement of financial sustainability is often challenging due to inexplicable policy support and guidelines. One systems engineering approach to make them more efficient is the upgrade of processing technologies and service efficiency [5–7]. Examples include optimized use of the excess heat for hydrothermal applications as well as providing expanded service for extra gate fees [8, 9]. The multifaceted decisions for improved financial performance often require effective stakeholder management and partnerships. Given that each city has different physical environments and socioeconomic contexts, it is more important to understand what could enhance and hinder financial success.

As a municipal infrastructure, biogas plants possess financial characteristics of strong asset specificity. When infrastructure (especially for centralized plants) has been installed with a huge investment, it seldom provides retrofitting flexibility in case of weak financial performance. From an economic aspect, the uncertainty issues make it confusing for the general public and sometimes even key stakeholders to determine the technology readiness stage of biogas plants nowadays. Consequently, it further undermines organizational behavior for stakeholder partnerships and collective decisions when managing more rigorous sustainability pursuits. Considering that the full economic/environmental potentials of biogas plants have not been met worldwide, it is necessary to illuminate some relevant improvements. This chapter provides vertical integration of experiential knowledge by practice. It covers the consideration of basic economic factors as to what stakeholder partnerships would biogas plants need. The overall content would prepare readers to open up more meaningful dialogs and debates for planning/managing biogas plants in the next decade.

11.2 Basic Technological Factors

In-depth discussions of technological factors are not the major focus of this chapter, but some prior knowledge is essential before discussing the economic aspects of the actual implementation of biogas plants. For example, socioeconomic factors (such as government policies and market prices of resource inputs/outputs) are crucial in the value chain. They could have direct impacts on the financial portfolios of biogas infrastructure, but it is of no use without the technological success of well-developed systems for biogas production (and outputs of other valuable products). It is often an ongoing learning by practice, which also reflects there is a long-term need for building local human assets and knowing how to advance more strategic technological designs and also stakeholder partnerships for biogas plants.

A wide range of technological factors can influence the efficiency and stability of anaerobic digestion processes. The most common parameters that operators can apply for process control are temperature and pH. Regarding temperature, the operating conditions can be classified into three regimes: psychrophilic, mesophilic, or thermophilic. The selection of temperature range needs to consider its system advantages/drawbacks as well as local climate conditions. There are operational trade-offs behind the options, where lower biogas production is often observed with daily fluctuations in temperature. For example, thermophilic anaerobic digestion processes (50–60 °C) are useful when pathogen removal is one of the process requirements. The thermophilic process asks more monitoring efforts, and there could be challenges in maintaining constant temperature if the gradient with surrounding environments is high. Potential mitigation measures include the bioaugmentation of more adaptive microbes and increased investment in reactor insulation [10, 11]. In general, microorganisms under psychrophilic digestion (10–20 °C) are more stable. However, psychrophilic conditions suffer from much slower rates of organics conversion and microbial growth, which results in the need for an extended biomass retention time and therefore requirements of large digester capacity. Given that methanogenesis occurs at neutral pH values between 6.5 and 7.5, it is particularly important to continuously maintain constant conditions without disturbance. Anaerobic co-digestion is a viable approach to increase the revenue of biogas plants by extra gate fees and energy output, but the higher organic loads would lead to excessive lowering of pH. In contrast, the existence of toxic substances (at excessive concentrations) from diverse waste streams may pose inhibitory effects on microbiological processes. Recent mitigation advancements include biotechnological applications of direct interspecies electron transfer mechanisms to promote conversion efficiency [12, 13].

Regarding the economic revenues, biogas production is not the end stage, and it is always desirable to be further converted into usable energy/bioproducts. Some examples include electricity generation by the gas engine, gas purification and injection into grids, upgraded as vehicle fuel, manufacturing applications for proteins/chemicals, and thermal source for heat and steam [14, 15]. However, it should be noted that direct flaring of biogas (i.e. without generating any profit from resource recovery) is common in some regions [16]. Another major output of anaerobic digestion is digestate which is a valuable biofertilizer considering its nutrient content and fertilization effects. For anaerobic digestion generally applying higher temperatures and longer retention times, it is also able to reduce odor and inactivate possible pathogens for reuse purposes. Depending on agriculture guidelines, the technological design of biogas plants needs to be in alignment with how local policy regulation recognizes digestate as a fertilizer. Additional processing to separate and further refine biogas/digestate is therefore sometimes necessary.

While biogas production and utilization depend on a wide range of technological factors, another important consideration for financial sustainability is scale. Centralized plants at the city-level municipality (such as wastewater treatment plants) tend to have more utilization options, though there are other financial concerns for retrofitting their fixed assets. Through systems integration (e.g. fermentation refinery), valued-added products could also be recovered in biogas plants, which can be referred to as an example of a circular bioeconomy. Regarding this, techno-economic assessment using machine learning applications could provide an efficient approach for quick evaluation [17]. Running a biogas plant necessarily involves much more knowledge nowadays, and technological understanding is a prerequisite behind exploring any decisions for financial success.

11.3 Economic Evaluation and Failures

The planning of biogas infrastructures can be more complex when it goes into scale and implementation. As a part of sustainable waste management, biogas production from anaerobic digestion can help waste treatment with less reliance on fossil fuels, thus reducing carbon emissions. Despite the recognition of technological maturity, the financial sustainability of actual implementation has been reported to face challenges (and failures). Built on past implementation experience, this section discusses the critical risks from economic aspects. It would clarify what cannot be ignored during long-term planning. The content can also lay the foundations for next-stage efforts by stakeholder partnership and co-governance.

11.3.1 Investment Risks for Fixed Assets

The infrastructure development for biogas plants requires high investments, and a long service lifespan of at least 20 to 30 years is often assumed. Due to its technological specificity for anaerobic digestion, the equipment and reactor settings are not easy to be retrofitted for other industrial purposes. This implies once the installments were confirmed, the economic returns would be diminished if it was not employed as planned functions and service lifespan. Regarding these, the uncertainty of spatial arrangement and human assets over time are two main economic risks that are often overlooked [18–20]. Human assets generally refer to expertise with a good understanding of infrastructure management and planning, to which they can keep pace with more rigorous progress of waste management goals.

The preliminary design of plant location commonly targets minimizing waste transport and balancing waste collection/re-utilization. In other words, there are economic dependencies of location selection and would encounter multilevel trade-offs in different geographic settings. In term of spatial arrangement, the asset requirements of biogas plant applying centralized planning is typically to be higher, since it entails extra collection costs and fixed facilities (of depots for storage and preprocessing). In contrast, to maximize the economic returns, the regional demand and potential processing needs for digestate reuse need to be quantified. The miscalculation of the local demand would substantially increase disposal costs, given that the thermal drying and posttreatment of digestate both demand high energy requirements. For projection of digestate generation, it also links to the uncertainty of whether the daily collection amount of organic waste is stable and sufficient to meet the capacity design. Depending on the policy agenda in different counties, the planning of a biogas plant often starts with clear goals that are static. However, without sufficient local recycling support (e.g. rewarding system and campaign), plant operators are unlikely to handle the system challenges arising from waste management in a larger context.

While optimal biogas production is often cited for research interests of the system upgrade, one major investment cost for bioenergy is a combined heat-power unit (CHP) (to which the conversion models can differ for climate regions and biogas generation rates). A mismatch planning of proper CHP devices, anaerobic digestion system, waste collection scheme, and reuse tend to damage the projected financial returns of biogas plants. Rather than providing a detailed guideline, the aforementioned discussion aims to illustrate the complexity and interlinks of economic factors which are essential to ensure the long-term operation as early infrastructure planning. It is about assisting stakeholders to understand ex ante decisions and minimize unexpected circumstances. Recently, researchers have started to push forward the economic frameworks of resolving uncertainty concerns and delivering more informed decision needs by stakeholders [9]. Rather than an analytical application for individual cases, the framework of forecast models with Monte Carlos analysis emphasizes the transferable and transformative needs (e.g. retrofitting for decarbonization).

11.3.2 Failures and Intervention

Waste management is of public interest, and bioenergy recovery has characteristics of positive externalities (e.g. reduction in greenhouse gas emissions, energy requirement, and odor control) [21]. However, biogas plants can also fall short as polluters, when the outputs are nonmarketable or not sufficiently valued by stakeholders. Through market mechanisms, the supply and demand equilibrium will bring the price discussion of goods or services together [22], which implies a logical way to legitimatize the financial sustainability of biogas plants. Regarding this, market failures of services and goods frequently exist in different areas. A perfect market is seldom naturally formed without close monitoring and intervention. As mentioned earlier, the value of product outputs (e.g. bioenergy and digestate reuse) depends on the supply chain relationship which involves the fundamental risk of information asymmetry and bounded rationality of stakeholders.

For example, the information asymmetry could take if a new recycling initiative (e.g. composting) was in parallel introduced by other interested parties. The disturbance on a local recycling network (no matter centralized or decentralized) is that those originally planned waste sources may become less available for waste collection. The competing relationship is not uncommon within the waste management sector and even in government circles. The development of an open platform may bring a solution so that an updated consolidation of quantitative findings can proceed and transfer to stakeholders. The support for more efficient waste management goes with an operating cost of building the institutional environment and an unavoidable responsibility by the government. Without such interventions, plant operators have limited capacity to regularly trace the waste flows at the city scale. Moreover, the social capacity to formulate and solve complex problems (in responding to new sustainability and nonmonetary objectives) would also be limited.

11.4 Stakeholders Partnership and Co-governance

This section illustrates why stakeholder engagement and effective co-governance could promote the financial sustainability of biogas plants. When carrying out the projection of economic profiles, the complexity could be subjected to a variety of decision scenarios. For example, biogas can be utilized in different ways. The application of a CHP device is a common approach for energy recovery, but the excess heat is not often fully utilized to generate economic returns. In contrast, anaerobic digestate has long been regarded more as a byproduct, but nowadays, it might probably emerge to be a major nutrient source for sustainable urban agriculture and local food security. It is clear that different waste sources and utilization pathways also exist when planning/upgrading a biogas plant. Given the different social contexts and technological scenarios, this section does not attempt to provide a comprehensive stakeholders analysis but rather restricts the discussion of common players in practices. In this regard, it would be sufficient to initialize the dialogs for more progressive stakeholder engagement in most cases.

11.4.1 Government

Government (at both local and national levels) is of great importance as it set waste management goals and initializes infrastructure planning which allows for the feasibility exploration of a biogas plant [23]. Particularly, the costs of electricity generation from biogas could be more expensive than that of traditional fossil fuels. Regarding this, financial subsidies are important to make plants' operations financially sustainable and thereby can stimulate continuous growth of the local environmental market. Thus, in some countries (e.g. the Netherlands), there are clear policies and budgets in place to provide long-term commitments for local recycling practices [24]. They are also the foundation to allow the development of long-term certainty among stakeholders. Environmental departments coordinate the monitoring of environmental concerns, the supervision of land use, and specific guidelines regarding resource reuse and waste collection schemes. Under current trends of climate challenges, more ambitious sustainability goals require clear policy support and proactive intervention from the government at all levels.

11.4.2 Consultant and Constructor

Depending on the tendering processing of goods/services, the consultant and constructor can be a single engineering firm or formed by a joint-venture partnership. In many cases of infrastructure projects, the specific adoption of advanced engineering technology needs to incorporate more technological support from multiple parties. During the early operation period after equipment installment, common problems (e.g. troubleshooting of low methane production and odor control) require further collaboration between the consultant and constructor [25, 26]. External consultancy companies could be further invited in cases to solve specific problems. Therefore, through the operation cycle, additional needs for subsidies and permits are common. Different forms of ownership of a biogas plant exist, and it depends on the management capacity of local government and thereby the procurement process.

11.4.3 Source of Waste Streams

Identification of co-digestion inputs is promising to increase biogas production and gain extra gate fees to support financial sustainability. Food industries can be a common supplier of organic waste. However, the introduction of their waste streams is not always easy. On the one hand, if the food industry already had an advanced valorization system for resource recovery [27, 28], the application of anaerobic digestion for bioenergy would imply lower economic incentives. On the other hand, the existing waste disposal practices and permits are guided by the local government. The diversion of waste streams to biogas plants needs to consider and manage relevant environmental regulations. Besides, possible concerns of stakeholders around the plants' location and other sensitive receivers need to be considered. For these reasons, the government has the role to oversee opportunities regularly and link up intermediary communication between waste suppliers and biogas plant operators.

11.4.4 Customers for Energy and Resource

The generated electricity from biogas can be sold to electricity supply companies. For centralized biogas plants of higher energy output, it would be required to work with grid operators for electricity transport and infrastructure connection [22, 29]. The financial investment for grid connection is high, which thereby needs to be assured by more effective projection and management of financial profiles. In many cases, such energy plans were not directly absorbed by plants, and loan support is needed from banks. Apart from energy recovery, the excess digestate has to be discharged after each operational cycle of loading new feedstocks. Despite its nutrient values, digestate sometimes needs to be disposed of, if a match of local demand was not in place and/or recycling guidelines are not available [30]. Digestate is used as fertilizer in general circumstances if the demand for arable land is sufficient for the country. In other cases, the digestate can be further processed (e.g. by pressing, drying, and pasteurizing the digestate) to satisfy customer needs outside the country.

11.5 Summary and Outlooks

The financial sustainability of biogas plants depends on a wide range of factors involving effective planning and stakeholder partnerships. Biogas industries and waste management are heading into the next decade of development. It will be imperative to further explore systemic interventions in more comprehensive ways. From the technological aspect, more work can be done to address the economic characteristics of positive externalities. For example, modular upgrades and decentralized networks of biogas plants might provide better infrastructural flexibility to access more waste types from different locations, which could reduce retrofitting expenses and enhance service lifespan. In contrast, research institutes worldwide can play more roles in providing more updated knowledge, whereas cities worldwide also need to build human assets in tackling more challenges under climate risks. In the last decade, our research center has concentrated on building the R&D ecosystem and delivering the scale-up implementation of environmental solutions in urban environments (including the implementation of twin test bedding in Singapore and Shanghai).

The findings from scale-up technological trials often quickly inform next-stage searching for potential improvement and stakeholder partnerships. Governments can also make good use of new knowledge as scientific underpinnings to facilitate their policy intervention (e.g. regulations, subsidies, and procurement planning). The fast-paced explorations and communication among stakeholders can encourage more collective involvement toward new initiatives for identifying critical changes to be made.

Acknowledgments

This research is supported by the National Research Foundation, Prime Minister's Office, Singapore, under its Campus for Research Excellence and Technological Enterprise (CREATE) program.

References

1. Lettinga, G. (2014). My anaerobic sustainability story. http://www.leaf-wageningen.nl/en/leaf.htm.
2. Jacobsen, B.H., Laugesen, F.M., and Dubgaard, A. (2014). The economics of biogas in Denmark: a farm and socioeconomic perspective. *International Journal of Agricultural Management* 3 (1029-2016-82300): http://dx.doi.org/10.22004/ag.econ.236887.
3. Igliński, B., Buczkowski, R., Iglińska, A. et al. (2012). Agricultural biogas plants in Poland: investment process, economical and environmental aspects, biogas potential. *Renewable and Sustainable Energy Reviews* 16 (7): 4890–4900. https://doi.org/10.1016/j.rser.2012.04.037.
4. Tsui, T.H. and Wong, J.W. (2019). A critical review: emerging bioeconomy and waste-to-energy technologies for sustainable municipal solid waste management. *Waste Disposal & Sustainable Energy* 1 (3): 151–167. https://doi.org/10.1007/s42768-019-00013-z.
5. Gebrezgabher, S.A., Meuwissen, M.P., and Oude Lansink, A.G. (2010). Costs of producing biogas at dairy farms in the Netherlands. *International Journal on Food System Dynamics* 1 (1012-2016-81162): 26–35. http://dx.doi.org/10.22004/ag.econ.91139.
6. Mao, L., Tsui, T.H., Zhang, J. et al. (2021a). Mixing effects on decentralized high-solid digester for horticultural waste: startup, operation and sensitive microorganisms. *Bioresource Technology* 333: 125216. https://doi.org/10.1016/j.biortech.2021.125216.
7. Tsui, T.H., Ekama, G.A., and Chen, G.H. (2018). Quantitative characterization and analysis of granule transformations: role of intermittent gas sparging in a super high-rate anaerobic system. *Water Research* 139: 177–186. https://doi.org/10.1016/j.watres.2018.04.002.
8. Mao, L., Tsui, T.H., Zhang, J. et al. (2021b). System integration of hydrothermal liquefaction and anaerobic digestion for wet biomass valorization: biodegradability and microbial syntrophy. *Journal of Environmental Management* 293: 112981. https://doi.org/10.1016/j.jenvman.2021.112981.
9. Tsui, T.H., Zhang, L., Zhang, J. et al. (2022a). Methodological framework for wastewater treatment plants delivering expanded service: economic tradeoffs and technological decisions. *Science of the Total Environment* 823: 153616. https://doi.org/10.1016/j.scitotenv.2022.153616.
10. Lee, J.T., Dutta, N., Zhang, L. et al. (2022b). Bioaugmentation of *Methanosarcina thermophila* grown on biochar particles during semi-continuous thermophilic food waste anaerobic digestion under two different bioaugmentation regimes. *Bioresource Technology* 360: 127590. https://doi.org/10.1016/j.biortech.2022.127590.
11. Lee, J.T., Lim, E.Y., Zhang, L. et al. (2022a). Methanosarcina thermophila bioaugmentation and its synergy with biochar growth support particles versus polypropylene microplastics in thermophilic food waste anaerobic digestion. *Bioresource Technology* 360: 127531. https://doi.org/10.1016/j.biortech.2022.127531.

12. Tsui, T.H., Wu, H., Song, B. et al. (2020). Food waste leachate treatment using an upflow anaerobic sludge bed (UASB): effect of conductive material dosage under low and high organic loads. *Bioresource Technology* 304: 122738. https://doi.org/10.1016/j.biortech.2020.122738.
13. Tsui, T.H., Zhang, L., Lim, E.Y. et al. (2021). Timing of biochar dosage for anaerobic digestion treating municipal leachate: altered conversion pathways of volatile fatty acids. *Bioresource Technology* 335: 125283. https://doi.org/10.1016/j.biortech.2021.125283.
14. Hakawati, R., Smyth, B.M., McCullough, G. et al. (2017). What is the most energy efficient route for biogas utilization: heat, electricity or transport? *Applied Energy* 206: 1076–1087. https://doi.org/10.1016/j.apenergy.2017.08.068.
15. Pöschl, M., Ward, S., and Owende, P. (2010). Evaluation of energy efficiency of various biogas production and utilization pathways. *Applied Energy* 87 (11): 3305–3321. https://doi.org/10.1016/j.apenergy.2010.05.011.
16. Tsui, T.H., Zhang, L., Zhang, J. et al. (2022b). Engineering interface between bioenergy recovery and biogas desulfurization: sustainability interplays of biochar application. *Renewable and Sustainable Energy Reviews* 157: 112053. https://doi.org/10.1016/j.rser.2021.112053.
17. Tsui, T.H., van Loosdrecht, M.C., Dai, Y., and Tong, Y.W. (2023). Machine learning and circular bioeconomy: building new resource efficiency from diverse waste streams. *Bioresource Technology* 369: 128445.
18. Lovrak, A., Pukšec, T., and Duić, N. (2020). A geographical information system (GIS) based approach for assessing the spatial distribution and seasonal variation of biogas production potential from agricultural residues and municipal biowaste. *Applied Energy* 267: 115010. https://doi.org/10.1016/j.apenergy.2020.115010.
19. Valenti, F., Porto, S.M., Dale, B.E., and Liao, W. (2018). Spatial analysis of feedstock supply and logistics to establish regional biogas power generation: a case study in the region of Sicily. *Renewable and Sustainable Energy Reviews* 97: 50–63. https://doi.org/10.1016/j.rser.2018.08.022.
20. Zirngast, K., Čuček, L., Zore, Ž. et al. (2019). Synthesis of flexible supply networks under uncertainty applied to biogas production. *Computers & Chemical Engineering* 129: 106503. https://doi.org/10.1016/j.compchemeng.2019.06.02.
21. Bergek, A., Jacobsson, S., and Sandén, B.A. (2008). 'Legitimation'and 'development of positive externalities': two key processes in the formation phase of technological innovation systems. *Technology Analysis & Strategic Management* 20 (5): 575–592. https://doi.org/10.1080/09537320802292768.
22. Guerin, T.F. (2022). Business model scaling can be used to activate and grow the biogas-to-grid market in Australia to decarbonise hard-to-abate industries: an application of entrepreneurial management. *Renewable and Sustainable Energy Reviews* 158: 112090. https://doi.org/10.1016/j.rser.2022.112090.
23. Qu, W., Tu, Q., and Bluemling, B. (2013). Which factors are effective for farmers' biogas use?–evidence from a large-scale survey in China. *Energy Policy* 63: 26–33. https://doi.org/10.1016/j.enpol.2013.07.019.
24. Winquist, E., Van Galen, M., Zielonka, S. et al. (2021). Expert views on the future development of biogas business branch in Germany, the Netherlands, and Finland until 2030. *Sustainability* 13 (3): 1148. https://doi.org/10.3390/su13031148.
25. Bruno, T., Gelderman, C.J., Lambrechts, W., and Semeijn, J. (2018). The promise of best value procurement: governance and (in) stability of specifications within an innovative biogas project. *Journal of Cleaner Production* 172: 1465–1475. https://doi.org/10.1016/j.jclepro.2017.10.251.
26. Ghanavati, H. (2018). Biogas production systems: operation, process control, and troubleshooting. In: *Biogas*, 199–219. Cham: Springer https://doi.org/10.1007/978-3-319-77335-3_8.
27. Nawaz, A., Li, E., Irshad, S. et al. (2020). Valorization of fisheries by-products: challenges and technical concerns to food industry. *Trends in Food Science & Technology* 99: 34–43. https://doi.org/10.1016/j.tifs.2020.02.022.

28. Otles, S., Despoudi, S., Bucatariu, C., and Kartal, C. (2015). Food waste management, valorization, and sustainability in the food industry. In: *Food Waste Recovery*, 3–23. Academic Press https://doi.org/10.1016/B978-0-12-800351-0.00001-8.
29. Karlsson, N.P., Halila, F., Mattsson, M., and Hoveskog, M. (2017). Success factors for agricultural biogas production in Sweden: a case study of business model innovation. *Journal of Cleaner Production* 142: 2925–2934. https://doi.org/10.1016/j.jclepro.2016.10.178.
30. Fuchs, W. and Drosg, B. (2013). Assessment of the state of the art of technologies for the processing of digestate residue from anaerobic digesters. *Water Science and Technology* 67 (9): 1984–1993. https://doi.org/10.2166/wst.2013.075.

12

Measuring the Resilience of Supply Critical Systems: The Case of the Biogas Value Chain

Raul Carlsson[1] and Tatiana Nevzorova[2]

[1] *Certification Development Unit, RISE Research Institutes of Sweden, Jönköping, Sweden*
[2] *Certification Development Unit, RISE Research Institutes of Sweden, Stockholm, Sweden*

12.1 Introduction

Different global events such as a pandemic, Brexit, international conflicts, and cyber security threats affect the industry and require to constantly review and optimize supply chains, raise capital commitments, sign new agreements, and seek new suppliers if needed. The industry is undergoing major changes through the energy crisis and climate change, which means that processes are set to change even without consideration of risk, resilience, and robustness. Another issue that the industry faces is the increased demand for material extraction including the scarcity of nonrenewable resources. This is also related to unethical and dangerous working conditions and child labor that require creating a strategic tool with a transparent and sustainable supply chain map. At the same time, there is a lack of simple methods to estimate/measure and ensure how resilient a supply system is. All these facts create a strong need in supporting the industry in the development of increased self-sufficiency and resilience in the event of a crisis. Therefore, this chapter aims to develop the methodology for measuring the resilience of supply critical systems.

The energy system (e.g. infrastructure systems throughout the energy supply chains) is considered to be one of the most complex and important critical infrastructure systems [1]. Energy supply is an essential part of modern societies and industries that facilitate economic

Biogas Plants: Waste Management, Energy Production and Carbon Footprint Reduction,
First Edition. Edited by Wojciech Czekała

growth and social welfare. Energy supply disruptions can lead to serious consequences for large segments of the population [2]. Therefore, it is important to build a resilient energy system that will be able to react to any external and internal disruptions from any kind of environmental catastrophes to technical failures and man-made accidents.

The biogas value chain is chosen as the practical case of the research because it can provide energy security to users while utilizing hazardous organic waste. Biogas can be seen as a resilient option that improves access to energy, jobs, and food in urban and rural areas [3]. Some studies explored different kinds of shocks that can affect biogas production. For example, since the productivity and stability of the anaerobic digestion process are highly dependent on fats, oils, and greases (FOG) composition, loading rate, and microbial community adaptation and the interactions between these three parameters [4, 5], accidental organic overloading is a common shock that affects anaerobic digester's biogas productivity (e.g. [6–8]). The increase in accidents on biogas stations (e.g. [9–11]) also stimulates to study different resilience measurements. At the same time, different studies analyzed biogas potential as a replacement fertilizer [12] or as a solution to poverty and energy security (e.g. [3]). However, there is limited availability of studies that create the universal model to measure the resilience throughout the whole value chain using biogas as a practical case looking at the biogas system from two perspectives – biogas as a resilience solution and resilience measures to occurred disruptions along the biogas value chain. Therefore, this research aims to fulfill this gap.

The rest of the chapter is structured as follows: Section 12.2 provides a background of the research. Section 12.3 presents the methodological details of the study. Section 12.4 introduces an extensive analysis of the biogas value chain as well as presents the full measurement scheme concept. Section 12.5 makes conclusions and provides some recommendations for future research.

12.2 Background

The term "resilience" can be defined as a "process that the observed system undergoes in response to a disruption quantified in terms of a measure of system performance and its evolution during the system response time after an event" ([1], p. 275). Even though the definitions of "resilience" may vary, there seems to be a consensus about the general characteristics of a resilient supply chain: namely to manage risks and survive disruptions. Resilience in this regard means being robust by being agile, which may sound slightly contradictory. Nevertheless, applying risk mitigation strategies seems to bring flexibility and increased performance [13]. A resilient company should be strong both in short term and long term. In the short term, this means surviving through the big external risks such as climate change and securing the logistic supply as well as the delivering of products to customers. In the long term, it means being able to protect the brand name and satisfy the customers as well as sustain the market position.

Biogas energy supply can become a solution because of its universal nature and multifunctionality. Since biogas has a high calorific value, it can be used for electricity, heating, and as a fuel for a vehicle after being upgraded and thus replacing scarce fossil fuels. One of the crucial benefits of biogas is the abundance of resources for its production. It can be produced from any kind of organic waste from the agricultural sector (e.g. livestock manure) that has the largest underutilized resources for biogas production [14], food and industrial sectors as well as from municipal wastewater facilities (i.e. untreated sewage). Utilizing anaerobic fermentation of animal wastes could fully or partially meet the energy needs of husbandry industries, while also producing organic fertilizers [15]. Therefore, biogas can improve waste management and contribute to a more circular economy by treating and disinfecting the waste and producing energy from it. This study aims to analyze the resilience of energy supply from two different dimensions: (i) resilience of energy production specifically of biogas energy itself where the biogas value chain experiences disruptions along the value chain; and (ii) resilience through locally produced energy (i.e. biogas), where biogas will become a solution to any kind of occurred disruptions.

The measurement concept of the resilience of the biogas value chain is divided into two parts: (i) resources and assets to analyze the resilience of physical resources necessary for the operation of the system to foreseeable individual or combined risks of stoppage/collapse, and (ii) management system to examine the preparedness of management system around the supply system to respond to elevated risk or handle outages/breakdowns. Under the methodology development, different scenarios for measuring the resilience of supply critical systems are developed, which are based on various well-defined internal disturbances where systems and system components are exposed to direct and indirect disturbances while their sensitivity to these disturbances is measured, as well as the consequences of these disturbances on the system's users. The system's stability is monitored, and its ability to cope with disturbances based on maintenance and repairs is analyzed to develop measurement methods. As a result, the created methodology aims to strengthen and simplify governance and monitoring toward better livelihoods in society as well as effectively distribute the responsibility for support capacity through individually allocated responsibility and independent certification of measurable ability to withstand and manage the crisis.

12.3 Methodology

The conducted study is based on two research methods. The first method that we provided is a quantitative analysis of biogas value chain data such as the amount of feedstock, energy production from biogas, and logistic consumption (e.g. transport fuels). The data were collected from the report written by Swedish Energy Agency and the Swedish Gas Association [16] (Tables 12.1 and 12.2). This information is combined with a risk calculation model needed to understand, analyze, and make concrete estimations of the potential risks of any kind of disruptions in the biogas value chain, especially with regard to disruptions connected with or independent from surrounding large-scale energy production systems, as well as to develop scenarios of feedback loop self-sustained circular biogas-based energy production.

Since the main aim of this chapter is to develop the methodology for measuring the resilience of supply critical systems that combines the sensitivity of assets and resources with organizational capacity for supply readiness, our second method includes a review of existing international standards (ISO) in the sphere of security, resilience, risk management, organizational resilience, societal security as well as any possible guidance for

Table 12.1 Production of biogas (GWh) and distribution by plant type in Sweden, 2021.

Facility type	Heating	Electricity	Transport fuel	Industrial use	Other	Total
Sewage treatment plant	194	10	419	1	0	624
Co-digestion facilities	67	9	1053	8	13	1150
Farm facilities	28	13	25	0	0	78
Industrial facilities	47	3	11	51	0	112
Other facilities	82	4	0	0	2	88

Source: Adapted from Swedish Energy Agency and Swedish Gas Association [16].

Table 12.2 Amount of digestate produced (kton wet weight) and its use as fertilizer in Sweden, 2021.

Facility type	Production of digestate	Use of digestate as fertilizer
Sewage treatment plant	597	232
Co-digestion facilities	1842	1810
Farm facilities	508	508
Industrial facilities	35	32
Total	**2981**	**2581**

Source: Adapted from Swedish Energy Agency and Swedish Gas Association [16].

ability assessment. The choice of reviewing ISO standards is determined by the role of standardization in measuring system resilience, which is partly a common conceptual apparatus that stands over time and between different social sectors and organizations, and partly a prerequisite for creating a common view of how resilience can be measured. By standardizing and sharing solutions and measures along the value chain, there is a possibility to create a common universal platform for safer and cheaper solutions to overcome any risks and disruptions.

12.4 Measurement Scheme

12.4.1 Introduction to the Measurement Concept

Although the risks and disruptions may vary for different industries since each industry has unique supply chain patterns and strategies, our measurement concept aims to find patterns and suggest a universal scheme that would help an organization to overcome and withstand risks and disruptions. The measurement concept of resilience is based on two parts (Figure 12.1):

(1) Management system that has four levels of resilience functionality based on the research written by Gasser et al. [1]. The first level is called "resist," where the organization has the capability to withstand risks and disruptions. At the second level, "restabilize," the organization absorbs the disruption and is able to reestablish key functionalities to the same level as they were before. At the third level, "rebuild," the disruption is severe, but the organization can recover and stabilize. The last level, "reconfigure," is the ability to adapt and conform to new conditions.
(2) Physical resources and assets are considered to match the same resilience levels as the management system in terms of designing them to resist predefined risks and disruptions, as well as being able to restabilize and rebuild after disruptions have occurred.

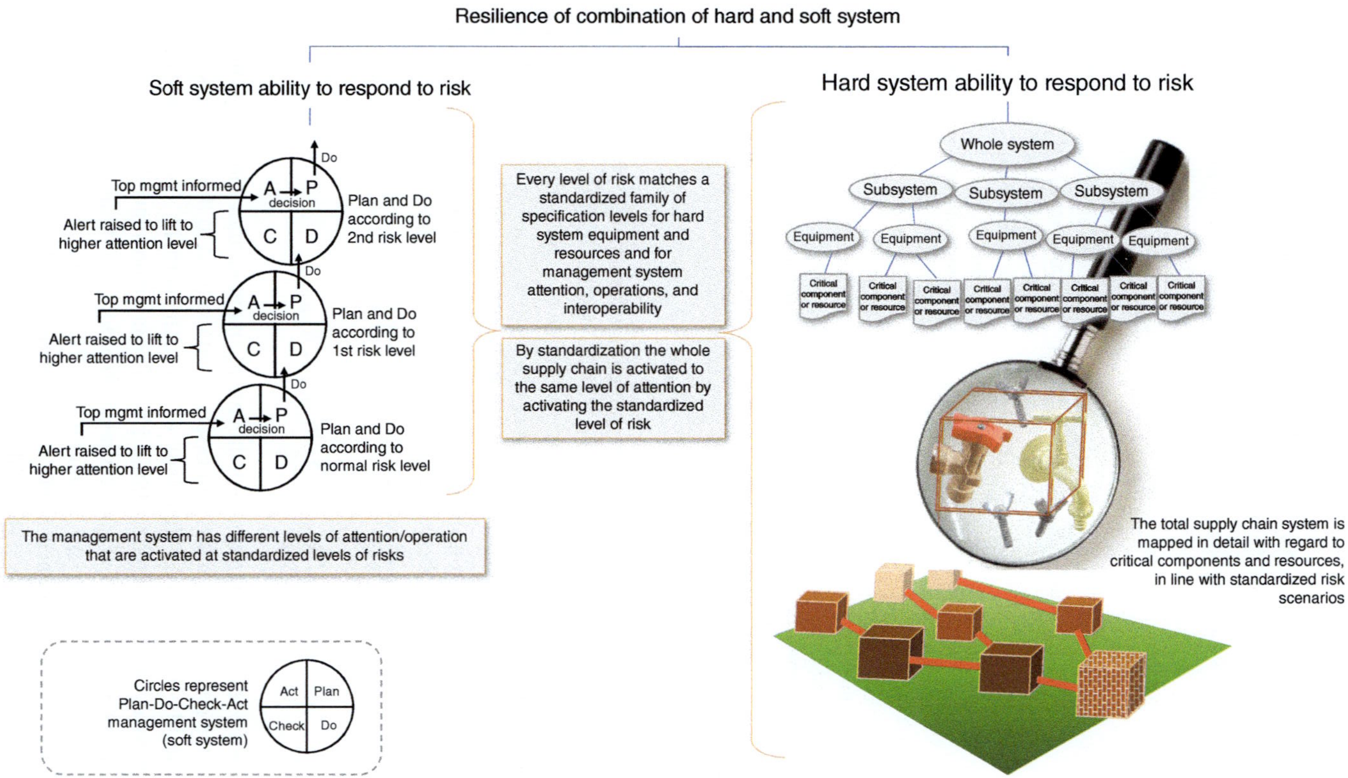

Figure 12.1 *The measurement concept of resilience (own study).*

Table 12.3 The resilience functions of the management system and physical resources and assets in relation to their levels of resilience functionality.

Levels of resilience functionality	Resilience functions of management system (R_{MsN})		Resilience functions of physical resources and assets (R_{RaN})	
Resist	Large value network	$0.9 < rf \leq 1$	Flexible technology and localization with several material and resources supply chain options	$0.9 < rf \leq 1$
Restabilize	Strong connections with supplier networks	$0.5 < rf \leq 0.9$	Availability of materials and resources for the most critical parts of the value chain system	$0.5 < rf \leq 0.9$
Rebuild	Strong finances of the owner and good documentation of the whole system	$0.25 < rf \leq 0.5$	Availability of materials and resources for the specific part of the value chain system	$0.25 < rf \leq 0.5$
Reconfigure	Good competence in the production system	$0 \leq rf \leq 0.25$	Efficient and effective production under regular circumstances	$0 \leq rf \leq 0.25$

Note: Rationale for each resilience functionality presented in the table is given in Section 12.4.5.2.

Assessing the resilience level of a management system or the physical resources and assets can be done in different ways. Section 12.4.2 presents ISO standards that guide risk and continuity ability assessments, which may combine to express the resilience of an organization with regard to specific scenarios. When considering that the total resilience of the organization (R_{TN}) is established from the bottom up, the management system and physical resources and assets are multiplicative. This means that without a management system capable of correctly utilizing the resources, the resources do not reach their full capacity and vice versa. Therefore, R_{TN} is formulated as follows:

$$R_{TN} = \sum_{k=1}^{4} R_{k,MsN} \times R_{k,RaN} \tag{12.1}$$

where R_{MsN} is the resilience functions for the management system and R_{RaN} is the resilience functions for the physical resources and assets when exposed to the specific scenario N.

The four levels of resilience functions of the management system and the physical resources and assets are presented in Table 12.3. The table presents the resilience functions as matching pairs at each level of the resilience function. The total resilience function R_{TN} is established from the bottom up, meaning that R_{TN} of, for example, "restabilize" must first have achieved full resilience at the lower levels, i.e. R_{MsN} and R_{RaN} both first have established both Reconfigure and Rebuild.

Subsequently, we base our measurements on the introduced resilience functions. Different risk scenario types represent one or several risks for disrupting a producing organization's ability to perform its task. Such scenario types may be logistics interruption, geopolitics, finances, raw materials, distribution, customer markets and partners, environment, e.g. river flood, energy supply, equipment, skills, etc. Both the management system and the physical resources and assets have one or several resilience functionality components which aim to withstand the risks of disruption introduced by the scenario. The measurement of resilience is based on these different scenarios where systems and system components are modeled to be exposed to impacts and shocks.

12.4.2 Measuring Management System Resilience

Figure 12.2 structures the relationships between ISO standards addressing the management of risk, resilience, and continuity of organizations. The boxes with dotted boundaries support functions for a resilient organization and address organization, behavior, and assessment of the capability of emergency organizations, while the light blue boxes are relating to preparations, risk assessment, management, and organizational collaboration to maintain the continuity of an organization. The black box represents the whole supply chain, from a specific organization's point of view. The ISO 31000:2018 [17] standard guides an organization to assess the risks for resilience, in particular how to assess the risk throughout the supply chain. Such risk assessment can be made either by one individual organization, which tries to get an overview of the whole supply chain, or jointly by all supply chain organizations.

At the core of Figure 12.2 are the standards ISO 22300:2021 [18], which provides a firm vocabulary for an organization's resilience, ISO 22316:2017 [19], which provides guiding principles for an organization's comprehensive resilience, and ISO 22301:2019 [20], which provides the required attributes of an organization to verifiably be a standardized resilience management system.

Based on the risk assessment, an organization may take many different paths toward how to achieve resilience, for example:

- to set the scope for sharing data about key supply chain risks [21],
- for relevant actors to quickly be able to attend and react to any approaching risks, to establish collaborative business relationship management [22],
- to establish a permeating continuity management system [23],
- to supply staff with training and equipment to handle incidents for worst-case scenarios [24],
- to prepare the general public community to react in case of disruptions and disasters caused by alerted risks [25–27].

For an organization to establish an overview of its security and resilience, as well as its capability to respond, they call for third-party review [28] or assessment [29]. Applying the combined set of standards, an organization may evaluate its capability for resilience by applying the concept of the four resilience functions and the general structure of the ability assessment standard (ISO 22325:2016). It should be noted that this standard is limited to evaluating emergency handling ability, but that this may be extended using the structure of the four resilience functions.

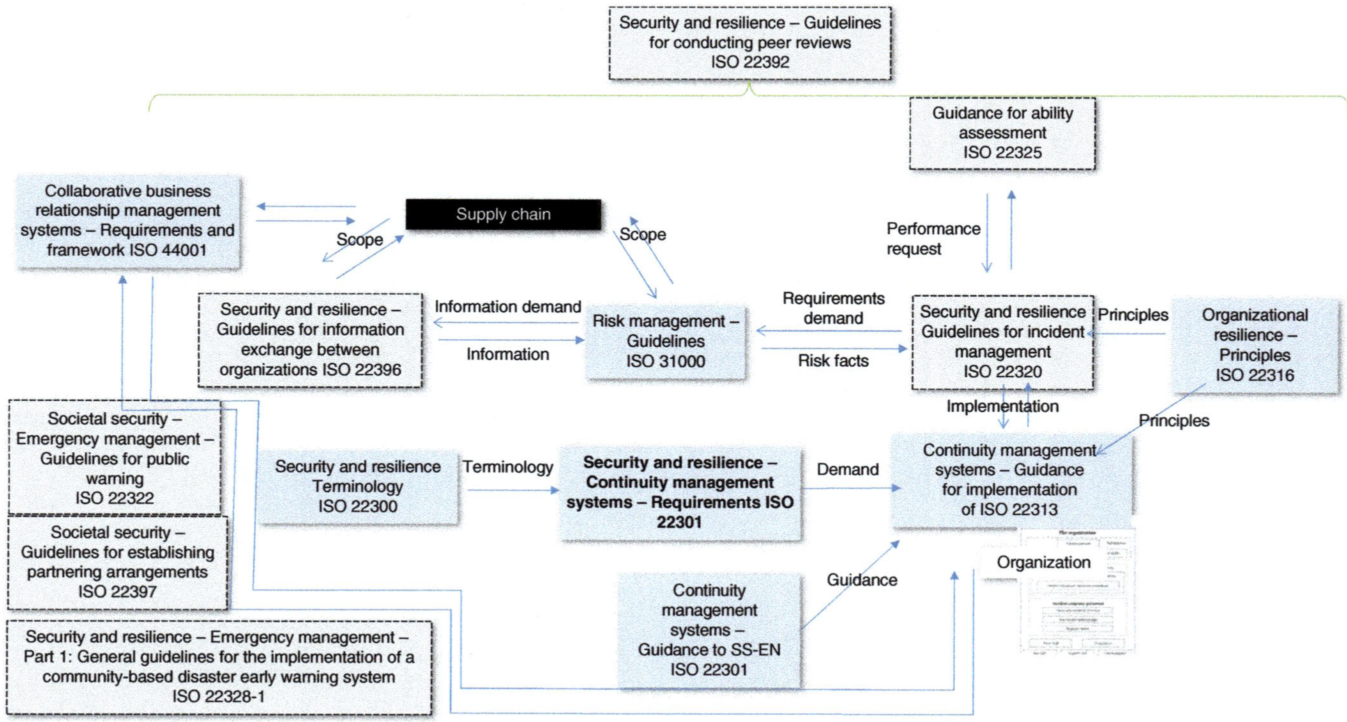

Figure 12.2 Relations between standards supporting the management of value chain resilience.

Quantitative parameters of resilience functions for management system

To measure the resilience of a management system, there is a need to quantify its different resilience levels, stating that:

- Resist: $0.9 < rf \leq 1$ (perfect resilience to defined scenario)
- Restabilize: $0.5 < rf \leq 0.9$ (good resilience to defined scenario)
- Rebuild: $0.25 < rf \leq 0.5$ (bad resilience to defined scenario)
- Reconfigure: $0 \leq rf \leq 0.25$ (no resilience to defined scenario)

12.4.3 Measuring the Resilience of Physical Resources and Assets

In analogy with the measurement of the resilience of management systems, the resilience of physical resources and assets is also based on the four resilience functionalities concerning predefined scenarios for risks for disruptions:

Resist: Physical resources and assets are designed to resist specific scenarios that cause different types of mechanical impacts, electromagnetic shocks, water flooding, etc. To achieve the intended level of resistance, the specifications of these resources and assets should be based on an appropriate choice of risk assessment methodology [30]. When disrupting impacts or shocks exceed the designed levels or when the events are of unpredicted types, such as disruptions of supply chains of critical spare parts or consumables, the functionality of the physical resources and assets are at risk to be disrupted. To resist this, a risk management system [17] needs to apply the correct risk assessment methods to identify these risks and to build up sufficient in-house supplies.

Restabilize: Since not all scenarios, impacts, or shocks can be predicted, a resilience management system also needs to have prepared for efficient restabilization after the disruption of the physical resources and assets. This means that even if a shock or impact has damaged the hardware, it has been designed so that its spare parts and consumables may be acquired from a wide range of sources, mended by accessible experts, or adapted from standardized and widely available stocks.

Rebuild: Depending on the level of technology and cause of disruption, it may be hard to in practice rebuild the capacity of the physical resources and assets. Rebuilding can only be done if expertise, assets, and sufficient adaptable components can be acquired in the way that they can reestablish the system so that it again can supply a sufficient level of function, service, or product, but not to the level as was before the disruption.

Reconfigure: If the disrupting forces destroy the physical resources and assets and no expertise is available to build them up, it will be up to the resilience of the soft system to reconfigure any functionality or purpose provided by the disrupted physical resources and assets. If the soft system does not have such resilience functionality that is able to reconfigure the system, it will no longer exist.

Quantitative parameters of resilience functions for physical resources and assets

As for the management system, we pragmatically chose to quantify the resilience functions for physical resources and assets as:

- Resist: $0.9 < rf \leq 1$ (perfect resilience to defined scenario)
- Restabilize: $0.5 < rf \leq 0.9$ (good resilience to defined scenario)

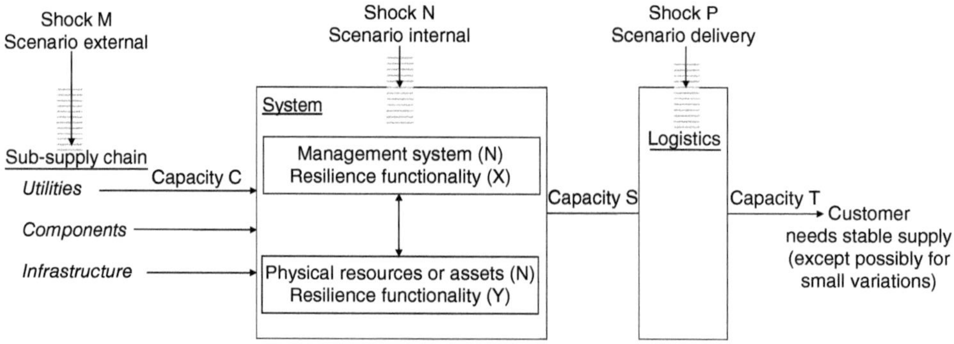

Figure 12.3 Model of a supply chain to exemplify calculation of total system resilience R_{Ts} (own study).

- Rebuild: $0.25 < rf \leq 0.5$ (bad resilience to defined scenario)
- Reconfigure: $0 \leq rf \leq 0.25$ (no resilience to defined scenario)

12.4.4 Total System Resilience

Based on the measurement and quantification of the management system and physical resources and assets presented in Sections 12.4.1, 12.4.2, and 12.4.3, we further present how to measure total system resilience (see Figure 12.3). A complex scenario of three independent disruptions (Shock M, Shock N, and Shock P) and three consecutively dependent levels (Supply chain, System, and Logistics) of a value chain that provides a stable supply to a customer are analyzed. Only the system is considered to have an internal structure where the resilience of the management system and the physical resources and assets are separate with separate resilience functionalities (X and Y). It is worth noting that both the management system and the physical resources and assets are considered with regard to the same scenario Shock N, whereas their resilience functionalities are considered with regard to different aspects of the shock, X and Y. This shows that an actual shock may lead to different impacts on two parts of the system or may even only affect one part, which may cause an indirect effect on another part.

In the simple case, we model that the customer's need is provided by the total available resources of the supply chain and that the resilience functions of the management system and the physical resources and assets match each other. The total system resilience for providing customer with a stable supply can be expressed as:

- Impact from Shock M on sub-supply chain: $0 \leq I_{SCM} \leq 1$
- Resilience of sub-supply chain to impact from Shock M: $R_{SCM} \leq I_{SCM}$
- Impact from Shock N on system's management system: I_{MsN}
- Resilience of system's management system to impact from Shock N: $R_{MsN} \leq I_{MsN}$
- Impact from Shock N on system's physical resources and assets: I_{RaN}
- Resilience of system's physical resources and assets to impact from Shock N: $R_{RaN} \leq I_{RaN}$
- Impact from Shock N on system: I_{RaN}

- Considering formula (12.1) the resilience of system to impact from Shock N:

$$R_{SN} = \sum_{k=1}^{4} R_{k,MsN} \times R_{k,RaN} \tag{12.2}$$

- Impact from Shock P on logistics: I_{LP}
- Resilience of logistics to impact from Shock P: $R_{LP} \leq I_{LP}$

The measurement methodology is developed to assess (i) the resilience performance and the resilience functionality of the total management system of each organization throughout the value chain, (ii) the resilience and the resilience function of the physical resources and assets, and (iii) the ability of the management system and the physical resources and assets to synchronize to achieve a synergetic total system resilience and resilience functionality:

$$R_{TsMNP} = R_{MsM} \times R_{RaN} \times R_{LP} \tag{12.3}$$

12.4.5 Applying the System Resilience Model to the Biogas Value Chain

This section elaborates on the resilience of the biogas value chain, considering variations of three locally independent scenarios of shocks directed to different parts of the value chain. We will investigate under which circumstances the biogas value chain may be considered resilient, produce a surplus of energy, and fail to be useful as a resilient source for energy production.

Figure 12.4 represents two parallel and independent simplified biogas production systems. The left part represents different feedstock resources (W and S) that goes to biogas production, either from the agricultural sector and the industrial sector (top left box) or from water and sewage utilities sector (bottom left box). The middle section represents the different biogas production systems such as anaerobic digestion and sewage treatment plant. The logistics of the solid fractions (F) and (Y) as well as the biogas components are shown in the right part of Figure 12.4, where G and C split up into GE or CE for energy production and the upgraded biogas of GT or CE uses as a transport fuel.

12.4.5.1 Analysis of Two Supply Chains Without Disruptions

Without disruptions, two alternative biogas value chains in (a) the agricultural and industrial sectors and (b) the water and sewage utilities sector (see Figure 12.4) can be expressed as follows:

(a) *Agriculture and industrial sectors*

$$B = W_{AI} \times e_{ADF} \times e_{FI} \tag{12.4}$$

$$E = W_{AI} \times e_{ADbE} \times e_{FI} \tag{12.5}$$

$$T = W_{AI} \times e_{ADbT} \times e_{TI} \tag{12.6}$$

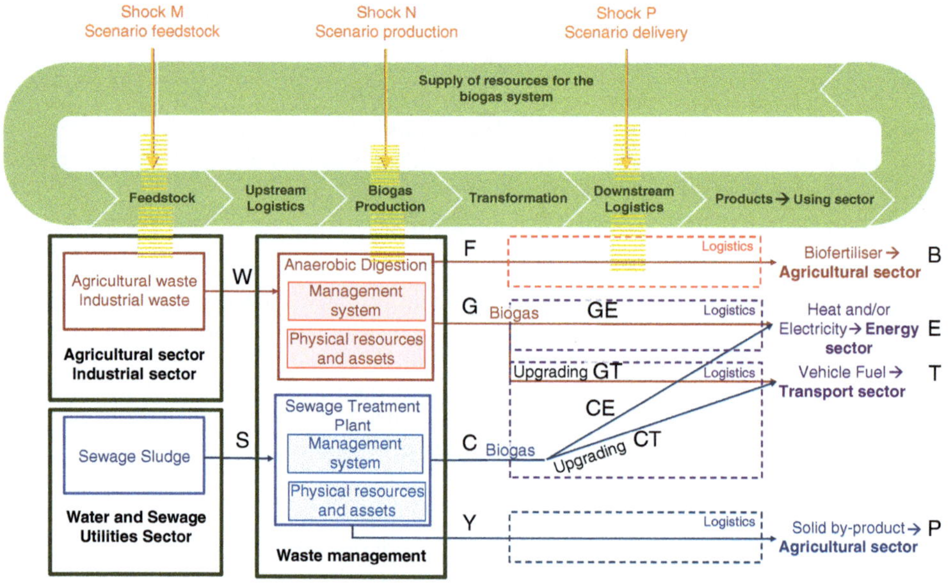

Figure 12.4 Two alternative separated biogas value chains agricultural and industrial sector (red arrows and boxes) and water and sewage utilities sector (blue arrows and boxes) (own study).

where:

- waste from agricultural and industrial sectors: W_{AI}
- efficiency of anaerobic digestion to fertilizer: e_{ADF}
- efficiency of anaerobic digestion to biogas for energy: e_{ADbE}
- efficiency of anaerobic digestion to biogas for transport: e_{ADbT}
- efficiency of fertilizer/solid byproducts logistics: e_{Fl}
- efficiency of energy sector logistics: e_{El}
- efficiency of transport sector logistics: e_{Tl}

(b) *Water and sewage utilities sector*

$$E = S_{Ss} \times e_{STbE} \times e_{El} \tag{12.7}$$

$$T = S_{Ss} \times e_{STbT} \times e_{Tl} \tag{12.8}$$

$$P = S_{Ss} \times e_{Wm} \times e_{Fl} \tag{12.9}$$

where:

- sludge from water and sewage utility sector: S_{Ss}
- efficiency of sewage treatment plant to biogas for energy: e_{STbE}
- efficiency of sewage treatment plant to biogas for transport: e_{STbT}
- efficiency of sewage treatment plant to solid byproducts: e_{Wm}

12.4.5.2 Disrupting Scenarios with Parametrized Resilience Functions

The resilience of two alternative biogas supply chains is analyzed under the circumstances of disrupting impacts related to climate and geopolitical issues. The ways when the management system can be prepared for such disruptions have an escalating scale nature and are presented in Table 12.3. In the "reconfigure" level, an organization is only well prepared to manage and operate the business in its good times having "good competence about production system" but may be unprepared for any disruption. Therefore, this level is considered to be the lowest level of resilience. For an organization to be able to "rebuild" its operation after a disruption, it needs strong finances of the owner and good documentation of the whole system. This perspective represents the second level of resilience of a management system. When disrupting impacts stretch over large sectoral and geographical areas, it may be expected that there is much competition for supplies and resources to make operations work. Therefore, strong connections with supplier networks are key to quickly "restabilizing" a business after a severe interruption. Organizations, which are able to provide this, will, therefore, be considered next to top-level resilient management systems. Organizations that fulfill all resilience levels from the bottom-up and have well-connected value networks are the ones most likely to stay in operation and business, since they have a large value network that can ensure good redundancy concerning both alternative supply chains and alternative customers.

The division of the levels for the resilience of physical resources and assets is similar to the levels of the resilience of the management system. The lowest level, "reconfigure," also constitutes an efficient and effective production under regular circumstances for a production system's physical resources and assets but not at the level of disruptions. The next level, "rebuild," is based on the good availability of materials and resources for the specific part of the value chain system. A production system with secured availability of materials and resources for the most critical parts of the value chain system is needed to fulfill the next "restabilize" level. The top level of resilience, "resist," is the level where the total production system and its resources are built on flexible technology and localization with several materials and resource supply chain options.

The parameters of resilience for the different impacts from shocks M, N, and P of Figure 12.4 are formulated as follows:

(a) *Agriculture and industrial sectors*

Resilience for Shock M disrupting agricultural and industrial sector: r_{AM}

Resilience for Shock N disrupting management system of anaerobic digestion: r_{AMsN}

Resilience for Shock N disrupting physical resources and assets of anaerobic digestion: r_{ARaN}

Resilience for Shock N disrupting total anaerobic digestion, in analogy with formula (12.2):

$$r_{AN} = \sum_{k=1}^{4} r_{k,AMsN} \times r_{k,ARaN} \tag{12.10}$$

Resilience for Shock P disrupting logistics: r_{LP}

(b) *Water and sewage utility sector*

Resilience for Shock M disrupting water and sewage utilities sector: r_{SM}

Resilience for Shock N disrupting management system of sewage treatment plant: r_{STMsN}

Resilience for Shock N disrupting physical resources and assets of sewage treatment plant: r_{STRaN}

Resilience for Shock N disrupting sewage treatment plant, in analogy with formula (12.2):

$$r_{STN} = \sum_{k=1}^{4} r_{k,STMsN} \times r_{k,STRaN} \tag{12.11}$$

Resilience for Shock P disrupting logistics: r_{LP}

12.4.5.3 Analysis of Two Supply Chains with Disruptions

With resilience functions for scenarios introduced, the two alternative biogas value chains in (a) the agricultural and industrial sectors and (b) the water and sewage utilities sector can be expressed as follows:

(a) *Agriculture and industrial sectors (compared with formulas* (12.4)–(12.6))*:*

$$B = W_{AI} \times r_{AM} \times e_{ADF} \times r_{AN} \times e_{Fl} \times r_{LP} \tag{12.12}$$

$$E = W_{AI} \times r_{AM} \times e_{ADbE} \times r_{AN} \times e_{Fl} \times r_{LP} \tag{12.13}$$

$$T = W_{AI} \times r_{AM} \times e_{ADbT} \times r_{AN} \times e_{Tl} \times r_{LP} \tag{12.14}$$

where:

- waste from agricultural and industrial sectors: W_{AI}
- efficiency of anaerobic digestion to fertilizer: e_{ADF}
- efficiency of anaerobic digestion to biogas for energy: e_{ADbE}
- efficiency of anaerobic digestion to biogas for transport: e_{ADbT}
- efficiency of fertilizer/solid byproducts logistics: e_{Fl}
- efficiency of energy sector logistics: e_{El}
- efficiency of transport sector logistics: e_{Tl}

(b) *Water and sewage utilities sector (compared with formulas* (12.7)–(12.9))

$$E = S_{Ss} \times r_{SM} \times e_{STbE} \times r_{STN} \times e_{El} \times r_{LP} \tag{12.15}$$

$$T = S_{Ss} \times r_{SM} \times e_{STbT} \times r_{STN} \times e_{Tl} \times r_{LP} \tag{12.16}$$

$$P = S_{Ss} \times r_{SM} \times e_{Wm} \times r_{STN} \times e_{Fl} \times r_{LP} \tag{12.17}$$

where:

- sludge from water and sewage utility sector: S_{Ss}
- efficiency of sewage treatment plant to biogas for energy: e_{STbE}

Table 12.4 Impacts and their probability.

	Climate change disruptions								Supply chain disruptions			
	I_{dry}				I_{flood}				I_{supp}			
	L1	L2	L3	L4	L1	L2	L3	L4	L1	L2	L3	L4
Impacts (I)	0.0001	0.02	0.5	1	0.01	0.2	0.6	1	0.1	0.3	0.8	1
Probability (r)	0.1	0.01	0	0.0001	0.1	0.01	0	0.0001	0.1	0.01	0	0.0001

- efficiency of sewage treatment plant to biogas for transport: e_{STbT}
- efficiency of sewage treatment plant to solid byproducts: e_{Wm}

The biogas production will be analyzed with resilience functionality according to Table 12.3 and formulas (12.10)–(12.17) by exposing them to combinations of the three following scenarios:

Disruptions due to climate change:

(1) Dry season, dry soil and dry growth, hence less biomaterial:
Impact on scenario dry: I_{dry}
Risk for scenario dry: r_{dry}
(2) Flooding, disrupts infrastructure, hence less gas production:
Impact on scenario flood: I_{flood}
Risk for scenario flood: r_{flood}

Supply chain disruptions due to geopolitics:

(3) Supply of technology:
Impact on scenario supply: I_{supp}
Risk for scenario supply: r_{supp}

Table 12.4 presents different scenario impacts with four levels of severity, ranging from disruptions of regular production, L1, to total disruption of societal productivity and function, L4. The likelihood of impacts is higher for lower level disruptions and lower for higher level disruptions. The actual figures are not based on real risk assessments but are introduced to represent the resilience model.

The results are presented in two different forms. In the first form, the system is exposed to different direct impacts, where the result is either the system is disrupted or not. In the second, the system is analyzed with regard to risk for disruptions, where the decreased productivity represents the probability of loss of biogas production capability. Our scenarios are built on the productivity figures in Table 12.5.

In the following different scenario models, the resilience functions, impacts, and probabilities are varied in different ways to show the result of applying the methodology. Sections 12.4.2 and 12.4.3 present quantitative parameters of resilience functions for both management systems and physical resources and assets. In the following calculations, these parameters are used for the total system resilience representing the resilience of the agricultural and industrial sectors, the water and sewage utilities sector, different types of energy, as well as for the logistics for the solid byproducts. The quantitative parameters presented in Table 12.3 are used as resilience functions for the management system and the physical resources and assets of the anaerobic digestion and the sewage treatment plant, respectively.

Table 12.5 Production and efficiency data for the biogas-producing waste management systems.

Agriculture and industrial sectors	Quantity	Unit
$B = W_{AI} \times e_{ADf} \times e_{FI}$ – flows of solid fraction	58	kton
$E = W_{AI} \times e_{ADbE} \times e_{EI}$ – biogas intended for energy production	11.6	GWh
$T = W_{AI} \times e_{ADbT} \times e_{TI}$ – biogas intended for energy production	21.9	GWh
P – flows of solid fraction from	0	kton
Waste from agricultural and industrial sectors: W_{AI}	30	kton
Efficiency of anaerobic digestion to fertilizer: e_{ADf}	1.92	kton/kton
Efficiency of anaerobic digestion to biogas for energy: e_{ADbE}	0.43	GWh/kton
Efficiency of anaerobic digestion to biogas for transport: e_{ADbT}	1.22	GWh/kton
Efficiency of fertilizer/solid byproducts logistics: e_{FI}	1	
Efficiency of energy sector logistics: e_{EI}	0.9	
Efficiency of energy for transport sector logistics: e_{TI}	0.6	
Water and sewage utility sector	**Quantity**	**Unit**
B	0	kton
$E = S_{Ss} \times e_{STbE} \times e_{EI}$	0.86	GWh
$T = S_{Ss} \times e_{STbT} \times e_{TI}$	1.17	GWh
$P = S_{Ss} \times e_{Wm} \times e_{FI}$	58	kton
Sludge from water and sewage utility sector: S_{Ss}	30	kton
Efficiency of sewage treatment plant to biogas for energy: e_{STbE}	0.032	
Efficiency of sewage treatment plant to biogas for transport: e_{STbT}	0.064	
Efficiency of sewage treatment plant to solid byproducts: e_{Wm}	1.9	

Note: The data in the table are based on the process model of Figure 12.4 and Tables 12.1 and 12.2.

Table 12.6 Statistical risk for loss of biogas production from impact under low system resilience.

	Climate change								Supply			
	Dry				Flooding							
	L1	L2	L3	L4	L1	L2	L3	L4	L1	L2	L3	L4
B	57.6	0.0	0.0	0.0	0.0	0.0	0.0	0.0	0.0	0.0	0.0	0.0
E	11.6	0.0	0.0	0.0	0.0	0.0	0.0	0.0	0.0	0.0	0.0	0.0
T	21.9	0.0	0.0	0.0	0.0	0.0	0.0	0.0	0.0	0.0	0.0	0.0

Resilience function parameter values: $r_{SM} = r_{AM} = 0.001$, $r_{1,AMsN} = r_{1,ARaN} = r_{1,STMsN} = r_{1,STRaN} = 1 \rightarrow r_{AN} = r_{SM} = 0.25$, $r_{LP} = 0.0001$.

The results of the analysis are presented in Tables 12.6–12.11. The left column of these tables represents the flows of solid fraction (B), biogas intended for energy production (E), and biogas production intended for transports (T). The performed parametrized calculations follow the principle laid out in works written by Carlson et al. [31] and Noda et al. [32]. The results of the calculations are presented in pairs so that Tables 12.6 and 12.7 show low resilience outcomes, Tables 12.8 and 12.9 demonstrate high resilience outcomes, and Tables 12.10 and 12.11 show a situation where the economic impact might be higher on the lower level impacts due to their higher risk of occurring.

Tables 12.6 and 12.7 show the result of applying the same impacts on the production of the biogas system. Both tables are filled with many values "0" (zero), which means that due to low resilience even the lowest level of impact will disrupt the biogas production.

Table 12.7 *Actual loss of biogas production from impact under low system resilience.*

	Climate change								Supply			
	Dry				Flooding							
	L1	L2	L3	L4	L1	L2	L3	L4	L1	L2	L3	L4
B	0.0	0.0	0.0	0.0	0.0	0.0	0.0	0.0	0.0	0.0	0.0	0.0
E	0.0	0.0	0.0	0.0	0.0	0.0	0.0	0.0	0.0	0.0	0.0	0.0
T	0.0	0.0	0.0	0.0	0.0	0.0	0.0	0.0	0.0	0.0	0.0	0.0

Resilience function parameter values: $r_{SM} = r_{AM} = 0.001$, $r_{1,AMsN} = r_{1,ARaN} = r_{1,STMsN} = r_{1,STRaN} = 1 \rightarrow r_{AN} = r_{SM} = 0.25$, $r_{LP} = 0.0001$.

Table 12.8 *Statistical risk for loss of biogas production from impact under high system resilience.*

	Climate change								Supply			
	Dry				Flooding							
	L1	L2	L3	L4	L1	L2	L3	L4	L1	L2	L3	L4
B	57.6	57.6	57.6	57.6	57.6	57.6	57.6	57.6	57.6	57.6	57.6	57.6
E	11.6	11.6	11.6	11.6	11.6	11.6	11.6	11.6	11.6	11.6	11.6	11.6
T	21.9	21.9	21.9	21.9	21.9	21.9	21.9	21.9	21.9	21.9	21.9	21.9

Resilience function parameter values: $r_{SM} = r_{AM} = 1$, $r_{1-4,AMsN} = r_{1-4,ARaN} = r_{1-4,STMsN} = r_{1-4,STRaN} = 1 \rightarrow r_{AN} = r_{SM} = 1$, $r_{LP} = 0.0001$.

Table 12.9 *Actual loss of biogas production from impact under high system resilience.*

	Climate change								Supply			
	Dry				Flooding							
	L1	L2	L3	L4	L1	L2	L3	L4	L1	L2	L3	L4
B	57.6	57.6	57.6	0.0	57.6	57.6	57.6	0.0	57.6	57.6	57.6	0.0
E	11.6	11.6	11.6	0.0	11.6	11.6	11.6	0.0	11.6	11.6	11.6	0.0
T	21.9	21.9	21.9	0.0	21.9	21.9	21.9	0.0	21.9	21.9	21.9	0.0

Resilience function parameter values: $r_{SM} = r_{AM} = 1$, $r_{1-4,AMsN} = r_{1-4,ARaN} = r_{1-4,STMsN} = r_{1-4,STRaN} = 1 \rightarrow r_{AN} = r_{SM} = 1$, $r_{LP} = 0.0001$.

However, the values in Tables 12.6 and 12.7 have different meanings and should therefore be understood differently. Table 12.6 is calculated using the probability of a disrupting impact matching the resilience value of the production system, in which the result for the lowest impact (L1) statistically will always be a full production as long as no other impact disturbs the production. However, the system has a very low resilience to all other forms of impacts, no matter how low level they have. Table 12.7 shows the same case, but with another type of calculation. The calculation shows the consequence of any actual impact. It does not take into consideration the probability of the impact. Therefore, all cells have the value "0" (zero). Hence, the resilience level is very low whenever this production system is hit by a disrupting impact. In summary, Tables 12.6 and 12.7 together show that the production system for biogas has a very low level of resilience. However, since the probability for even the lowest level of impact is considered very low, it is lead to the conclusion that the system

Table 12.10 Statistical risk for loss of biogas production from impact under moderate system resilience.

	Climate change								Supply			
	Dry				Flooding							
	L1	L2	L3	L4	L1	L2	L3	L4	L1	L2	L3	L4
B	57.6	57.6	57.6	57.6	0.0	0.0	57.6	57.6	0.0	0.0	57.6	57.6
E	11.6	11.6	11.6	11.6	0.0	0.0	11.6	11.6	0.0	0.0	11.6	11.6
T	21.9	21.9	21.9	21.9	0.0	0.0	21.9	21.9	0.0	0.0	21.9	21.9

Resilience function parameter values: $r_{SM} = r_{AM} = 0.001$, $r_{1-3,AMsN} = r_{1-3,ARaN} = r_{1-3,STMsN} = r_{1-3,STRaN} = 1 \rightarrow r_{AN} = r_{SM} = 0.75$, $r_{LP} = 0.001$.

Table 12.11 Loss of biogas production from actual impact under moderate system resilience.

	Climate change								Supply			
	Dry				Flooding							
	L1	L2	L3	L4	L1	L2	L3	L4	L1	L2	L3	L4
B	57.6	0.0	0.0	0.0	0.0	0.0	0.0	0.0	0.0	0.0	0.0	0.0
E	11.6	0.0	0.0	0.0	0.0	0.0	0.0	0.0	0.0	0.0	0.0	0.0
T	21.9	0.0	0.0	0.0	0.0	0.0	0.0	0.0	0.0	0.0	0.0	0.0

Resilience function parameter values: $r_{SM} = r_{AM} = 0.001$, $r_{1-3,AMsN} = r_{1-3,ARaN} = r_{1-3,STMsN} = r_{1-3,STRaN} = 1 \rightarrow r_{AN} = r_{SM} = 0.75$, $r_{LP} = 0.001$.

can provide biogas under regular conditions. However, this depends on how correct the data is, and whether it is even realistic that biogas production will not be affected by any disrupting impact over a foreseeable time.

Tables 12.8 and 12.9 show the results of applying the same impacts on the production of the biogas system. All cells in Table 12.8 and most cells in Table 12.9 hold the value "0" (zero), which means that due to high resilience almost no level of impact will disrupt the biogas production. Similarly to Tables 12.6 and 12.7, Tables 12.8 and 12.9 also have different meanings and should be understood differently. Table 12.8 is calculated using the probability of a disrupting impact matching the resilience value of the production system, which demonstrates that even for the highest impact (L4) statistically there will always be full production as long as no other impact disturbs the production. Table 12.9 shows the same situation, but the calculation indicates the consequence of any actual impact. It does not take into consideration the impact probability. Therefore, only those cells where the impact is higher than the resilience will have been disrupted, i.e. it shows the value "0" (zero). Hence, the resilience level is very high, and the system withstands almost any disruptive impact. In summary, Tables 12.8 and 12.9 together show that this biogas production system has a very high level of resilience.

Tables 12.10 and 12.11 also demonstrate the series of results having applied a series of impacts on the biogas production system. In comparison to Table pairs 12.6–12.7 and 12.8–12.9, the patterns in the Tables 12.10 and 12.11 are less easy to interpret. For example, Table 12.10 shows that only for the two lowest level impacts (L1 and L2) the biogas production has been disrupted for both flooding and supply chain, since the stronger impacts occur statistically more seldom for flooding and supply chain disruptions. Table 12.11 may show that the production of biogas will operate well under high impact disruptions and be disrupted under low impacts. However, Table 12.11 shows that the system has very low resilience to any of the impact levels whenever such an impact occurs. It can only withstand the lowest level of dry climate impact.

Reflecting on the results presented in Tables 12.6–12.11, the methodology can be used to assess both the actual resilience of the design and operation of one physical supply chain and the overall production resilience of several similarly designed and operating systems under risk of similar impacts, or one production system over a longer period under such similar impacts.

12.5 Conclusion and Recommendations

The chapter aimed to develop the methodology for measuring the resilience of supply critical systems that combine the sensitivity of assets and resources with organizational capacity for supply readiness. The biogas value chain has been studied from two perspectives: as a resilience solution to any kind of disruptions and as a practical case to study its resilience ability to occurred disruptions. The measurement concept of resilience is based on two interconnected parts, the management system, and physical resources and assets.

As a result, our analysis showed that any methodology for measuring the resilience of supply systems needs to include the necessary concepts related to resilience. This means that the supply system requires to be clearly understood and scoped. It is recommended that the scoping is made with a preempting methodology, meaning that it should have the ability to follow the causality of all supply flows all the way to the original source. This, in turn, would help map up and react to any type of risks for disrupting scenarios globally. It may also be recommended to apply the system used to scope and detail a product system in the methodology of life cycle assessment [33]. In addition, methodological tools to identify, assess, quantify, prepare for, and manage risks for critical disruptions are needed. The assessment and structuring of standards for resilience in Section 12.4.2, continuity management, risk assessment, and supply chain cooperation provided much guidance on this aspect. However, the described methodology (particularly in Section 12.4.1) highlighted that handling and preparing for risks and resilience related to the hard systems, i.e. the physical resources and assets, is separated from handling and preparing for risks and resilience related to the soft systems of supply chains and organizations, i.e. the management systems. If the resilience of hard and soft systems matches, it is likely that investments made in, for example, highly resisting machinery and equipment show no significance unless the supply of spare parts has been secured by the management system.

Another important aspect of the results underlay the fact that it is not possible to correctly measure the resilience of the total supply system, especially when considering the hard predictability of exact levels and probabilities of different impacts from different scenarios, as well as the exact quantification of the resilience functionality level of subsystems and components of a supply chain. However, by setting up a model and by elaborating and testing with different combinations of values for these parameters, it is possible to find which aspects may be critical under severe and unlikely occasions, and which economic appraisals may be applied to establish a sufficiently resilient supply of critical resources.

Therefore, applying our methodology to the biogas system, the system that produces a versatile form of energy for a wide range of supply critical applications, such as production, heat, and transport with the ability to localize the energy production, we have shown that the created methodology is of direct practical value. It should be stressed that once the resilience measurement of a supply system based on the presented methodology is established, the model can be maintained and improved over time to continually produce better decision support and guidance to its end users and stakeholders.

References

1. Gasser, P., Lustenberger, P., Cinelli, M. et al. (2021). A review on resilience assessment of energy systems. *Sustainable and Resilient Infrastructure* 6 (5): 273–299. https://doi.org/10.1080/23789689.2019.1610600.
2. Willis, H.H. and Loa, K. (2015). *Measuring the Resilience of Energy Distribution Systems*. Santa Monica: RAND Corporation. Measuring the Resilience of Energy Distribution Systems|RAND (accessed 3 February 2023).
3. D'Aquino, C.A., Pereira, B.A., Sawatani, T.F. et al. (2022). Biogas potential from slums as a sustainable and resilient route for renewable energy diffusion in urban areas and organic waste management in vulnerable communities in São Paulo. *Sustainability* 14: 7016. https://doi.org/10.3390/su14127016.
4. Long, J.H., Aziz, T.N., and Ducoste, J.J. (2012). Anaerobic co-digestion of fat, oil, and grease (FOG): a review of gas production and process limitations. *Process Safety and Environment Protection* 90 (3): 231–245. https://doi.org/10.1016/J.PSEP.2011.10.001.
5. Salama, E.-S., Saha, S., Kurade, M.B. et al. (2019). Recent trends in anaerobic co-digestion: fat, oil, and grease (FOG) for enhanced biomethanation. *Progress in Energy and Combustion Science* 70: 22–42. https://doi.org/10.1016/j.pecs.2018.08.002.
6. Berninghaus, A.E. and Radniecki, T.S. (2022). Shock loads change the resistance, resiliency, and productivity of anaerobic co-digestion of municipal sludge and fats, oils, and greases. *Journal of Cleaner Production* 362: 132447. https://doi.org/10.1016/j.jclepro.2022.132447.
7. Kougias, P.G., Treu, L., Campanaro, S. et al. (2016). Dynamic functional characterization and phylogenetic changes due to long chain fatty acids pulses in biogas reactors. *Scientific Reports* 6 (1): 28810. https://doi.org/10.1038/srep28810.
8. Silvestre, G., Rodríguez-Abalde, A., Fernández, B. et al. (2011). Biomass adaptation over anaerobic co-digestion of sewage sludge and trapped grease waste. *Bioresource Technology* 102 (13): 6830–6836. https://doi.org/10.1016/j.biortech.2011.04.019.
9. Heezen, P.A.M., Gunnarsdottir, S., Gooijer, L., and Mahesh, S. (2013). Hazard classification of biogas and risks of large scale biogas production. *Chemical Engineering and Technology* 37–42. https://doi.org/10.3303/CET/1331007.
10. Moreno, V.C., Papasidero, S., Scarponi, G.E. et al. (2016). Analysis of accidents in biogas production and upgrading. *Renewable Energy* 96: 1127–1134. https://doi.org/10.1016/j.renene.2015.10.017.
11. Travnícek, P., Kotek, L., Junga, P. et al. (2018). Quantitative analyses of biogas plant accidents in Europe. *Renewable Energy* 122: 89–97. https://doi.org/10.1016/j.renene.2018.01.077.
12. Koszel, M. and Lorencowicz, E. (2015). Agricultural use of biogas digestate as a replacement fertilizers. *Agriculture and Agricultural Science Procedia* 7: 119–124. https://doi.org/10.1016/j.aaspro.2015.12.004.
13. Rezapour, S., Zanjirani Farahani, R., and Pourakbar, M. (2017). Resilient supply chain network design under competition: a case study. *European Journal of Operational Research* 259 (3): 1017–1035. https://doi.org/10.1016/j.ejor.2016.11.041.

14. Jørgensen P.J. (2009). Biogas – green energy. Process. Design. Energy supply. Environment. PlanEnergi and Researcher for a Day – Faculty of Agricultural Sciences, Aarhus University 2009. 2nd edition. https://www.lemvigbiogas.com/BiogasPJJuk.pdf.
15. Nevzorova, T. and Kutcherov, V. (2019). Barriers to the wider implementation of biogas as a source of energy: a state-of-the-art review. *Energy Strategy Reviews* 26: 100414. https://doi.org/10.1016/j.esr.2019.100414.
16. Swedish Energy Agency and Swedish Gas Association (2021). Produktion av biogas och rötrester och dess användning år 2021 [In Swedish]. biogasstatistikrapport_2021_webb.pdf (energigas.se) (accessed 30 November 2022).
17. ISO 31000:2018 (2018). Risk management – Guidelines.
18. ISO 22300:2021 (2021). Security and resilience – Vocabulary.
19. ISO 22316:2017 (2017). Security and resilience – Organizational resilience – Principles and attributes.
20. ISO 22301:2019 (2019). Security and resilience – Business continuity management systems – Requirements.
21. ISO 22396:2020 (2020). Security and resilience – Community resilience – Guidelines for information exchange between organizations.
22. ISO 44001:2017 (2017). Collaborative business relationship management systems – Requirements and framework.
23. ISO 22313:2020 (2020). Security and resilience – Business continuity management systems – Guidance on the use of ISO 22301.
24. ISO 22320:2018 (2018). Security and resilience – Emergency management – Guidelines for incident management.
25. ISO 22322:2015 (2015). Societal security – Emergency management – Guidelines for public warning.
26. ISO 22328-1:2020 (2020). Security and resilience – Emergency management – Part 1: General guidelines for the implementation of a community-based disaster early warning system.
27. ISO 22397:2014 (2014). Societal security – Guidelines for establishing partnering arrangements.
28. ISO 22392:2020 (2020). Security and resilience – Community resilience – Guidelines for conducting peer reviews.
29. ISO 22325:2016 (2016). Security and resilience – Emergency management – Guidelines for capability assessment.
30. ISO/IEC 31010:2019 (2019). Risk management – Risk assessment techniques.
31. Carlson, R., Roos, S., Noda, H., and Takahashi, R. (2008). Damage cost calculation method for energy apparatuses. *Proceedings of 8th International Conference on EcoBalance* 14: 1402.
32. Noda, H., Takahashi, R., Kobayashi, T. et al. (2011). Development of evaluation model for substation damage. *IEEE Transactions on Power Delivery* 26 (3): 1920–1926. https://doi.org/10.1109/TPWRD.2011.2116161.
33. ISO 14044:2006 (2006). Environmental management – Life cycle assessment – Requirements and guidelines.

13

Theory and Practice in Strategic Niche Planning: The Polish Biogas Case

Stelios Rozakis[1], Katerina Troullaki[1], and Piotr Jurga[2]

[1] *BiBELab, Department of Chemical and Environmental Engineering, Technical University of Crete, Chania, Greece*

[2] *Department of Bioeconomy and Systems Analysis, Institute of Soil Science and Plant Cultivation, Pulawy, Poland*

13.1 Introduction

13.1.1 The Promising Potential of Biogas Transition in Central Eastern European Countries

The European Green Deal and related policies that aspire to the ambitious goal of climate-neutral Europe by 2050, represent a challenge for Central Eastern European (CEE) Countries, especially those with high GHG intensity of electricity, namely Estonia, Poland, Czech Republic, Bulgaria and Romania. Although decreasing since 1990, GHG intensity of electricity generation in these countries still exceeds 300 g CO_2e/kWh [1]. Despite their adaptation to the market economy and the spectacular economic growth of the last decades, fossil energy sources, predominantly coal, and significant industrial sector require the mobilization of their full potential in order to attain European average. Moreover, the expectation and upcoming obligation of carbon footprint limits in food products urges countries with important export-oriented food industry to improve their performance increasing the value created by carbon negative energy carriers. An important asset toward this endeavor is the size of the agricultural sector and its available resources,

Biogas Plants: Waste Management, Energy Production and Carbon Footprint Reduction,
First Edition. Edited by Wojciech Czekała

provided that sustainable energy production will be prioritized. Biogas based on plant and animal residues management is a crucial activity in this respect.

Since the accession of CEE countries to the EU, a take-off in agricultural biogas has been observed, however its deployment is observed uneven. Legal framework adoption was a necessary but not sufficient step to enable agricultural biogas energy production, if not coupled with appropriate financial support systems to make production economically efficient for investors. Unlike Czech Republic and Slovakia, where financial support has been high and persistent, Poland and Hungary seem to be deeply underdeveloped considering their biogas potential. In fact, most CEE countries have hardly exploited their potential; e.g. Poland, which is deemed to have a comparable potential with Germany, avails about 1% of its actual installed power.

Currently though, compelling reasons motivate laggard countries to act. The soaring energy prices in late 2021 followed by the ongoing major conflict in Ukraine bring about tremendous geopolitical impacts. Due to high dependence of Central and Eastern Europe on Russian supplies of oil and natural gas, what used to be theoretical energy security threats [2] are now harsh reality obliging for immediate action. While fossil fuel stocks are being drastically depleted and the use of nonrenewable energy sources restricted due to ecological considerations, well-being expectations of contemporary societies contribute to a surging demand for energy. In the face of the imminent disruption of the world's energy balance, much emphasis is placed on the transition to renewable energy sources (RES). In parallel, alternative configurations of energy generation, especially short supply chains and distributed production, as well as activities favoring efficient local energy use and away from energy-intensive production, are of paramount importance to minimize replacing present dependence with another one. In this respect, untapped biogas production potential can be exploited by means of prompt and significant investment, where state-of-the-art, economically-justified projects should be a major priority.

Detailed analysis of agricultural biomass and agro-food waste [3] reveals the feasibility of a very ambitious and politically critical objective. According those analyses, there is evidence that the Polish agricultural biogas sector (without urban waste) can reach over 6600 MW of installed electric power, the equivalent of two nuclear power plants (6000 MW in total) planned to be built in Poland by 2035. As the total power of agricultural biogas plants is currently less than 130 MW (2% of potential) compared with nearly 6 GW in Germany, perspectives of development are tremendous. "*Poland has over 1.5 mln ha bigger agricultural surface (than Germany) and has AD technologies focused mainly on biowaste from the agro-food industry and agricultural biomass (i.e., different straws). So it seems that the target of 6.6 GW of electric power in the Polish biogas sector is real to reach*" [3].

Biogas in Poland is obtained mainly from energy crops and partially from animal manures. Only a small part of the agro-food industry's biowaste is treated in the anaerobic digestion (AD) process; most is composted. After a period of growth between 2011 and 2016, there has been stagnation, mainly due to legislative barriers and unstable public support [4]. Chodkowska-Miszczuk et al. [5] provide detailed information on the Polish biogas plants substrate mix and value creation before and after RES policy change in 2016, observing adaptive business decisions both in input and output, without in most cases altering the basic configuration. A clear trend of increased use of food industry waste is observed, either at no expenses or even receiving a disposal fee, which has allowed for financial sustainability of biogas businesses, despite inconsistent public support. The waste input increases the potential feedstock volume considerably, allowing for ambitious objectives, such as the significant substitution of imported gas.

There are innovations in Poland which demonstrate that opting for a "zero waste" circular bioeconomy may bring positive outcomes for waste transformation per se, facilitating the transition from the linear to the circular economy [6]. In the context of circular sustainable biobased economy waste management is supposed to turn waste into reusable materials. The biogas technology can proceed to biowaste cascading, producing high-value products – biogas and biofertilizer. In addition to the current know-how of AD for a large spectrum of substrates and interactions to optimize output efficiency, correlation between the manure microbial dynamics and biogas yield is the focus of genetic engineering and metagenomics research. The potential of such techniques is enormous; some talk about scientific breakthroughs changing the game in biological methane production with future high-yielding bioreactors [7].

The aforementioned conditions have triggered the development of the biogas sector in CEE countries. However, shifts in norms and education of the stakeholders are also needed, as well as the support of investors and policymakers throughout the value chain. This constitutes a transition evolving through friction, barriers, and conflicts, and, because of the magnitude of the endeavor, eventually ventures a system transformation. The objective of such a transition is twofold: being compatible with social concerns and environmental sustainability, as well as being economically sustainable itself. To this end, interdisciplinary sustainability-oriented research is a central vector for unlocking the potential of biogas toward a sustainable energy transition.

13.1.2 State-of-the-Art Research for Navigating Sustainability Transitions

Studying sustainability transitions has been the focus of the homonymous research field over the last two and a half decades. Sustainability transitions (ST) research stems from the field of innovation studies and combines ideas from evolutionary economics, history, and sociology of technology, sociology of innovation, and institutional theory (among others) [8]. In this strand of research, innovations for sustainability are seen as embedded in "socio-technical systems," i.e. systems that are not merely technological but comprise multiple, interdependent elements: technologies, market and industry structures, politics and policies, culture, user practices and expectations, supply and distribution chains, and broader infrastructures. Underlying ST studies is the recognition that today's major environmental and social challenges cannot be addressed by incremental innovations; rather radical shifts are needed, i.e. transitions, to new kinds of socio-technical systems [8].

In this context, sustainability transitions are "*long-term, multi-dimensional, and fundamental transformation processes through which established socio-technical systems shift to more sustainable modes of production and consumption*" ([9], p. 956). Transitions are therefore not linear; they are co-evolutionary processes that consist of interdependent developments across a range of dimensions. A central concept in transition research is the socio-technical regime, which represents the "deep structure" of existing socio-technical systems. Regimes are resilient, and thus favor incremental rather than radical changes, along established pathways of development (path dependence or lock-in). On the contrary, radical innovations originate in niches: "protected spaces," in which alternative socio-technical configurations emerge and develop away from the selection pressures of the prevailing regime [10].

13.1.3 Chapter Organization

In this chapter we overview and apply sustainability transition research concepts coupled with insights from other disciplines to finally set up the process of strategic planning for the case of sustainable biogas development. This strategic planning approach is illustrated in Polish conditions, where biogas is still in niche status, using the perception of stakeholders to articulate strategies for its development as a pivotal sector toward a circular sustainable bioeconomy in the country.

The chapter comprises five sections. Following the overview of the context favoring biogas development in CEE countries and basic concepts for studying a sustainable biogas transition, another four sections are comprised. In Section 13.2, the main conceptual frameworks in ST research for the analysis and governance of sustainable innovation are presented. Next, ST concepts and frameworks combined with insights from other disciplines are applied in the case of biogas production, detailed in Section 13.3. Section 13.4 presents the process of strategic niche planning. The case study of the Polish biogas sector resulting in strategic propositions and the conclusive comments complete the chapter.

13.2 Main Conceptual Frameworks for Studying Sustainability Transitions

The aforementioned foundational concepts of "regime," "niche" and "landscape," are the basis of two widely used conceptual frameworks of ST research, the multi-level perspective (MLP) and the Strategic Niche Management (SNM), which we present below. Two more frameworks that are not directly influenced by the concepts of "niche" and "regime" are presented, namely the Technological Innovation Systems (TIS) and Transition Management (TM). These founding frameworks of ST studies [8, 9] take a systemic perspective to address the multi-dimensional nature of sustainability transitions and the nonlinear, co-evolutionary dynamics of structural change [11].

Besides their analytical value, ST frameworks have been widely used as strategic tools to navigate sustainability transitions. Much of the work on governing transitions acknowledges its normative nature, "recognizing that transitions cannot be solely governed from a top-down perspective and that a plurality of actors, not just governments are involved" [8]. In particular SNM and TM focus on governing sustainable socio-technical transitions through particular processes. The TIS approach has also been developed to inform policy-making. As transitions research is gaining ground in policy contexts, it has been positioned as a transdisciplinary science [8]. Particularly work in TM has applied action-oriented methods to co-create transitions knowledge.

13.2.1 Strategic Niche Management (SNM)

SNM [10, 12] is a governance-oriented framework, which focuses on proactively developing niches as protected spaces for emerging sustainable innovations. Niches are conceptualized as "cosmopolitan" spaces constituted of two interdependent levels: sequences of local projects/experiments (local level) linked together by networks and intermediary organizations (global niche level) [13]. These intermediaries aggregate the knowledge generated by projects on the ground and systematize it into transferrable forms to help new projects be established, which in turn adopt and localize the knowledge into their contexts. Through the replication of projects, niches grow and under the right conditions they may influence, compete with or displace existing, less sustainable regimes.

The SNM literature has studied the conditions under which niches gain momentum with the potential to trigger wider systemic changes, and has identified three core elements contributing to successful niche development: visions and expectations, networks and learning:

- When **visions and expectations** are precise, achievable and aligned among niche actors, they effectively guide innovation activities and may attract attention and support from external actors.
- **Networks** contribute when they are broad and deep, i.e. when they consist of many different actors who commit substantial resources to support niche activities. The participation of powerful actors particularly may provide increased "legitimacy and resources to niche-innovation networks" [11].
- Two types of **learning** are important for niche advancement: (i) first-order learning, which is about practical, instrumental knowledge for improving the innovation (e.g. technical design, infrastructure requirements and organizational models), and (ii) second-order learning, which is about reflexively questioning, re-framing and re-orienting the innovation, its role and value.

According to SNM, therefore, niche innovations develop and become influential to the extent that expectations, networks and learning become robust enough to trigger wider regime changes. These three areas of activity are interdependent and co-evolve in either virtuous or vicious cycles constituting a dynamic niche-development trajectory [14].

13.2.2 Multi-Level Perspective (MLP)

The niche perspective, while valuable, has been considered insufficient to explain transition trajectories as it only emphasizes internal niche processes. Scholars have argued that external processes need to be also analyzed to understand the nonlinear, structural changes involved in transitions [15]. To this end, the multi-level perspective [16, 17] contextualizes SNM providing a deeper understanding of how innovations develop over time.

The MLP explains sustainability transitions as resulting from co-evolutionary changes in socio-technical systems at three analytical levels: niches (where radical innovations originate), regimes (which comprise the established practices and rules that stabilize the incumbent system), and an exogenous landscape (the wider socio-technical context which influences regime and niche dynamics, and which evolves through long-term processes). These three levels interact dynamically potentially leading to destabilization of existing systems and transitions to more sustainable ones: niche innovations build-up internal momentum, landscape developments put pressure on and destabilize the regime, and destabilization of the regime creates windows of opportunity for further expanding niche influence.

Contrary to the SNM, the MLP is a middle-range theory; it focuses on the meso-level (regime) because transitions are defined as shifts from one regime to another [11]. The systemic nature of transitions and the dynamic co-existence of stability and change in socio-technical systems are central to this approach [8].

13.2.3 Transition Management (TM)

TM [18] is a policy-oriented framework, combining ideas from complex systems theory and governance studies. TM scholars have developed a prescriptive strategy, suggesting that policy makers can govern ongoing transitions into more sustainable paths through a sequence of steps [18], shortly: (i) Strategic activities including vision development and identification of potential transition pathways, (ii) Tactical activities including concrete plans and agendas, support coalitions and investment commitments. (iii) Operational activities, such as experimental, demonstration and implementation activities, and (iv) Reflexive activities including evaluation and monitoring, that leads to adjustments and articulation of best-practices. TM conceives management as a reflexive governance process involving a variety of actors from science, policy, businesses and civil society in so-called transition arenas, and fostering cooperation rather than competition among them. Such cooperation has met significant barriers when TM was applied in actual policy contexts [9]. Besides national policy-making processes, interest for TM has increased for regional and local levels as well.

13.2.4 Technological Innovation Systems (TIS)

Another approach frequently used to analyze innovations is the TIS framework [19], combining ideas from industrial economics and innovation systems theory. A TIS consists of actors, technologies, and institutions. Carlsson and Stankiewicz [20] introduced the TIS concept as a "*network of agents interacting in a specific economic/industrial area under a particular institutional infrastructure and involved in the generation, diffusion and utilization of technology*" [20]. Later the concept of "functions" that need to run smoothly for the innovation system to perform well was introduced [21]. Thus, the development of a new technology is conceived as a result of seven system functions: (i) knowledge development and diffusion, (ii) entrepreneurial activities, (iii) market formation, (iv) guidance of the search, (v) resource mobilization, (vi) legitimation and (vii) positive externalities.

TIS has been applied to analyze the development of innovative niches, such as biogas [22]. In short, the aim of this approach is to identify strengths and weaknesses in the performance of evolving innovations, as well as drivers and barriers in their wider institutional context that needs to be developed hand in hand with the innovative technology. TIS has in fact replaced the narrow concept of market failures with the broader concept of system failures. From their beginning, TIS analyses were intended to inform policy-making, which is why the identification of drivers and barriers to innovation, as well as technology specific policies is a typical outcome of TIS studies [23].

In Section 13.3, a combination of ST concepts and frameworks are applied to analyze the development of the biogas innovation in the EU, with a focus on CEE countries.

13.3 Studying Biogas from a Sustainability Transitions Perspective

13.3.1 Landscape, Regime, and Niche Dynamics

Historically biogas plants had been used to address agricultural issues, such as waste management by transforming manure into enhanced fertilizer. Nowadays, they are important vectors in the energy transformation process in multiple ways: they simultaneously constitute waste treatment sites implementing circular economy ecological transition, as well as cogeneration production sites providing organic fertilizer to the fields, and enhancing local development and energy transition [24]. Anaerobic co-digestion for biogas is an innovative sector crossing several socio-technical regimes namely the energy regime (end products), the agricultural regime (supply of livestock and plant biomass), and the waste regime (substrate inputs and digestate output).

AD should not only be understood as a waste management system and source of renewable energy but also as a pollution-abatement technology [25]. As agricultural use of manure is regulated by the EU (Directives 2001/81/EC and 91/676/EEC) to prevent nitrate leaching and ammonia emissions, manure disposal has become an economic and environmental challenge for many farmers. Conversion of manure in biogas plants through AD could be a sustainable solution, reducing emissions to the soil and atmosphere and providing additional benefit to livestock farms, as the AD byproducts (solid and liquid digestates) could be used as fertilizers in crop production [26]. From a human health perspective, it is interesting that AD can destroy the antibiotics in manure [27]; this way it can provide a solution to the problem of massive antibiotics usage in animal production, which is the main reason for the creation of "super-bacteria."

Another regime related to the biogas niche, i.e. transport, is brought up by Huttunen et al. [28]. Biomethane is a technologically feasible second generation fuel that can aid to achieve the RE objectives for 2023–2030 and fulfill the obligation to use RE in transport [29, 30]. Besides its use in transport, biomethane can be pumped into the gas grid. Such use is prioritized – thus becomes economically feasible – due to the political imperative of energy independence from the imported gas. Such landscape pressures in addition to the climate change urgency are creating windows of opportunity for biogas niche development. "Network anchoring" by recruiting actors from an existing regime shape the direction of niche development. "Technological anchoring" appears when technology matures to meet regime needs. Elzen et al. [31] identify adaptation of legislation and regulation, market formation and societal attitudes as "institutional anchoring" negotiated between niches and regimes.

In Poland, different landscape pressures, such as the demand for RES and agricultural diversification, have prompted the agriculture and electricity regimes to interact symbiotically. Legislation for feed-in tariffs introduced a minimum compensation for grid-tied RES systems, including biogas plants ("institutional anchoring" into the electricity regime). The biogas actor network expanded to include the waste industry, energy suppliers and industrial biogas plant operators (i.e. "network anchoring" into the electricity regime). The legal adaptations that followed, as well as the investment funding for setting up RES facilities have been major factors for the increase in biogas plants ("institutional anchoring" into the electricity and agriculture regimes).

The intensive development of agriculture in Poland causes an increase in the supply of organic waste. Given the large production volumes, not only agriculture but also the agrifood industry can supply biogas plants with feedstock [32]. Currently the sector experiences increasing professionalization driven by the agrifood industry requirements for waste disposal [33]. The number of biogas plants is expected to double, and the expansion of the gas network along with infrastructure to accommodate biomethane, has already begun. The strategy for renewable energy deployment, including the increase of biowaste share, is effectively supported by the Energy Policy of Poland (PEP2040). The legal framework is in place, the sales support system functions efficiently and new biogas plants will benefit from preferential sales regimes until 2035 [34]. More recently the concept of energy cooperatives has been included in the Renewable Energy Sources Act (2019) to allow for decentralization of the energy supply system and to better meet the needs of local consumers.

Observing the multi-sector interactions in biogas development, it is evident that – as ST literature explicates – overlapping regimes influence the success of commercial biogas activities. Therefore, working across regime boundaries is critical for innovation development [11, 31]. Moreover, as sustainability transitions tend to follow slower paths toward market integration [35, 36], they may rely on state intervention through appropriate policies for long-term sustainability.

13.3.2 Policy Coherence for Niche Development

Precisely because of innovative niches balancing on different regime boundaries, policy conflicts often occur impeding sustainability transitions. Analysis of policy coherence may focus on one domain (internal coherence) by comparatively assessing different goals (horizontal assessment) and/or by comparing goals with instruments (vertical assessment) [37]. In addition, consistency between various domains identifies to external policy coherence. It can be assessed by examining the meaning and implications of relevant sections of documents arbitrarily using a scoring system in a top-down evaluation. On the other hand, policy coherence can be understood as the consistency of expressed policy goals, instruments and other policy-related signals from the perspective of actors affected by the policies. "*The analysis of policy coherence aims to point out where policies give inconsistent signals to actors, leading to different interpretations on what kind of action is promoted*" [28]. The bottom-up perspective on policy coherence proposed by Huttunen et al. [28] is focusing on local and regional actors' perceptions.

In addition to the aforementioned internal and external dimensions, temporal coherence is also important to consider, i.e. that policies are consistent and predictable over time. Competing interests, whatever the transition under consideration, moderate decision makers' targets to avoid unnecessary coherence problems searching trade-offs to optimize synergistic effects of policies [38]. For the biogas case, four systems are identified, namely waste management, agriculture/food production, energy production, and biogas for transport use, whose policies (potentially in the case of transport) influence the biogas sector development. Relevant policies per identified system are presented in Table 13.1.

The innovation process can be presented by analyzing the components that influence the acceleration of innovation development and its support by policy instruments. In the case of the Polish biogas sector, lack of facilitated regulations for the application of digestate for agrotechnical purposes – cumbersome legislative path or lack of long-term support system and uncertainty related to subsidies are identified as the main policy incoherence.

Table 13.1 *Current Polish and EU-level laws that influence the biogas sector development.*

Policy themes	Policy instruments			
Energy	Renewable Energy Sources Act (478/2015)		The Energy Law Act – grid access (627/62/2001)	"Agroenergy program – biogas investment subsidies/National Fund for Environmental Protection and Water Management" Environmental Protection Act (348/54/1997)
Agriculture	Act of 10 July 2007 on fertilizers and fertilizing (1033/147/2007)	Animal byproducts Regulation (EC 1069/2009)	Measures required for agri-environmental payments and related terms	Nitrate Action Programme (243/2020)
Waste and Environment	Waste Act (21/2013)	Government Decree on Landfills (523/2013)	The Act on the Protection of the Environment (627/62/2001)	Act on Providing Information on the Environment and Environmental Protection, Public Participation on Environmental Impact Assessment (1227/199/2008)
Transport	Act on Biocomponents and Liquid Biofuels (1199/169/2006)	Act on Tax on Civil Law Transactions (959/86/2000) – car tax	Act on Local Taxes and Charges (31/9/1991) – vehicle tax	Act on Excise Duty (11 March 2009) – fuel taxes

Exemplary knowledge and development of the biogas sector is supported by external policy coherence, through R&D European and national programs as well. The lack of general incentives to invest in the biogas sector, an unstable support system, and changing legislation limit the policy goals set by the Renewable Energy Sources Act. Temporal policy coherence is related to the existence of policy goals which have encouraged the increased utilization of biowastes.

A matrix has been constructed in the similar case of Finnish biogas comprising the main TIS components/functions (vertically) plotted against the main biogas development areas (horizontally) [28]. The occurring coherence (positive or negative) between the system functions and the four areas of development can thus be identified. The identification of coherence and incoherence for each of the mentioned system functions, in connection with important areas for biogas sector development, can contribute to the removal of barriers and strengthening the elements enabling the development of the analyzed sector.

13.3.3 Transition Pathways

Dietz et al. [39] distinguish transition pathways toward the bioeconomy implying different factors mobilized that may interact and relate to each other at some extent completed by a fifth one added by Lovec and Juvančič [40] (Figure 13.1). Fossil fuel substitution is the first (pathway 1) accompanied by productivity development in the primary sector (pathway 2). The third pathway refers to biorefining creating value in the markers and it can be entitled "new, more efficient biomass use." The fourth pathway (4) concerns low bulk, high-value applications enhancing complex interactions between value chains (e.g. food and health). Public and private action that generates value-providing ecosystem services identifies to the last (5) pathway.

The do-use-interact (DUI) model includes improvements in productivity, practice and commercialization-oriented innovation, and is just as important as the Science-Technology-Innovation (STI) model, which is based on innovation adoption, R&D, and technology transfer, and typically characterizes high-value bio-industrial applications [40]. DUI is more suitable for less complex pathways such as 1, 2, and 5, as it highlights the significance of practice and interaction within the firm, the supply chain and with competitors. Informal knowledge exchange may suit better sectors with mature and standardized technology and practices, such as forestry, agriculture, food, paper, wood and energy, including conventional bioeconomy sectors, and particularly biogas. Social network analysis of the niche sector can provide evidence of DUI versus STI configurations in particular contexts or countries.

13.3.4 Social Network Analysis

The growing interest in sustainability transitions has triggered significant progress in the field. Besides comparing and combining insights generated by different frameworks within the ST literature, scholars have started blending coming from other scientific disciplines. One example of the latter trend is the combined application of SNM with Social Network Analysis (SNA) [41], a widely-applied technique which brings into analysis the value of social connections.

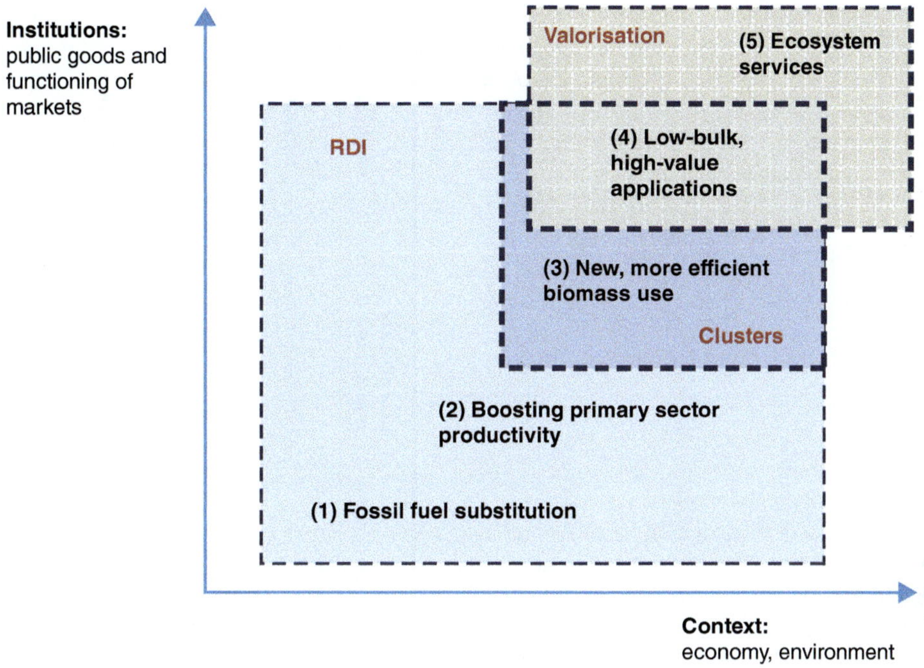

Figure 13.1 Contexts (axes), specific drivers (in bold), and transition pathways. Source: Lovec and Juvančič [40]/MDPI/Public Domain CC BY 4.0.

According to SNM, networking is one of the three processes that directly influence the development of a niche. It is assumed that a larger actor network and more intensive networking activities, result in more experimentation and learning, therefore increased likability for an innovation to gain momentum and establish itself in the market. However, other network characteristics may also be important, such as the nature, quality, and morphology of interactions in different parts of the network. In addition, indirect links between networking and innovation performance exist according to SNA literature. Network connections and structures may also influence the other two SNM processes, i.e. learning and formation of expectations.

A primary SNA of the biogas niche actor network, gives additional insights (Table 13.2). Six different levels of relationships between actors have been identified: (i) cooperation in terms of knowledge building (indicated as "information" in the table), (ii) incidental contact between the actors ("service"), (iii) mere financial relationship ("money"), (iv) input exchange ("substrate," "equipment," "land," "material"), (v) output exchange ("digestate," "heat"), and (vi) institutional contact ("co-operation," "subcontracting," "regulation"). More in-depth analyses can give further insights by measuring various network indicators, i.e. "density" (measure of cohesiveness), "distance" (measure of the efficiency of diffusion within the network) and "degree centrality" (number of direct ties an actor has within the network denoting opportunities to access diverse types of knowledge) [42].

Table 13.2 Network of actors in the biogas sector linked with different relationship types.

Configuration of stakeholder relationships	Farmer	Biogas plant company	Neighboring farms	Bank/private investor	Public funding agencies	Univ/research lab
Farmer	x	Land; substrate	Information	x	x	x
Biogas plant company	Money; heat; digestate	x	Digestate	Information	Information	x
Neighboring farms	x	Substrate	x	x	x	x
Bank/private investor	x	Money		x	Information	x
Public funding agencies	x	Money	x	Money; information	x	x
Univ/ research lab	x	Information		x	x	x
Surrounding households	Information	x		x	x	x
Local administrative entity	x	x		x	x	x
Manuf: Dyn Biogas	x	Equipment; information		Information	Information	Cooperation; information
Maintenance company	x	Service, information		x	x	x
Public authority (e.g. Ministry)	x	Regulation; information		x	Information	Information
Agrifood industry	x	Material; money	x	x	x	Information

Configuration of stakeholder relationships	Surrounding households	Local administrative entity	Manuf: Dyn Biogas	Maintenance company	Public authority (e.g. Ministry)	Agrifood industry
Farmer		x	Information	Information; money	x	x
Biogas plant company	Heat	x	Money; information	Money	Information; money	Information
Neighboring farms	Information	Information	x	x	x	x
Bank/private investor	x	x	x	x	x	x
Public funding agencies	x	x	x	x	Information	x
Univ/ research lab	x	x	Information	x	x	x
Surrounding households	x	x	x	x	x	x
Local administrative entity	x	x	Information	x	x	x
Manuf: Dyn Biogas	Information	Information; money	x	Money	Information, cooperation	Information
Maintenance company	x	x	Subcontracting; information	x	x	x
Public authority (e.g. Ministry)	x	x	Regulation; information	x	x	x
Agrifood industry	x	x	x	x	x	x

13.4 Strategic Niche Planning for Sustainable Transitions

13.4.1 Methodological Steps

Considering biogas activity in CEE countries as a niche sector according to the multi-level perspective approach, sustainable transition to exploit the full potential of biogas in a frame of circular bioeconomy depends on dynamic co-evolution within the landscape and incumbent regimes. The understanding of these dynamics can be supported by examining transition drivers and constraints at different levels in such a way that generate sector level enabling strategies.

This section details the process of strategic planning indicating the potential of the technology in question to trigger a shift in the regime, and at the same time examining pressures coming from the landscape. The Strength, Weaknesses, Opportunities, and Threats (SWOT) framework provides insights on the effective conditions to be met for the transition to become possible: the internal condition revealing strengths and weaknesses of the niche technology, and the external condition described as opportunities and threats originating in the related landscape and regime(s) encompassing the sector. Relevant literature and inquiry of stakeholder network reveals the driving forces and barriers to such a development. These can consequently be elaborated and classified in internal and external factors plotted against the opportunities and threats axis. Exploitation of such information completed by current policy coherence overview can lead to explicit strategic actions. These latter are then evaluated against objectives and performance indicators in order to select the most appropriate ones.

Firstly, a stakeholder analysis is undertaken, by means of an academic and gray literature to detect the most important agents involved in the biogas niche sector development, such as firms, NGOs, research institutes, business support organizations, trade organizations, and public institutions. A questionnaire composed of three sections (i) general information, (ii) expectations on developments regarding current and future technologies and their sustainability, and (iii) informal knowledge sharing, and formal knowledge exchange – can complete the picture. Based on the data collected (especially those of section b) context-specific drivers and barriers to the development of the niche sector are pointed out. These are then transformed in SWOT analysis homogenous elements of strengths, weaknesses, opportunities, and threats.

Next, stakeholder focus groups assess the validity of the questionnaire content, preferably evaluating the importance of the SWOT matrix elements. The SWOT elements are then transposed in all possible pairs in a threats-opportunities-weaknesses-strengths (TOWS) matrix. Finally, policy coherence analysis examines synergies and conflicting provisions, and policy measures to consider when elaborating strategic actions to governance promoting sustainability. The steps of analysis are schematically presented in Table 13.3.

In Table 13.4, the development of a new technology, as a result of the positive fulfillment of seven functions [43] (first column) is matched with the different dimensions required from the strategist to reduce corresponding uncertainty [44] (second column). These correspond to the supply chain, related stakeholders, and requirements to be fulfilled, thus translated in seven categories to classify driving factors and barriers [45] (last column).

The identified driving factors and barriers are then characterized by their position in the internal or external environment thus populating the SWOT matrix [46]. Threats and

Table 13.3 Sequence of steps for strategy design and elaboration.

Strategy crafting steps	Outcome
1. Surveying stakeholders	Setting up the network structure – Definition of agents-actors
2. Multi-level perspective view	Detect drivers and barriers to the niche development
3. Investigating SWOT factors	Identify elements of SWOT matrix PESTEL (political, economic, social, technological, environmental, and social dimensions) frame for external factors
4. Conducting AHP analysis	Pairwise comparisons to measure Factor intensity and Group intensity
5. Formulating strategies	Combine strategies using the TOWS interactions Consider most powerful facilitators and obstacles
6. Policy coherence analysis	Top-down versus bottom-up detection of policy incoherencies Primary policy suggestions Strategy value Prioritize strategic actions
7. Prioritizing strategic actions	Assess relative importance of strategic actions Strategy value
8. Resume strategic actions	Synthesis of strategies outcome with policy coherence analysis to articulate actions, operational objectives, and strategic goals that serve each subsector deployment or niche development, scaling up and integration to the socio-technical landscape

Table 13.4 Matching classification based on TIS with foresight dimensions and drivers rationale.

TIS functions		Foresight dimensions	Driving factors and barriers	
[43]		[44]	[45]	
Resource mobilization	Human, financial, networking capital	Biomass availability and trends	Supply chain	Raw material
Market formation	Nursing, bridging, and mass markets	Market acceptance		Customer/market
Influence on the direction of search	Incentives and/or pressures for the organizations to enter it	Regulatory and policy framework	Stakeholders	Financing (public–private)
Legitimation	Social acceptance and compliance with relevant institutions			Legislation and policies
Entrepreneurial experimenta-tion	To tackle with uncertainty in technologies, applications, and markets	Development and horizon scanning of the emerging technology	Requirements	Collaboration
Knowledge development and diffusion	Types and sources of knowledge (R&D, DUI)			Innovation management
Development of positive externalities	generation of positive external economies > new entrants			Sustainability

Table 13.5 Questions to be discussed using the TOWS structure.

TOWS matrix	Opportunities (external, positive)	Threats (external, negative)
Strengths (internal, positive)	*Strength and Opportunity (S&O) Strategies* How can we use strengths to maximize the identified opportunities?	*Strengths and Threats (S&T) Strategies* How can we use strengths to minimize the effect of threats?
Weaknesses (internal, negative)	*Weakness and Opportunity (W&O) Strategies* How can we minimize the weakness using the identified opportunities?	*Weakness and Threats (W&T) Strategies* How can we minimize the weaknesses to avoid the identified threats?

opportunities pertaining to the present and the future environment while the audit of strengths and weaknesses focuses on the internal resources of the enterprise. Recent works enrich SWOT analysis for strategy formation with multi-criteria evaluation of the elements of the matrix [45, 47–50] ranking them according to their importance, so that they later capacitate combinations in order to attribute priorities to the strategic propositions for action.

Then the question is how to develop more specific actions and tactics to achieve the project's objectives. For this purpose, a conceptual framework for situational analysis stemming from the SWOT matrix is proposed in the business management literature [51]. The TOWS matrix (Table 13.5) provides the framework to indicate potential strategies to effectively and efficiently pursue the enterprise's organizational objectives and mission based on all conceivable combinations of its external and internal factors.

The TOWS matrix facilitates interaction of the four sets of SWOT variables resulting in four conceptually distinct alternative strategies, tactics and actions [51]. (i) The WT Strategy (mini-mini): In general, the aim of the WT strategy is to minimize both weaknesses and threats. Actions need to be devised with the intent of either overcoming weaknesses or diminishing threats over time. (ii) The WO Strategy (mini-maxi) aims to identify opportunities in the external environment that can help overcome organizational or technical weaknesses. (iii) The ST Strategy (maxi-mini): This strategy addresses threats in the external environment by optimally using its strengths with great restraint and discretion. (iv) The SO Strategy (maxi-maxi): Any combination of strengths and opportunities most probably leads to multiplicative impacts.

The decision makers start explicating specific tactics to support concerted actions that can serve broader strategies, resuming the lower level proposition to articulate strategies at a higher level of abstraction. Table 13.6 shows an example of relationship building which subsequently may guide strategic actions. In this matrix, a "+" indicates a relationship between a strength of the company and an external opportunity, while a "0" indicates the absence of a relationship. For instance, strength S3 in Table 13.6 can be matched with several opportunities. Similarly, more than one strength can be utilized to exploit opportunity O2. Similar tables may be used to analyze the other three strategy quadrants (W&O, S&T, and W&T). Relationships may vary in terms of their potential, thus each should be carefully assessed. Still, this matrix is proposed as "a relatively simple way of recognizing promising

Table 13.6 Interaction grid Strengths-Opportunities: illustrative example.

Strength Opportunity	S1	S2	S3	S4
O1	0	+	+	+
O2	+	+	+	0
O3	+	0	+	0
O4	0	0	+	0

strategies that use the company's strengths to take advantage of opportunities in the external environment" [51].

Proceeding to the classification of developed strategies, actions related with the same area/domain are grouped. These classified groups are then matched to different aspects/domains according to their functionality. Multi-criteria techniques are widely used to assess ex ante the importance of various strategic propositions. Analytic hierarchy process (AHP) is used to assess the local priorities lp(f) and ranks of the individual factors per group. Alternative simpler methods can be used such as Simos procedure [52] when the pairwise comparisons required by AHP are difficult to obtain or when SWOT matric elements are numerous resulting in inconsistencies. Based on the weights of individual factors and the importance of SWOT vectors lead to global priorities and ranks among all factors. Alternatively, an ex post approach is suggested in Falcone et al. [53]: after thorough discussion of TOWS based strategies and their subsequent validation by expert opinion, a weighting process can follow employing a Likert scale to denote the relevance of each strategic proposition. The weighted average of the group allows to rank the strategic propositions from the least to the most relevant.

From a policy perspective, the SWOT-AHP technique is a valuable method for strategy design. However, a limitation is that it is not suitable for policy engineering (e.g. optimizing the economic value of subsidies for different consumer income ranges). For this purpose the analysis could be further pursued beyond the previous phase following the guidelines by Patidar et al. [54]. The objectives to be achieved by the niche sector are first defined. Next, among the developed strategies, the most essential, urgent, and impactful, considering the defined objectives, are selected. Sustainable Performance Measures (PM) evaluate the niche sector in economic, environmental, and social dimensions. The evaluators prepare an incidence matrix comprising the strategies and their affected PMs and assess each strategy's possible outcomes based on their knowledge and experience. This procedure can pinpoint the shortlisted strategies that satisfy the maximum number of PMs.

Alternatively, at this stage impact assessment can be undertaken by means of more or less sophisticated tools. Indicatively, Olszewski et al. [55] develop a decision support system by integrating economic, spatial and demographic data, and processing them with spatial information modeling methods and game theory to assess the impacts of various policy rationales. Relevant models can also be consulted. Indicatively, Rozakis et al. [56] denote conditions beyond economic profitability to be fulfilled for further sector development, and Mamica et al. [34] investigate the impact of different policy measures configurations on the economic performance of typical biogas facilities. Finally, the evaluation of the possible outcome(s) of each strategy is done based on evaluators' knowledge and experience. Evaluators prepare an incidence matrix comprising the strategies and their affected PMs. From this incidence matrix, strategies can be shortlisted by identifying the ones that satisfy the maximum number of PMs.

13.4.2 Case Study: Biogas Sector in Poland

In the case of sustainable biogas development strategic planning is then implemented for Polish conditions, where biogas is still in niche status, using the perception of stakeholders to articulate strategies for its development as a pivotal sector toward a circular sustainable bioeconomy. In the case of biogas, the actor network is quite diverse consisting of investors, technology developers, consultants, public authorities, funding institutions, private farmers, farmer groups, NGOs, and research organizations. Table 13.2 has depicted the network of actors and their relationships in the sector. According to the classification by Brunnhofer et al. [45] as shown in the first column of Table 13.7, we proceed to detail driving factors and barriers for biogas development considering the multiple techno-economic levels in Poland detailed in Section 13.3.1. The elements of the table are mostly harvested from the academic and gray literature (i.e. Ref. [63]).

In a second phase the outcome of the first phase is examined by means of in-depth discussions with experts on the important factors for the sector development. This process results in additional drivers and barriers characterized in categories (economic, social, ecological, policy, business model, industry attitude) to complete the picture. Then the identified drivers and barriers are categorized and transformed into SWOT groups merged and condensed to the shortest feasible number of SWOT factors per group (Tables 13.8 and 13.9).

One should search for major external forces affecting the activity by considering homogeneous categories such as political, economic, social, technological, environmental, and legal issues. Issues collected may represent both opportunities and threats. Finally, unbiased and reliable information is required so that we make a judgment on whether an external factor affects the project in question positively or negatively.

The TOWS matrix of the biogas case study is mapped in Table A.1 (Appendix) where grids appear for each one of the four quadrants (SO, ST, WO, WT). After careful consideration of all possibilities, the grid cells can be used to point out feasible combinations, imagine tactics, and plan subsequent actions. For instance in the first (strengths-opportunities) quadrant, opportunity O1, generally described as "resource availability," can be exploited through S1 (diverse feedstock treatment capability) maximizing synergy if "support development of specialized labs certifying AD from various sources Reference." The plot of strengths and weaknesses against opportunities and threats is quite dense, thus resulting in numerous propositions for action. In total circa 15–20 unique actions are proposed in each quadrant summing up in a decent number of actions that require organization so that they make sense.

Based on those, a synthesis is formulated in tabular form for the classification of proposed actions where those pertaining to the same domain are grouped (Table 13.10). Domains are characterized according to their functionality, namely "knowledge development, influence on the demand-side, entrepreneurial experimentation, command and control measures and incentives on the supply, legitimation, resource mobilization and positive externalities" (second column in Table 13.10). Within these aspects, strategic actions are specified in classes as shown in the third column.

Table 13.7 Driving factors and barriers for biogas development.

Driving factors	Description
Raw material	Sufficient feedstock and Solution to problems caused by manure surplus [57, 58] Large quantities of agrifood waste
Customer/market	Serves multiple end uses Energy security [59]
Financing (public–private)	Support for energy production from renewable energy sources Support to research and demo projects from the EU
Legislation and policies	Energy self-sufficiency Emergency global trends regarding renewable sources
Collaboration	High-yielding bioreactors supported by metagenomics demo projects [7]
Innovation management	Process heterogeneous input feedstocks [60] "Flexible and standardized design and size of digester" [60]
Sustainability	"Framing AD in terms of the water-energy-food nexus" [61]
Barriers	
Collaboration	Coordination and collaboration increase capacity building, institution development, and management Training of labor force for construction, operation, and repair Persisting lack of reliable professionals in the country so that biogas plant operators must learn from their mistakes Limited availability and capacity of the power grid in rural areas Problems with digestate management
Customer/market	The market is still quite small, difficult, and unstable Implications of co-production – problems encountered by the production of fertilizer granules and their export in South Europe [59] "Heavy metal build-up and the risk of spreading pathogens" [62] Lack of specialist companies on the market few designers, general contractors, construction companies, and technologists specializing in the design, construction, and operation of ABP Lack of expert advice
Financing (public–private)	Lack of a national substantive expert background for investors interested in the construction of ABP
R&D expertise	High investment costs
Legislation and policies	Time-consuming and complicated administrative procedures. Legal instability, Need to adapt to constantly changing legal regulations Legal problems, e.g. how to classify a digestate
Raw material	
Risk averse culture	Resistance by the local community Lack of social awareness of the production and use of biogas Lack of an organized system of reliable knowledge transfer result in increased investment risk and cause local communities and officials at various levels to experience unnecessary concern
Technology	Quality check of biogas and follow up services Equipment for the construction of a biogas plant has to be imported from abroad Lack of domestic producers of certain technological devices

Table 13.8 *Internal factors derived from the drivers/barriers list.*

Code	Strengths	Description
S1	Capabilities for a spectrum of waste substrate	1.1 Local accessibility of waste from various sources 1.2 Application potential of various waste types
S2	Income generation from various outflows	2.1 Wide range of products in el/heat 2.2 Satisfying demand for base/peak electricity
S3	Capacity for positive externalities	3.1 Potential of local energy use 3.2 Creation of new workplaces 3.3 Air quality improvement by reducing odors from direct use of waste (tamburini 3–5) 3.4 Regulation of soil nutrients by reducing spreading of organic load and digestate 3.5 Global climate regulation by saving GHG emissions
S4	Technological advantages	4.1 Mature technology available 4.2 Networking R&D centers active throughout the country 4.3 Fermenting chambers tightly sealed preventing odor
	Weaknesses	**Description**
W1	Technology state-of-the-art	1.1 Problems with access to power supply 1.2 Technical constraints to handle heterogeneity of biological waste 1.3 Breakdowns and high repair costs
W2	Plant establishment issues	2.1 Significant costs for grid connection 2.2 Demand of considerable capital for investment 2.3 Lack of expertise to deal with the administration
W3	Inherent issues to technology	3.1 Heat unused directly become waste 3.2 Bulky raw material increasing transport costs
W4	Lack of consensus	4.1 Local community protests

13.5 Strategic Propositions and Concluding Comments

Based on the analysis of the TWOS matrix, policy recommendations are provided below. Instead of ranking the elements of the SWOT matrix in terms of importance, for the sake of this illustrative case study those that are supported by literature references have been selected. Proposed actions are grouped according to the classification of Table 13.10. Also, the classification proposed by Lovec and Juvančič [40] in *enabling* (supportive supply-side and demand-side measures) and *regulatory* (requirements, standards and certificates) measures is applied. In this regard, categories of enabling measures include R&D, education, networking and resource mobilization; while regulatory categories include legislation, legitimation, incentives and demand pull – cooperation: brokerage, clusters, and rural development.

13.5.1 Research and Development

In this category, the highest priorities identified are "Enhance biorefinery R&D for cascading use of waste," "Fund R&D for max biogas production efficiency for various plant capacities," "Develop mass produced standard technology equipment flexible," "Development of

Table 13.9 *External factors derived from the drivers/barriers list.*

Code	Opportunities	Description
O1	Resource availability	1.1 Unexploited potential (numerous farms circa or above 100 LSU) 1.2 Financial support/funding from domestic and EU sources 1.3 Great potential for generating biodegradable waste from industry
O2	Technology progress	2.1 Fast development of the technology 2.2 Experience/knowledge transfer from neighbor countries 2.3 Emerging innovative technologies for biogas production
O3	Defossilization momentum and climate policy	3.1 Feed-in price for 15 years by current legislation 3.2 iLUC restriction clause in EU energy/climate policies
O4	Agricultural policy synergy	4.1 Compatible with Polish CAP national strategic plan 4.2 Compatible with regional/rural development
O5	Societal imperatives	5.1 Raising environmental awareness/deficit of organic fertilizers 5.2 Growing demand by dairy industry for low GHG milk production
O6	Energy independence Geopolitical concerns	6.1 Energy self-sufficiency objective 6.2 Fossil fuel prices fluctuation
O7	Waste policy synergy	7.1 Stricter regulations for manure/slurry 7.2 Increasing cost of disposing food waste 7.3 Financial stability of food industry/free of charge or willing to pay
	Threats	Description
T1	Input availability	T1.1 No guarantee of stable feedstock supplies T1.2 Geographically scattered input supplies
T2	Resource issues	T2.1 Decrease of agricultural land for food-prices volatility T2.2 Loan high cost by banking sector or investment funds
T3	Legislative issues	T3.1 Unstable policy (i.e. installation subsidy, feed-in tariffs for electricity) T3.2 Constantly changing legal regulations T3.3 Long process for issuing permits/bureaucratic formalities
T4	Market issues	T4.1 Lack of experienced professionals in the country T4.2 Small, difficult, unstable market T4.3 Development of biogas competitors
T5	Policy conflicts	T5.1 Incentives/programs to reduce waste in the agricultural and food sectors T5.2 Legal restrictions on fertilizers

reliable equipment," "Exploit unused heat to increase the electricity efficiency," and "Technology improvement" (i.e. co-digestion to tackle the high concentration of ammonium nitrogen in the case of poultry litter substrate that challenges the process of AD and biogas production from a research, economic, and environmental standpoint [32]). In order to minimize the gap among EU member countries, tailor made financial support to encourage scientific and technological collaborations among value chain actors in the CEE countries are of paramount importance. Such preference needs to be targeted preferably founded in Strategic Research and Innovation Agenda eventually promoting coordinated R&D activities following success stories in leading countries [64].

Table A.1 Populated TOWS with exhaustive strategic actions.

SO quadrant in TOWS	S1: feedstock treatment capability	S2: capacity for income generation	S3: social advantages	S4: tech advantages
O1. resource availability	Develop certified labs for efficient anaerobic digestion from various sources	Support for state-of-the-art storage to increase peak electricity capacity	Environmental credits to finance new jobs with reduced employer charges	Support alternative business model tailor made to resources
O2. technology progress	Training for business model expertise	Standardize infrastructure construction and equipment	R&D for prevent odor technology improvement	Fund clustering with established firms to implement do-use-interact approach
O3. defossilization and climate policy	Higher iLUC percent allowed in final biobased products	Green certificates and/or increased quota in auction calls for biogas	Promote biogas business models that max climate mitigation performance	Adopt index-linked feed-in tariffs to promote use of risky substrates – hybrid feed-in tariffs differentiating for electricity and biomethane
O4. agr policy synergy	Land use policy for agricultural/urban zones	Credits to farmers for decreasing direct use of liquid manure in the fields	Support for establishment grant in low population density areas	Include digestate from biogas plants in organic fertilizer subsidy
O5. societal imperatives	Enforce tests for epidemiologically clean and antibiotic free digestate	Finance consumer awareness for low GHG dairy products	Disseminate biogas contribution to sustainable circular bioeconomy	Promote training for operation/maintenance technician jobs in the periphery
O6. energy independence geopolitical concerns	Explicit self-sufficiency targets critical raw materials (i.e. phosphates)	Incite food industry to acquire equity to attain energy independence objectives	Campaign for independence from imported natural gas Inform the public on the potential of biogas/biomethane comparing with nuclear power plant capacity	Develop DSS for the strategic performance evaluation of the state-of-the-art technology
O7. waste policy synergy	Stricter command and control measures for waste treatment from the industry	Tax credit to industry to compensate expenses for waste anaerobic digestion	Enhance biorefinery R&D for cascading use of waste	Fund R&D for max biogas production efficiency for various plant capacities

Table A.1 (continued)

WO quadrant in TOWS	W1. feedstock concerns	W2. establishment issues	W3. inherent tech issues	W4. consensus
O1. resource availability	Improve funding terms for small distributed production	Training biogas brokers to facilitate administrative and grid connection task	Funding for heat infrastructure and operating expenses	Involve neighboring farms as suppliers or customers of the activity
O2. technology progress	R&D to diversify raw material to minimize distance of feedstock transport	Develop mass produced standard technology equipment flexible	Development of reliable equipment R&D to exploit heat to increase the electricity efficiency	
O3. defossilization and climate policy	Support for biomass production in abandoned/marginal land	Simplify formalities due to urgent action for climate	Support heat use for substitution of Imported gas	Climate awareness campaign to dominate over nimby attitudes
O4. agr policy synergy	Manure management in cross compliance requirements	Subsidy to establishment costs for on-farm biogas	Credit for satisfying heat demand by local sources	Information campaign on impacts to rural development
O5. societal imperatives	Eligible LCA studies on environmental footprint of biogas activity	Promote technical education in circular economy technologies	Undertake studies for evaluating savings in houses from biogas heat	Promote citizen participation and co-creation of biogas location decisions
O6. energy independence geopolitical concerns	Subsidy to fossil fuels for transport for transport related to energy supply	Establish energy independence fund for base load energy generation from domestic sources	Expand gas pipeline network in the rural areas	Benefit from geopolitical momentum to raise sensitivity and emotional attitudes
O7. waste policy synergy	R&I funding to cope with substrate heterogeneity	Synergies related to investment support with waste management policy	Harmonize feedstock substrate fee with digestate delivery	Translate circular economy benefits to financial inflows to local communities

T1. input availability	Market power: diversify suppliers	Market power: high prices for feedstock to ensure facility operation	Market power: monopolistic competition related to the location	Support alternative business model development to exploit potential
T2. resource issues	Distributed production linked to livestock activity	Networking with venture capital for innovative configurations	Increased establishment funding for short biobased chains	Build capacity for fund-raising through submission of competitive demo projects
T3. legislative issues	Simplification of formal requirements for establishment permit	Special tariffs for peak demand electricity from biogenic sources	Environmental credits to finance new jobs with reduced employer charges	Adjustment of legislation on technology transfer offices in higher education and state research institutes
T4. market issues	Support for professional associations to organize courses of lifelong education	AKIS brokers to consult on diversification of biogas business plan configurations	Market power: monopolistic competition related to the location	Support for creation and activities of biobased energy clusters
T5. policy conflicts	Craft strategy to enhance various policies coherence	Special tariffs for peak demand electricity from biogenic sources	Incentives for vertical integration of energy module in food industry	Constant update of legal quality requirements to scientific advances

Table A.1 (*continued*)

Weaknesses-threats quadrant in TOWS	W1. tech state-of-the-art	W2. establishment issues	W3. inherent tech issues	W4. consensus
T1. input availability	Incentives for vertical integration of energy module in food industry	Prioritize small size on-farm biogas configuration	Locate next to important heat/cooling demand sites to compensate for input transport expenses	Distributed production linked to livestock activity improving odor situation
T2. resource issues	Support for biomass production in abandoned/marginal land	Direct subsidies for establishment expenses in biogas	Finance public electricity network expansion	Flexible credit for district heating infrastructure with local source heat energy
T3. legislative issues	Science based legislation updating of quality requirements of byproducts	Simplify formalities due to urgent action for climate	Flexible outsourcing contracts to decrease maintenance costs for small biogas only	Institutionalize public participation in the decision process for locating RES
T4. market issues	AKIS brokers to consult on diversification of biogas business plan configurations	Support of regional biobased energy clusters	AKIS brokers to consult on diversification of biogas business plan configurations	Training of cooperatives for capacity building of local labor force
T5. policy conflicts	Involve all stakeholder to craft strategy to enhance various policies coherence	Fast track investment subsidies for laggard regions	Targeted policies to fill efficiency gaps of state-of-the-art technology	Fast track investment subsidies for laggard regions with low population density

Table 13.10 *Strategic propositions suggested in the TOWS grid (Table A.1) grouped for different aspects and classifications.*

S#	Domain	Classification	Strategies	Factor				Description
				S	W	O	T	
1	*Knowledge development and diffusion*	R&D funding schemes	S_1O_1 S_3O_2 S_3O_7 S_4O_7 W_1O_2 W_1O_7 W_2O_2 W_3O_2 W_3O_2	3	3	5	2	*R&D* Develop certified labs for efficient anaerobic digestion from various sources – to cope with substrate heterogeneity – diversify raw material to minimize distance of feedstock transport – Prevent odor technology improvement – Enhance biorefinery R&D for cascading use of waste – Fund R&D for max biogas production efficiency for various plant capacities – Develop mass produced standard technology equipment flexible – Development of reliable equipment – exploit unused heat to increase the electricity efficiency – Surcharge or subsidy for biogas producers who utilize biodegradable waste from industry (processing of raw material that is produced anyway) – promotion of cooperation – Financial support for investments that accompany the biogas plant, e.g. to use the produced heat or energy for self-consumption purposes *EDU* Promote training for operation/maintenance technician jobs in the periphery – Support for professional associations to organize courses of lifelong education – Promote technical education in circular economy technologies – Adjustment of legislation on technology transfer offices in higher education and state research institutes – Training for business model expertise – Promotion of the role and importance of biogas plants in a zero-emission energy system *Info* DSS for the strategic performance evaluation of state-of-the-art tech – Undertake studies for evaluating savings in houses from biogas heat – Creation of an IT tool for information exchange and partnerships on biowaste production and collection
		Educational policies	S_4O_5 W_2O_5 S_1T_4 S_4T_3 S_1O_2 S_3O_3					
		Informational instruments	S_4O_6 W_3O_5 S_1O_1					

Table 13.10 *(continued)*

S#	Domain	Classification	Strategies	Factor S	Factor W	Factor O	Factor T	Description
2	*Influence on the direction of search*	Overarching visions: energy security, sustainable bioeconomy	S_1O_6 S_2O_6 S_3O_6 W_2O_6 W_3O_3	4	4	3	3	*Security* Explicit self-sufficiency targets critical raw materials (i.e. phosphates) – Campaign for independence from imported natural gas – Establish energy independence fund for base load energy generation from domestic sources – Support heat use for substitution of Imported gas
	Incentives to enter		W_4O_5 S_4T_2					
		Expectations	S_2O_1 S_2O_5 S_1T_2 W_2T_1 S_3T_5					*Expectations* Promote citizen participation and co-creation of biogas location decisions – Build capacity for fund-raising through submission of high TRL demo projects
		Generate demand	W_2T_5 W_3T_1					*Generate demand* Support for state-of-the-art storage to increase peak electricity capacity – promote consumer awareness for low GHG food products Incentives for vertical integration of energy module in food industry – Locate next to important heat/cooling demand sites to compensate for input transport expenses – Fast track investment subsidies for laggard regions – Prioritize small size on-farm biogas – Distributed production linked to livestock activity
3	*Entrepreneurial experimentation* Testing new technologies, applications, and markets	Brokerage for suitable business Models and targeted legal interventions	S_4T_1 S_1O_2 S_3O_3 S_4O_1 S_4O_2 W_2O_1	4	2	3	4	*Brokerage* Support alternative business model development to exploit potential – Promote biogas business models that max climate mitigation Training biogas brokers to facilitate administrative and grid connection task AKIS brokers to consult on diversification of biogas business plan configuration – Support biogas producers with tools for scaling up the production of digestate as biofertilizer and promotion opportunities

		Develop networking through clustering	S_2T_2 S_2T_4 S_4T_4 S_2O_2 W_2T_4 S_1T_1					*Networking-clusters* Networking with venture capital for innovative configurations – Fund clustering with established firms to implement do-use-interact approach – Support for creation and activities of biobased energy clusters – Standardize infrastructure construction and equipment – Support of regional biobased energy clusters – boost market power: diversify suppliers
		Improve legislation	W_3T_3 S_1T_3 W_2O_3					*Improve legislation* Simplify formalities due to urgent action for climate Flexible outsourcing contracts to decrease maintenance costs for small biogas Simplification of formal requirements for establishment permit
S#	Domain	Classification	Strategies	S	W	O	T	Description
4	*Market formation* Factors driving new market formation	Command and control measures on waste management	S_1O_5 W_3O_7 S_3O_1 S_1O_7 S_2O_3 S_2T_1 S_2O_4	3	3	6	3	*Command and control measures – niche markets* Enforce tests for epidemiologically clean and antibiotic free digestate Stricter command and control measures for waste treatment from the industry Green certificates and/or increased quota in auction calls for biogas Market power: high prices for feedstock to ensure facility operation
		Efficiently designed subsidy support	W_1O_6 S_2T_3 W_3O_4 W_2O_4 S_3T_2					*Credits-subsidies* Compensate feedstock substrate fee with digestate delivery – Environmental tax credits to finance new jobs with reduced employer charges – Credits to farmers for decreasing direct use of liquid manure in the fields – Subsidy for transport related to feedstock for energy Special tariffs for peak demand electricity from biogenic sources Credit for satisfying heat demand by local sources – Subsidy to establishment costs for on-farm biogas – Increased establishment funding for short chains

Table 13.10 (continued)

S#	Domain	Classification	Strategies	Factor S	Factor W	Factor O	Factor T	Description
	Legitimation Social acceptance Compliance with institutions and society	Evidence based multiple dividends Participative procedures – cooperative decisions	S_1O_3 W_1O_5 S_3O_5 S_3O_6 W_4O_3 W_4O_4 W_4O_6 S_4T_5 W_4T_4 W_4O_1	3	2	5	2	Higher iLUC percent allowed in final biobased products – Eligible LCA studies on environmental footprint of biogas activity – Disseminate biogas contribution to sustainable circular bioeconomy – Constant update of legal quality requirements to scientific advances Inform the public on the potential of biogas/biomethane comparing with nuclear power capacity – Climate awareness campaign to dominate over nimby attitudes – geopolitical momentum to raise sensitivity and emotional attitudes Training of cooperatives for capacity building of local labor force – Involve neighboring farms as suppliers or customers of the activity – Information campaign on impacts to rural development
	Resource mobilization Financial and human factors and other assets	Manage food conflict Regular waste supply Labor market	S_1O_4 S_2O_7 S_4O_3 W_1O_1 W_1O_3 S_3T_3	4	1	4	1	*Land* Land use policy for agricultural/urban zones – Support for biomass production in abandoned/marginal land *Waste source* Tax credit to industry to compensate for waste anaerobic digestion – Adopt index-linked feed-in tariffs to promote use of risky substrates – hybrid feed-in tariffs differentiating for electricity and biomethane *Rural labor potential* Improve funding terms for small distributed production – Environmental credits to finance new jobs with reduced employer charges
	Positive externalities	Harmonize related policies Benefits to other actors	S_3O_4 S_4O_4 W_1O_4 W_3O_6 W_2O_7 W_4O_7 S_1T_5 W_2T_3 W_3T_2	3	5	2	3	*Rural Development* Establishment grants in low population density areas Translate circular economy benefits to financial inflows to local communities Finance public electricity network expansion in rural areas *Agriculture* Include digestate from biogas plants in organic fertilizer subsidy Manure management in cross compliance requirements *Waste-climate* Investment support related to waste management policy Climate – Simplify formalities due to urgent action for climate *Policy coordination* Craft strategy to enhance various policies coherence

13.5.2 Education Activity – Enhance Brokerage

Enriching the curricula to strengthen the formal education system favoring direct contact to RES companies by means of discursive meetings and in situ visits in order to benefit from the professional knowledge shared by entrepreneurs. Chodkowska-Miszczuk [5] highlights the need to focus on place-based education, introducing the narrative experience to promote the acquisition of knowledge.

Support alternative business model development to exploit potential – Promote biogas business models that max climate mitigation. Agricultural knowledge and information systems (AKIS) brokers consult on diversification of biogas business plan configuration including training biogas brokers to facilitate administrative and grid connection task. The transition studies framework that describes the eventual shaping of sustainable markets comprises three critical processes: enabling exchange practices, proving the system, and constructing the narrative, generating traded, demonstrated, and expected value respectively. The value output from each process serves as input to the other two processes illustrated In the case of market-shaping processes for biogas in Sweden. Public and private agents have engaged key actors in a multitude of activities that have built up the market-shaping processes. Ottosson et al. [65], Lazarevic and Valve [66] investigate potentially disruptive technology, and observe the diversity of niche configurations that all make use of the same technological core to generate economic value examining the Finnish biogas production. Various combinations of the resources recovered and the outputs that remain mere externalities, result in some niche configurations that are not as sustainable as others.

13.5.3 Networking-Clusters

Clusters can enhance early transition pathways to bioeconomy improving productivity and practice-oriented innovation, a feature of the do-use-interact model [40]. Consequently, transaction costs are decreased with untraded dependencies, the proximity of actors necessary for tacit knowledge triggers innovation, whereas mentoring, role models, learning, communication, and commercialization that arise from the biocluster setting boost income and employment. From a multi-level perspective, bioclusters contain a mix of niche players and established regime actors, and often coordinate a constellation of niche activities in a region. "A biocluster can thus be defined as a protected place where innovations are (temporarily) shielded from the mainstream selection pressures, nurtured through experimentation and learning, and eventually become empowered" [67]. Concurrent to clustering, networking novel conceptual tools beyond value chain comprise the "biomass-based value web," an extension of the value chain concept with the aim to capture the links within and between value chains that arise from the cascading and joined use of biomass [68]. Adoption of the value web approach may enforce the innovation network to meet future opportunities and challenges of the biogas sector.

13.5.4 Resource Mobilization

Selecting suitable substrates with sustainability criteria in mind is crucial and should be considered in any new incentive scheme. In this respect, the food versus fuel debate – and the associated land use change issues – is extremely relevant, especially since the last geopolitical developments. From this perspective, dedicated energy crops, such as maize production for biogas, are not acceptable; rather generating value from byproducts should be the focus.

In addition, by co-digesting different biowastes, synergistic effects can be created that result in total biogas production higher than the sum of separate substrate digestions. Using dairy waste (i.e. cheese whey), brewing pomace and other materials (e.g. biochar as a process improver or biostimulant) are excellent ways to increase biogas productivity. However, the logistics to transfer these resources to biogas plants should be considered, as for instance transferring resources over long distances can be uneconomic [27].

13.5.5 Elaborate Legislation

Speaking the language of transitions research, key challenges undermining the biogas niche development refer either to a lack of policy support or to the changing and unstable institutional and regulatory framework. Technical knowledge and workforce qualifications, on the contrary are not conceived as a real threat in the biogas case. The harmonization of legislative and financial frames conditions the regular flow of public aid, whereas precise criteria for biogas plants operations, including substrate structure, attenuate tensions between biogas plants and local entities. In other words, the regulations governing the operation of biogas plants should be better specified, and monitoring or management instruments applicable at local level should be elaborated.

13.5.6 Legitimation

Consultation, dialogue and interactions among stakeholders may be achieved by engaging in local community roles and participation in local cultural events, as well as involving the local community in the operation of the biogas plant [24, 69], e.g. by organizing visits to the plant. Interactions can be mutually beneficial to businesses and to the local society. For biogas plant operators, having the approval of the local community can result in improved access to raw materials and greater opportunities for distributing their products, such as energy, heat, and organic fertilizer. In turn, local communities can benefit from biogas plants in both material and immaterial ways [24]. The opportunity to use locally produced heat and organic fertilizer at preferential price, additional income from directing raw materials for biogas production, assistance in local initiatives, renovations, revitalization, and implementation of public services represent material impacts; while gaining cooperative attitudes, shaping de facto ecological awareness and attaining energy literacy constitute immaterial benefits [24].

13.5.7 Incentives for Market Penetration

The role of subsidies and the adoption of an appropriate incentive scheme is fundamental for the biogas market development. The duration of subsidization should consider the lifetime of a plant, and should be carefully defined, so as not to undermine productivity gains by discouraging producers from engaging in learning activities and establishing economies of scale.

The lack of mechanisms to support the sale of green energy may be the major refraining factor for potential investors. Through an analysis of the Polish socio-economic environment, Mamica et al. [34] suggest standardizing regulations and implementing long-term support mechanisms for the sale of energy generated from agricultural biogas. They also recommend an increased use of thermal energy in biogas production to increase the system's effectiveness. Efforts should also be made to adjust the regulations and support systems to the needs of the various actors. Locating agricultural biogas plants in energy cluster systems, as a way of ensuring energy flexibility in the event of shortages of other RES, such as wind or solar, should be promoted to a greater extent.

Appropriate actions should be accompanied by appropriate infrastructure (e.g. new fueling stations) and the adoption of a bonus-malus scheme for new natural gas vehicles (NGVs). The link between waste management practices and human behavior should also be taken into consideration, and actions to this end, such as increasing citizens' awareness of the potential of biogas production, should be pursued.

13.5.8 Demand Pull Actions and Rural Development

Biogas plants are often "first movers" [24], meaning that they are the first power plants in rural areas, and in case the experience is not good, they can also turn out to be the last. One of the most important risks refers to the competition for resources with the primary agricultural production (e.g. increased competition for agricultural land, driving up the farmland prices). However, they constitute nuclei of innovation in rural areas and they help establishing a network of links with local actors generating synergies to provide various ecosystem services [70].

The potential magnitude of the Polish biogas sector and the spectrum of final products that can be derived from biogas plants renders biogas a pivotal element of the evolving energy market. Last but not least the potential processing of biowaste to valuable products, as well as the net positive impacts to rural development, contribute to the role of biogas plants in shaping ecological and social awareness. It is not exaggerated to say that they are the RES entities characterized by the greatest social significance. Compared to other RES installations, they require more permanent human labor and closer cooperation with the local environment.

13.6 Conclusion

The chapter in hand attempts to analyze the process of crafting strategies to navigate a sustainable transition toward a circular bioeconomy at the sectoral niche level. This is founded in multi-level perspective concepts allowing for thorough understanding of the co-evolution of the niche and the incumbent system. Such kind of analysis can support strategic planning in order to enable actions for the promotion of niche activities. For this purpose, methods and ideas drilled from corporate strategic management, sustainability transitions as well as policy analysis are extensively applied resulting in concrete strategic propositions supplying fertile background material for strategies short-listing and further specification in policy measures. The above approach is illustrated in the case of biogas in Poland fueled by abundant literature and original information extracted through contact with stakeholders. This is a promising and vibrant sector capable to transform the energy and agricultural regimes constituting a regime per se in the foreseeable future due to the paramount potential of Poland in this respect.

References

1. EEA Report (2021). Trends and projections in Europe 2021. *European Environment Agency Report No 13/2021*, Denmark. https://www.eea.europa.eu/publications/trends-and-projections-in-europe-2021
2. Bielski, S., Marks-Bielska, R., Zielińska-Chmielewska, A. et al. (2021). Importance of agriculture in creating energy security—a case study of Poland. *Energies* 14 (9): 2465. https://doi.org/10.3390/en14092465.
3. Dach J., Siebielec, G., Mazurkiewicz, J., Pochwatka, P. (2022). Anaerobic digestion for renewable energy, carbon sink and organic fertilizers as an integral part of bioeconomy development. Report Horizon 2020 Bioeastsup Project. https://bioeast.eu/download/6-study-bioenergy-pdf
4. Chodkowska-Miszczuk, J. (2019). Institutional support for biogas enterprises – the local perspective. *Quaestiones Geographicae* 38 (2): 137–147. https://doi.org/10.2478/quageo-2019-0018.
5. Chodkowska-Miszczuk, J., Martinát, S., and van der Horst, D. (2021). Changes in feedstocks of rural anaerobic digestion plants: external drivers towards a circular bioeconomy. *Renewable and Sustainable Energy Reviews* 148: 111344. https://doi.org/10.1016/j.rser.2021.111344.
6. Kaszycki, P., Głodniok, M., and Petryszak, P. (2021). Towards a bio-based circular economy in organic waste management and wastewater treatment – The Polish perspective. *New Biotechnology* 61: 80–89. https://doi.org/10.1016/j.nbt.2020.11.005.
7. Achinas, S. and Euverink, G.J.W. (2020). Rambling facets of manure-based biogas production in Europe: a briefing. *Renewable and Sustainable Energy Reviews* 119: 109566. https://doi.org/10.1016/j.rser.2019.109566.
8. Köhler, J., Geels, F.W., Kern, F. et al. (2019). An agenda for sustainability transitions research: state of the art and future directions. *Environmental Innovation and Societal Transitions* 31: 1–32. https://doi.org/10.1016/j.eist.2019.01.004.
9. Markard, J., Raven, R., and Truffer, B. (2012). Sustainability transitions: an emerging field of research and its prospects. *Research Policy* 41 (6): 955–967. https://doi.org/10.1016/j.respol.2012.02.013.
10. Kemp, R., Schot, J., and Hoogma, R. (1998). Regime shifts to sustainability through processes of niche formation: the approach of strategic niche management. *Technology Analysis & Strategic Management* 10 (2): 175–198. https://doi.org/10.1080/09537329808524310.
11. Geels, F.W. (2011). The multi-level perspective on sustainability transitions: responses to seven criticisms. *Environmental Innovation and Societal Transitions* 1 (1): 24–40. https://doi.org/10.1016/j.eist.2011.02.002.
12. Schot, J. and Geels, F.W. (2008). Strategic niche management and sustainable innovation journeys: theory, findings, research agenda, and policy. *Technology Analysis & Strategic Management* 20 (5): 537–554. https://doi.org/10.1080/09537320802292651.
13. Seyfang, G., Hielscher, S., Hargreaves, T. et al. (2014). A grassroots sustainable energy niche? Reflections on community energy in the UK. *Environmental Innovation and Societal Transitions* 13: 21–44. https://doi.org/10.1016/j.eist.2014.04.004.
14. Seyfang, G. and Longhurst, N. (2013). Desperately seeking niches: grassroots innovations and niche development in the community currency field. *Global Environmental Change* 23 (5): 881–891. https://doi.org/10.1016/j.gloenvcha.2013.02.007.
15. Geels, F. and Raven, R. (2006). Non-linearity and expectations in niche-development trajectories: Ups and downs in Dutch biogas development (1973–2003). *Technology Analysis and Strategic Management* 18 (3–4): 375–392. https://doi.org/10.1080/09537320600777143.

16. Geels, F.W. (2002). Technological transitions as evolutionary reconfiguration processes: a multi-level perspective and a case-study. *Research Policy* 31 (8-9): 1257–1274. https://doi.org/10.1016/S0048-7333(02)00062-8.
17. Geels, F.W. (2005). Processes and patterns in transitions and system innovations: Refining the co-evolutionary multi-level perspective. *Technological Forecasting and Social Change* 72 (6): 681–696. Cited 550 times. https://doi.org/10.1016/j.techfore.2004.08.014.
18. Loorbach, D. (2010). Transition management for sustainable development: a prescriptive, complexity-based governance framework. *Governance* 23 (1): 161–183. https://doi.org/10.1111/j.1468-0491.2009.01471.x.
19. Markard, J., Hekkert, M., and Jacobsson, S. (2015). The technological innovation systems framework: response to six criticisms. *Environmental Innovation and Societal Transitions* 16: 76–86. https://doi.org/10.1016/j.eist.2015.07.006.
20. Carlsson, B. and Stankiewicz, R. (1991). On the nature, function and composition of technological systems. *Journal of Evolutionary Economics* 1 (2): 93–118. https://doi.org/10.1007/BF01224915.
21. Hekkert, M.P., Suurs, R.A.A., Negro, S.O. et al. (2007). Functions of innovation systems: a new approach for analysing technological change. *Technological Forecasting and Social Change* 74 (4): 413–432. https://doi.org/10.1016/j.techfore.2006.03.002.
22. Negro, S.O., Suurs, R.A.A., and Hekkert, M.P. (2008). The bumpy road of biomass gasification in the Netherlands: explaining the rise and fall of an emerging innovation system. *Technological Forecasting and Social Change* 75 (1): 57–77. https://doi.org/10.1016/j.techfore.2006.08.006.
23. Jacobsson, S. and Bergek, A. (2011). Innovation system analyses and sustainability transitions: contributions and suggestions for research. *Environmental Innovation and Societal Transitions* 1 (1): 41–57. https://doi.org/10.1016/j.eist.2011.04.006.
24. Chodkowska-Miszczuk, J. (2021). A new narrative for sustainability: exploring biogas plants as 'first movers' in raising energy awareness. *Australian Journal of Environmental Education* 1–16. https://doi.org/10.1017/aee.2021.17.
25. Walsh, J.J., Jones, D.L., Edwards-Jones, G., and Williams, A.P. (2012). Replacing inorganic fertilizer with anaerobic digestate may maintain agricultural productivity at less environmental cost. *Journal of Plant Nutrition and Soil Science* 175 (6): 840–845. https://doi.org/10.1002/jpln.201200214.
26. Valentinuzzi, F., Cavani, L., Porfido, C. et al. (2020). The fertilising potential of manure-based biogas fermentation residues: pelleted vs. liquid digestate. *Heliyon* 6 (2): e03325. https://doi.org/10.1016/j.heliyon.2020.e03325.
27. Pochwatka, P., Kowalczyk-Juśko, A., Sołowiej, P. et al. (2020). Biogas plant exploitation in a middle-sized dairy farm in Poland: energetic and economic aspects. *Energies* 13 (22): 6058. https://doi.org/10.3390/en13226058.
28. Huttunen, S., Pivimaa, P., and Virkamäki, V. (2014). The need for policy coherence to trigger a transition to biogas production. *Environmental Innovation and Societal Transitions* 12: 14–30. https://doi.org/10.1016/j.eist.2014.04.002.
29. Fallde, M. and Eklund, M. (2015). Towards a sustainable socio-technical system of biogas for transport: the case of the city of Linköping in Sweden. *Journal of Cleaner Production* 98: 17–28. https://doi.org/10.1016/j.jclepro.2014.05.089.
30. Piechota, G. and Igliński, B. (2021). Biomethane in Poland—current status, potential, perspective and development. *Energies* 14 (6): 1517. https://doi.org/10.3390/en14061517.
31. Elzen, B., van Mierlo, B., and Leeuwis, C. (2012). Anchoring of innovations: assessing Dutch efforts to harvest energy from glasshouses. *Environmental Innovation Societal Transitions* 5: 1–18. https://doi.org/10.1016/j.eist.2012.10.006.
32. Piwowar, A. (2020). Agricultural biogas-an important element in the circular and low-carbon development in Poland. *Energies* 13 (7): 1733. https://doi.org/10.3390/en13071733.

33. Iwaszczuk, N., Szyba, M., Iwaszczuk, A., and Yakubiv, V. (2019). Production of agricultural biogas from waste – an element of socially responsible actions in the food sector. *Acta Innovations* 33: 52–62. https://doi.org/10.32933/ActaInnovations.33.5.
34. Mamica, Ł., Mazur-Bubak, M., and Wróbel-Rotter, R. (2022). Can biogas plants become a significant part of the new Polish energy deal? Business opportunities for Poland's biogas industry. *Sustainability* 14 (3): 1614. https://doi.org/10.3390/su14031614.
35. Karanikolas, P., Vlahos, G., and Sutherland, L.-A. (2015). Utilising the multi-level perspective in empirical field research: methodological considerations. In: *Transition Pathways Towards Sustainability in Agriculture: Case Studies from Europe* (ed. L.-A. Sutherland, I. Darnhofer, G.A. Wilson, and L. Zagata), 53–68. Wallingford, UK: CABI https://doi.org/10.1079/9781780642192.0051.
36. Sutherland, L.-A., Zagata, L., and Wilson, G.A. (2015). Conclusions. In: *Transition Pathways Towards Sustainability in Agriculture: Case Studies from Europe* (ed. L.-A. Sutherland, I. Darnhofer, G.A. Wilson, and L. Zagata), 205–220. Wallingford, UK: CABI https://www.cabidigitallibrary.org/doi/book/10.1079/9781780642192.0000.
37. Nilsson, M., Zamparutti, T., Petersen, J.E. et al. (2012). Understanding policy coherence: analytical framework and examples of sector–environment policy interactions in the EU. *Environmental Policy and Governance* 22: 395–423. https://doi.org/10.1002/eet.1589.
38. Nilsson, M. and Weitz, N. (2019). Governing trade-offs and building coherence in policy-making for the 2030 Agenda. *Politics and Governance* 7 (4): 254–263. https://doi.org/10.17645/pag.v7i4.2229.
39. Dietz, T., Börner, J., Förster, J.J., and von Braun, J. (2018). Governance of the bioeconomy: a global comparative study of national bioeconomy strategies. *Sustainability* 10: 3190. https://doi.org/10.3390/su10093190.
40. Lovec, M. and Juvančič, L. (2021). The role of industrial revival in untapping the bioeconomy's potential in central and eastern Europe. *Energies* 14 (24): 8405. https://doi.org/10.3390/en14248405.
41. Caniëls, M.C.J. and Romijn, H.A. (2008). Strategic niche management: towards a policy tool for sustainable development. *Technology Analysis and Strategic Management* 20 (2): 245–266. https://doi.org/10.1080/09537320701711264.
42. van den Bergh, J.C.J.M., Truffer, B., Kallis, G. Environmental innovation and societal transitions: introduction and overview (2011) *Environmental Innovation and Societal Transitions*, 1 (1), pp. 1–23.; https://doi.org/10.1016/j.eist.2011.04.010
43. Bergek, A., Jacobsson, S., Carlsson, B. et al. (2008). Analyzing the functional dynamics of technological innovation systems: a scheme of analysis. *Research Policy* 37 (3): 407–429. https://doi.org/10.1016/j.respol.2007.12.003.
44. Ladu, L. and Quitzow, R. (2017). Bio-based economy: policy framework and foresight thinking. In: *Food Waste Reduction and Valorization* (ed. P. Morone, F. Papendiek, and V. Tartiu). Springer https://doi.org/10.1007/978-3-319-50088-1_9.
45. Brunnhofer, M., Gabriella, N., Schöggl, J.-P. et al. (2020). The biorefinery transition in the European pulp and paper industry – a three-phase Delphi study including a SWOT-AHP analysis. *Forest Policy and Economics* 110: 101882. https://doi.org/10.1016/j.forpol.2019.02.006.
46. Kamp, L.M. and Bermúdez Forn, E. (2016). Ethiopia's emerging domestic biogas sector: current status, bottlenecks and drivers. *Renewable and Sustainable Energy Reviews* 60: 475–488. https://doi.org/10.1016/j.rser.2016.01.068.
47. D'Adamo, I., Falcone, P.M., Gastaldi, M., and Morone, P. (2020). RES-T trajectories and an integrated SWOT-AHP analysis for biomethane. Policy implications to support a green revolution in European transport. *Energy Policy* 138: 111220. https://doi.org/10.1016/j.enpol.2019.111220.
48. Gottfried O., D. De Clercq, E. Blair, X. Weng, C. Wang, 2018. SWOT-AHP-TOWS analysis of private investment behavior in the Chinese biogas sector, *Journal of Cleaner Production*, 184: 632–647. https://doi.org/10.1016/j.jclepro.2018.02.173

49. Mukeshimana, M.-C., Zhen-Yu, Z., Munir, A., and Irfan, M. (2021). Analysis on barriers to biogas dissemination in Rwanda: AHP approach. *Renewable Energy* 163: 1127–1137. https://doi.org/10.1016/j.renene.2020.09.051.
50. Rudolf, J. and Udovč, A. (2022). Introducing the SWOT scorecard technique to analyse diversified AE collective schemes with a DEX model. *Sustainability* 14 (2): 785. https://doi.org/10.3390/su14020785.
51. Weihrich H. The TOWS matrix–a tool for situational analysis (1982) *Long Range Planning*, 15 (2), pp. 54–66https://doi.org/10.1016/0024-6301(82)90120-0
52. Figueira, J. and Roy, B. (2002). Determining the weights of criteria in the ELECTRE type methods with a revised Simos' procedure. *European Journal of Operational Research* 139 (2): 317–326. https://doi.org/10.1016/S0377-2217(01)00370-8.
53. Falcone, P.M., Tani, A., Tartiu, V.E., and Imbriani, C. (2020). Towards a sustainable forest-based bioeconomy in Italy: findings from a SWOT analysis. *Forest Policy and Economics* 110: 101910. https://doi.org/10.1016/j.forpol.2019.04.014.
54. Patidar, R., Agrawal, S., and Pratap, S. (2018). Development of novel strategies for designing sustainable Indian agri-fresh food supply chain. *Sadhana – Academy Proceedings in Engineering Sciences* 43 (10): 167. https://doi.org/10.1007/s12046-018-0927-6.
55. Olszewski, R., Pałka, P., Wendland, A., and Majdzińska, K. (2021). Application of cooperative game theory in a spatial context: an example of the application of the community-led local development instrument for the decision support system of biogas plants construction. *Land Use Policy* 108: 10548. https://doi.org/10.1016/j.landusepol.2021.105485.
56. Rozakis, S., Bartoli, A., Dach, J. et al. (2021). Policy impact on regional biogas using a modular modeling tool. *Energies* 14 (13): 3738. https://doi.org/10.3390/en14133738.
57. Raven, R.P.J.M. (2004). Implementation of manure digestion and co-combustion in the Dutch electricity regime: a multi-level analysis of market implementation in the Netherlands. *Energy Policy* 32: 29–39. https://doi.org/10.1016/S0301-4215(02)00248-3.
58. Raven, R.P.J. (2007). Niche accumulation and hybridisation strategies in transition processes towards a sustainable energy system: an assessment of differences and pitfalls. *Energy Policy* 35: 2390–2400. https://doi.org/10.1016/j.enpol.2006.09.003.
59. Raven, R.P.J.M. and Geels, F.W. (2010). Socio-cognitive evolution in niche development: comparative analysis of biogas development in Denmark and the Netherlands (1973–2004). *Technovation* 30: 87–99. https://doi.org/10.1016/j.technovation.2009.08.006.
60. Marks, S., Dach, J., Morales, F.J.F. et al. (2020). New trends in substrates and biogas systems in Poland. *Journal of Ecological Engineering* 21 (4): 19–25. https://doi.org/10.12911/22998993/119528.
61. Hoolohan, C., Soutar, I., Suckling, J. et al. (2018). Stepping-up innovations in the water-energy-food nexus: a case study of anaerobic digestion in the UK. *The Geographical Journal* 185: 1–15. https://doi.org/10.1111/geoj.12259.
62. Raven, J.A., Cockell, C.S., and De La Rocha, C.L. (2008). The evolution of inorganic carbon concentrating mechanisms in photosynthesis. *Philosophical Transactions of the Royal Society B: Biological Sciences* 363 (1504): 2641–2650. https://doi.org/10.1098/rstb.2008.0020.
63. Urbaniec, M. (2015). Towards sustainable development through eco-innovations: drivers and barriers in Poland. *Economics and Sociology* 8 (4): 179–190. https://doi.org/10.14254/2071-789X.2015/8-4/13.
64. Negro, S.O. and Hekkert, M.P. (2008). Explaining the success of emerging technologies by innovation system functioning: the case of biomass digestion in Germany. *Technology Analysis and Strategic Management* 20 (4): 465–482. https://doi.org/10.1080/09537320802141437.
65. Ottosson, M., Magnusson, T., and Andersson, H. (2020). Shaping sustainable markets—A conceptual framework illustrated by the case of biogas in Sweden. *Environmental Innovation and Societal Transitions* 36: 303–320. https://doi.org/10.1016/j.eist.2019.10.008.
66. Lazarevic, D. and Valve, H. (2020). Niche politics: biogas, technological flexibility and the economisation of resource recovery. *Environmental Innovation and Societal Transitions* 35: 45–59. https://doi.org/10.1016/j.eist.2020.01.016.

67. Hermans, F. (2018). The potential contribution of transition theory to the analysis of bioclusters and their role in the transition to a bioeconomy. *Biofuels, Bioproducts and Biorefining* 12 (2): 265–276. https://doi.org/10.1002/bbb.1861.
68. Scheiterle, L., Ulmer, A., Birner, R., and Pyka, A. (2018). From commodity-based value chains to biomass-based value webs: the case of sugarcane in Brazil's bioeconomy. *Journal of Cleaner Production.* 172: 3851–3863. https://doi.org/10.1016/j.jclepro.2017.05.150.
69. Chodkowska-Miszczuk, J., Martinat, S., and Cowell, R. (2019). Community tensions, participation, and local development: factors affecting the spatial embeddedness of anaerobic digestion in Poland and the Czech Republic. *Energy Research and Social Science* 55: 134–145. https://doi.org/10.1016/j.erss.2019.05.010.
70. Tamburini, E., Gaglio, M., Castaldelli, G., and Fano, E.A. (2020). Biogas from agri-food and agricultural waste can appreciate agro-ecosystem services: the case study of Emilia Romagna region. *Sustainability* 12 (20): 8392. pp. 1–15; https://doi.org/10.3390/su12208392.

14

Social Aspects of Agricultural Biogas Plants

Wojciech Czekała

Department of Biosystems Engineering, Poznań University of Life Sciences, ul. Wojska Polskiego 50, 60-627 Poznań, Poland

14.1 Introduction

Waste management is one of the most important environmental protection issues [1]. The amount and variety of waste increase yearly, making it more challenging to manage [2]. The problem is global and occurs everywhere on Earth. Activities related to the idea of the circular economy allow, to some extent, solving the problem or reducing its severity [3]. Therefore, such activities should be undertaken, including waste management and renewable energy production.

Agricultural biogas plants are installations that produce renewable energy in an environmentally friendly manner. Profit from this type of installation comes primarily from the sale of the products of the anaerobic digestion process. In contrast, environmental benefits come mainly from reducing fossil fuel consumption and the possibility of using waste to produce biogas [4]. The exploitation of biogas plants can bring environmental, financial, and social benefits [5, 6]. The social aspects are often crucial regarding the feasibility of building and operating a biogas plant. Actions led by the local community can delay and, in exceptional cases, even wholly block a planned biogas plant construction project. Social resistance and protests can be critical in starting the construction of a biogas plant, although the installation brings numerous benefits. The study aim to discuss agricultural biogas plants' social benefits and acceptability.

Biogas Plants: Waste Management, Energy Production and Carbon Footprint Reduction,
First Edition. Edited by Wojciech Czekała

14.2 The Benefits of Agricultural Biogas Plants for Society

14.2.1 Biogas Plant as a Renewable Energy Production Facility

Burning fossil fuels, carbon emissions, and reduction of biodiversity are just some of the causes of increasing climate change. This problem was recognized in the last century, and its severity increases yearly. Because these changes adversely affect the environment and society, measures are being taken to reduce them. The municipal and industrial sectors need increasing electricity and heat [7]. Accordingly, various energy sources are being used, among which renewable energy sources play a significant role. However, their resources are limited, and the use of such fuels adversely affects the environment. Therefore, alternative sources of energy generation are being sought, where renewable energy sources play a unique role [8].

All biomass energy carriers can be converted into biofuels [9]. The group includes solid, liquid, and gaseous biofuels. One type of plant that produces renewable energy is the agricultural biogas plant, where anaerobic digestion occurs, and its main product is biogas [10]. After the purification of the gas mixture, biomethane is produced and characterized by similar parameters to natural gas. Another direction of biogas processing is the generation of electricity and heat through cogeneration. Regardless of the direction of biogas utilization, its products are classified as being generated by renewable means.

14.2.2 Reducing the Negative Impact of Waste on the Environment

Environmental issues are gaining importance as awareness of sustainability and environmental protection increases [11]. Rational waste management and renewable energy production are among the most critical topics. Biogas plants use a variety of raw materials to produce energy. Agricultural biogas plants are sometimes seen as installations that primarily use raw materials from targeted crops, such as corn, and natural fertilizers generated on farms, mainly manure, and slurry. Some installations use these types of raw materials, but increasingly a variety of wastes are being used for energy production [12].

Biodegradable waste are characterized by properties that allow it to be subjected to recovery processes, including biological treatment processes, which include composting and anaerobic digestion. The discussed group of waste is characterized primarily by a high content of organic matter and susceptibility to decomposition. Consequently, these wastes are suitable for biogas production [13]. However, this requires particular documents and permits, and once obtained, the biogas plant is treated as a waste management site. Sometimes, waste are managed illegally, an example being liquid waste discharge directly into rivers or water reservoirs. In contrast, solid waste can be dumped without proper permits or left in the forest. It should therefore be emphasized that proper management of biodegradable waste will, to some extent, reduce or even eliminate the risk of contamination of the local environment [14]. It is possible if a biogas plant accepts the waste at favorable prices and manages it. The presence of a biogas plant also provides an opportunity to reduce the risk of surface and groundwater contamination from nuisance and improperly applied waste on the soil surface or stored in unsuitable locations. Another aspect related to waste management is the topic of gaseous emissions [15]. These problems are often cited as the most troublesome.

14.2.3 Create Markets for Substrates Used in Biogas Production

Various substrates for agricultural biogas production are large [16]. They differ in many properties, including their state of aggregation, dry weight, organic matter content, pH, chemical composition, and impurities. Due to the need to ensure the profitability of an investment such as a biogas plant and in line with circular economy efforts, wastes are increasingly being used to produce energy in biogas plants [5, 6]. Therefore, the anaerobic digestion process allows the production of biogas and the management of many organic wastes.

For a biogas plant, the most crucial challenge is obtaining the right amount of feedstock, which is the basis for the proper operation of the plant [17]. Bearing in mind that a typical 1 MW biogas plant requires several tens of Mg of feedstock per day, appropriate logistical measures are required to obtain the necessary amount of feedstock for the digesters. It is recommended that substrates, especially those with lower biogas yields, be transported from distances of up to 20 km and preferably less than 10 km. It concerns the cost of purchasing substrates accounting for up to 40% of the cost of operating a biogas plant. Therefore, looking for readily available substrates with a high energy yield and a relatively low price is necessary.

A good solution is to cooperate with local farmers and businesses, allowing for a systematic supply of raw materials. Another advantage of sourcing raw materials locally is the reduction in expenses associated with their transportation [18]. In particular, the cost of transporting such a quantity of substrates significantly affects the balance of plant operation. It also has a positive impact on the environment through, among other things, lower fuel consumption and reduced emissions of vehicles transporting substrates to the biogas plant. Such cooperation provides farmers and companies with a systematic income, as contracts are often contracted for a year or several years.

14.2.4 Integration with Agro-Industrial Plants

The variety of substrates used for biogas production is enormous. Despite this, new substrates that can be processed in biogas plants are still being sought. These activities are aimed at increasing the profitability of the plant's operation, which should promote environmental improvement resulting in rational waste management. One solution is to combine biogas production with a functioning agro-industrial plant [19]. This idea is already being applied in practice, resulting in increased interest in the construction of biogas plants by agri-food companies [20].

Among the establishments interested in such a solution are fruit and vegetable processing plants, dairies, distilleries, breweries, bakeries, or fish and meat processing plants. Each plant produces daily waste, which must be managed following the law. Inputs to biogas plants cooperating with an agri-food manufacturer are usually efficient substrates. It can be blood, fat-rich substrates, gastrointestinal contents, and slaughterhouse entrails. For a distillery, it is a stock, and for a sugar mill, beet pulp.

Because they are produced systematically, and their properties are similar and specific to each other, biogas plant owners can be interested in obtaining them. An even better solution may be to build a biogas plant next to a functioning agri-food enterprise [21]. It saves money in the long run that the plant would otherwise have to spend on fees associated with managing the waste generated. In addition, the benefit comes from the reduced transportation of the generated waste, as such facilities are built on the plant's premises or close to it. Another advantage for a company with biogas plants is using electricity from cogeneration. This is the right step to reduce the steadily increasing cost of energy purchases. An additional benefit is using the generated heat for the plant's needs. Given the benefits presented, it is expected that investments in biogas plants will, over time, have a positive effect on many areas of the agri-food company's operations.

14.2.5 Production and Use of Electricity

In Poland, the most popular way to use of biogas produced in agricultural biogas plants is to process it in a cogeneration. Cogeneration allows for energy production in two forms: electricity and heat [22]. Biogas should be purified before being directed to the engine. Part of the electricity generated (usually several percent) is consumed for the biogas plant's needs, while the remainder is fed into the electricity grid and sold to the local power system operator.

When the engine is running, a generator is driven, from which electrical energy, after appropriate parameterization, is injected into the power grid. Electricity at the generator's output has different parameters from the current flowing in the power grid. For this reason, the voltage and frequency of the electricity produced must be adjusted before it is injected into the grid. This process occurs at a transformer station, where the current is sent directly to the consumers. For each unit of electricity produced, the biogas plant receives money. The amount depends primarily on the price of energy on the market and the support system for the generated energy classified as renewable in force in the country. Regardless of the support system, revenues from the sale of electricity are the largest for biogas plants operating in cogeneration. One interesting solution is to use the generated electricity on the premises of a given company [23]. For example, a biogas plant operating at an agri-food plant can support a given enterprise in the context of energy supply. A similar situation can be a solution for waste management facilities, where the electricity generated can support, for example, municipal waste processing machinery.

14.2.6 Production and Use of Heat

The second type of energy produced by cogeneration is heat. Heat is produced by the combustion of biogas in an internal combustion engine due to cooling the engine and exhaust gases. The heated water is approximately 90 °C and can be used as an energy source. The most rational solution is to heat the digesters, where the anaerobic digestion occurs under thermophilic conditions. This is because this process requires more heat than anaerobic digestion under mesophilic conditions.

Nevertheless, in either case, regardless of the temperature at which the process is conducted, the vast majority of the heat requires further development. For the use of heat from cogeneration, the owner of a biogas plant can receive additional financial benefits independent of the profit from selling electricity. Therefore, it is in his interest to develop as much heat as possible, benefiting from its sale and rational management.

Such a large amount of heat is produced in biogas plants that it can and should be efficiently managed, including outside the biogas plant [24]. The most obvious solution is to feed it into the local district heating network. However, the problem is the distance of such a network from the biogas plant, which should be relatively close, preferably less than 1 km. For greater distances, transmission losses may limit the profitability of the investment. In exceptional cases, a connection can be made to these properties when the district heating network is far away and relatively close to it is located in multi-apartment buildings. This solution helps ensure a stable heat supply, and the local community reduces the expenses associated with its purchase. Heat supply to nearby homes can be free or much cheaper than coal or gas heating.

14.2.7 Possibility of Biomethane Production

There are several basic directions for biogas management. As mentioned earlier, one of the most common options is using biogas in cogeneration. Another possibility is the injection of the produced and purified biogas into pipeline networks. In recent years, there has been an increase in the production and use of biomethane, which, due to its properties and multiple directions of use, allows for the decarbonization of areas of the electricity or transportation sectors, among others.

Biogas upgrading is the most common method of producing biomethane [25]. This method involves purifying the biogas produced in anaerobic digestion from carbon dioxide. The produced biomethane is characterized by parameters similar to those of natural gas. For this reason, the fuel in question is often referred to as renewable natural gas, the main component of which is methane. Due to its high methane content, biomethane is an excellent alternative to cold gas. Like natural gas, biomethane can also produce heat and electricity and plays a vital role in transportation. It should also be mentioned that the increase in the number of biogas and biomethane plants can benefit the development of gas networks, both locally and nationally.

In the current geopolitical situation, topics concerning energy supply assume particular importance. Biogas plants play a unique role in this context. First, producing electricity, heat, and biomethane provides opportunities for local use. It should be emphasized that if biogas plants have adequate substrates, the supply of energy or biomethane to the grid will also be systematic and stable. This type of plant is a significant advantage over other renewable energy installations.

14.2.8 Local Fuel in Developing Countries

Another option for utilizing the generated biogas is meal preparation [26]. This solution is often practiced in developing countries and requires pipelines to connect the digesters of the biogas plant to the meal preparation site [27]. Cookstoves and ovens powered by biogas operate similarly to conventional ones powered by natural gas. The cookstoves can even have several burners characterized by different fuel consumption, which is related to the biogas pressure, and the amount of gas entering is controlled by a nozzle. Among the advantages of this solution is the possibility of adapting conventional equipment to use biogas. Another advantage is the ability to regulate the flow of biogas, allowing, among other things, simmering. Among the most significant disadvantages of this solution are the investment costs associated with purchasing the digester, cookstoves/oven, and gas supply system. Therefore, control and safety measures are crucial for this type of solution.

14.2.9 Production of Valuable Fertilizer

One way to increase yield per unit area is through rational fertilization [28]. Natural fertilizers are associated with livestock production, primarily manure, slurry, and artificial fertilizers. Considering the increase in prices and the possible adverse environmental impact of artificial fertilizers, mainly nitrogen fertilizers, attention is increasingly turning to organic fertilizers [29]. One of the alternatives is the use of digestate, which is a byproduct of the anaerobic digestion process. Due to its numerous advantages, the product in question could become another attractive source of revenue for biogas plants.

The digestate, a byproduct of the anaerobic digestion process in a biogas plant, is a perfect organic fertilizer, and its properties have been repeatedly analyzed and discussed in the literature. On the one hand, the fertilizing properties of the digestate ensure the maintenance of soil fertility, and on the other hand, the digestate is competitively priced compared to mineral fertilizers. Therefore, as a valuable fertilizer, farmers can successfully use digestate, mainly on farms near biogas plants. Depending on the capacity of the biogas plant and the amount of waste received, the amount of digestate generated is several tens of thousands of Mg per year for a 1 MW biogas plant [30].

14.2.10 Creating New Jobs for the Local Community

Like most energy production facilities, agricultural biogas plants need qualified personnel to operate them. For this reason, energy production enterprises are a potential place to hire more people. For practical reasons, these should be people who live relatively close to the installation and have the appropriate qualifications. Staffing at the biogas plant should start with the person in charge of the entire installation, as they are responsible for all activities performed at the site. In addition, there is a need for personnel dealing with the reception and management of substrates for biogas production. An important place of work falls in the biogas plant to the person responsible for inspecting the technical condition of machinery and installations responsible for the proper operation of the entire plant. A biogas company should employ a person who supervises and monitors the anaerobic digestion process [31]. It is worth mentioning that external companies are providing such services. In addition to substantive and technical services, essential issues concern formal and administrative services. An important element in a biogas plant is working with the documents collected by the plant and those required by government authorities. In conclusion, it should be noted that agricultural biogas plants, due to their nature of work, create attractive jobs, especially for the local community.

14.2.11 Development of Nearby Infrastructure and Companies

Like most manufacturing enterprises, a biogas plant needs constant cooperation with other companies. Such action leads to local economic development, which creates favorable conditions for the development of companies. During the construction phase, the plant can benefit from construction companies responsible for implementing the project. In addition, before the construction phase, it is necessary to involve companies in preparing the documentation, primarily on environmental aspects. During the operation phase of the biogas plant, it is necessary to establish cooperation with other companies. First, the companies involved in transporting substrates and waste from the place of their generation or storage to the biogas plant. Second, it is necessary to ensure the collection and management of digestate.

Another issue is the commercial use of heat, a byproduct of cogeneration power generation. When it is impossible to supply heat to residents, other solutions should be sought to use it efficiently. The heat generated by cogeneration engines can also be used for processes that require large amounts of this type of energy. A rational solution would be to use the heat in belt dryers. Among the raw materials most often deprived of water at this type of installation operating at biogas plants are sawdust, the solid fractions of the digestate or biomass used to produce solid biofuels. An alternative solution requiring significant amounts of heat is drying crops. For example, immediately after harvesting, they require grain to be redried. Providing this service to nearby farmers or plants producing briquettes, pellets, or fertilizer is another solution to benefit the local community [32].

14.2.12 Tax Revenues to the Budget of Local Government Units

An essential element is the payment of taxes by biogas plants. Revenue-generating enterprises are required to remit specific amounts as taxes. Part of these fees should go to the local government units where the biogas plant is located. The local administration can allocate this money for investments that benefit residents of the area near the biogas plant. Good examples include infrastructure development, environmental protection, or using the money for educational knitting. Agricultural biogas plants can therefore be an essential element in local authorities' budget, generating numerous benefits for the community living in the administrative area.

14.3 Social Acceptability of Agricultural Biogas Plants

Some aspects of the local community resulting in the operation of biogas plants were discussed in Section 14.2. Despite the many benefits directly related to the construction and operation of biogas plants, these installations often face public dissatisfaction. This resistance does not only apply to biogas plants. Similar problems often arise with installations that process waste, produce biofuels, or produce energy. If constructing a biogas plant raises concerns or doubts among some public members, it is the investor's job to explain to interested residents that biogas plants are installations that positively impact the environment and not the other way around. The sooner an agreement is reached between the investor and the residents, the sooner the biogas plant can be built, bringing benefits to society. Selected arguments put forward by some residents living in areas where a biogas plant is planned and are discussed in Sections 14.3.1–14.3.5.

14.3.1 Fear of Something New

The public's reluctance to use biogas plants in many cases may be due to a lack of awareness about the operation and benefits of biogas plants. Fear of something new can undoubtedly be a reason why some residents are not satisfied with the investment. The public often expresses opinions about a biogas plant or any other installation without knowing how it works.

14.3.2 Concerns About Unpleasant Odors

It is, by far, one of the most common arguments raised by residents of nearby communities. Unpleasant odors are primarily associated with transporting substrates to the biogas plant and their eventual storage before being directed to the digester. It should be noted, however, that in some well-managed biogas plants, substrates are not stored, as they go immediately to the digesters. Ideally, all the raw materials from which biogas is produced should be used immediately. In this case, emissions are kept to a minimum. For the owner of a biogas plant, this solution is advisable because, in this case, he/she does not need to have part of the permits and additionally reduces the expenses arising from the storage of substrates and preparation for their use. Another factor in favor of using raw materials as quickly as possible is their biodegradation, which results in, among other things, unpleasant odors. Biomass undergoing decomposition reduces its energy value, translating into losses for the biogas plant. In summary, properly functioning biogas plants significantly reduce animal manure and organic waste odor nuisance.

14.3.3 Concerns About Contamination of Soils and Groundwater When Using Digestate as Fertilizer

For years, natural fertilizers such as manure and slurry have been used in agriculture for fertilizer purposes. This is because they are abundant in minerals, manure, and organic matter. Another factor determining the popularity of the fertilizers in question is their availability in agricultural areas. Due to increasingly stringent environmental requirements, numerous restrictions on using the fertilizers in question are noted. It is related, among other things, to the possibility of uncontrolled leakage into the soil, surface water, and groundwater and the emission of odors or gases, including methane, which is a greenhouse gas. Digestate may prove to be a solution to these problems. Digestate is an organic fertilizer characterized by a significant content of organic matter and macro- and micronutrients [33]. Its quantity is generated at a biogas plant site, and its composition is closely related to the substrates used for biogas production and the course of the technological process. Numerous studies indicate that digestate can be successfully used as a fertilizer. Because the dry matter content of the digestate usually does not exceed 10%, the digestate can be applied to soils similarly to a slurry. An indifferent or slightly alkaline reaction has a beneficial effect on the pH of soils. To reduce or eliminate the risks associated with the application of digestate, it is recommended to control its parameters, especially concerning the content of heavy metals and microbiological contaminants.

14.3.4 Concerns About Declining Property Values Around Biogas Plants

One of the frequently cited objections related to the operation of biogas plants is concerns about the decline in the value of property located around the biogas plant. This topic is often compared to the situation occurring near landfills. One of the more frequently cited concerns is the potential for biogas plants to generate odors, as explained in Section 14.3.2 [34]. However, regulations strictly indicate the requirements for siting biogas plants to limit the adverse effects on the public.

14.3.5 Concerns About the Destruction of Access Roads

Another argument raised by the public is the fear of increased truck and tanker traffic bringing substrates to the methane digestion process. The fact is that feedstocks are delivered to biogas plants almost every day. However, it should be emphasized that these are usually two to four trips per day for typical plants. This number should be considered marginal in the context of vehicle traffic in each municipality.

With adequately functioning biogas plants, the symptoms mentioned above are somewhat unjustified. An action that allows solving many problems and misunderstandings is certainly public consultations. During them, investors can be asked questions about all aspects, whether social, environmental, economic, or legal. In addition, the community should be informed about many benefits of building and operating a biogas plant. Indeed, attention should also be paid to training aimed at representatives of local government units that impact the creation or blocking of investments such as biogas plants.

In contrast, care should be taken to ensure proper technological supervision of biogas plants. In this context, it is vital to maintain and control the basic parameters of the process, as well as for qualified people to solve any problems that arise on an ongoing basis. It would guarantee a properly operating plant, benefiting investors and residents.

14.4 Conclusion

Biogas plants are unique installations where valuable products such as biogas, biomethane, electricity, heat, or digestate can be obtained from biodegradable waste. In addition to the numerous economic and environmental benefits, the benefits to the local community cannot be overlooked. In contrast, it is essential to mention the public's concerns about building and operating a biogas plant. It should be emphasized that these installations are safe for the environment in the case of a well-designed and built biogas plant and with appropriate technological support. The public living near them can benefit in numerous ways. Biogas plants can and should function as multidirectional enterprises. This is because they are waste management sites where biogas, electricity, heat, and valuable fertilizers are produced. The installations should become an impetus for developing and improving living conditions for the local community. It should be emphasized that, among other things, thanks to biogas plants, as well as other RES installations, climate policy goals can be achieved. Regardless of any reservations, an investor preparing an investment should cooperate with local authorities and the local community. This is because it is a matter of conducting activities to clarify all the doubts pointed out and demonstrate the numerous benefits of having a biogas plant in a municipality.

References

1. Puntillo, P. (2022). Circular economy business models: toward achieving sustainable development goals in the waste management sector-empirical evidence and theoretical implications. *Corporate Social Responsibility and Environmental Management* 1–14. https://doi.org/10.1002/csr.2398.
2. Arena, M., Azzone, G., Grecchi, M., and Piantoni, G. (2021). How can the waste management sector contributeto overcoming barriers to the circular economy? *Sustainable Development* 29 (6): 1062–1071. https://doi.org/10.1002/sd.2202.
3. Loizia, P., Neofytou, N., and Zorpas, A.A. (2019). The concept of circular economy strategy in food waste management for the optimization of energy production through anaerobic digestion. *Environmental Science and Pollution Research* 26: 14766–14773. https://doi.org/10.1007/s11356-018-3519-4.
4. Awadalla, O.A., Atawy, W.A., Bedaiwy, M.Y. et al. (2023). Anaerobic digestion of lignocellulosic waste for enhanced methane production and biogas-digestate utilization. *Industrial Crops and Products* 195: 116420. https://doi.org/10.1016/j.indcrop.2023.116420.
5. Czekała, W., Pulka, J., Jasiński, T. et al. (2023a). Waste as substrates for agricultural biogas plants: a case study from Poland. *Journal of Water and Land Development* 56 (I–III): 1–6. https://doi.org/10.24425/jwld.2023.143743.
6. Czekała, W., Nowak, M., and Piechota, G. (2023b). Sustainable management and recycling of anaerobic digestate solid fraction by composting: a review. *Bioresource Technology* 375: 128813. https://doi.org/10.1016/j.biortech.2023.128813.
7. Andrei, M., Thollander, P., and Sannö, A. (2022). Knowledge demands for energy management in manufacturing industry – a systematic literature review. *Renewable and Sustainable Energy Reviews* 159: 112168. https://doi.org/10.1016/j.rser.2022.112168.
8. Islam, M.K., Hassan, N.M.S., Rasul, M.G. et al. (2022). Green and renewable resources: an assessment of sustainable energy solution for Far North Queensland, Australia. *International Journal of Energy and Environmental Engineering* 1–9. https://doi.org/10.1007/s40095-022-00552-y.
9. Hakeem, I.G., Sharma, A., Sharma, T. et al. (2022). Techno-economic analysis of biochemical conversion of biomass to biofuels and platform chemicals. *Biofuels, Bioproducts & Biorefining* https://doi.org/10.1002/bbb.2463.
10. Calbry-Muzyka, A., Madi, H., Rüsch-Pfund, F. et al. (2022). Biogas composition from agricultural sources and organic fraction of municipal solid waste. *Renewable Energy* 181: 1000–1007. https://doi.org/10.1016/j.renene.2021.09.100.
11. Setoguchi, A., Oishi, K., Kimura, Y. et al. (2022). Carbon footprint assessment of a whole dairy farming system with a biogas plant and the use of solid fraction of digestate as a recycled bedding material. *RCR Advances* 15: 200115. https://doi.org/10.1016/j.rcradv.2022.200115.
12. Ferreira, L.O., Astals, S., and Passos, F. (2022). Anaerobic codigestion of food waste and microalgae in an integrated treatment plant. *Journal of Chemical Technology and Biotechnology* 97 (6): 1545–1554. https://doi.org/10.1002/jctb.6900.
13. Otieno, E.O., Kiplimo, R., and Mutwiwa, U. (2023). Optimization of anaerobic digestion parameters for biogas production from pineapple wastes codigested with livestock wastes. *Heliyon* 9 (3): e14041. https://doi.org/10.1016/j.heliyon.2023.e14041.

14. Cusenza, M.A., Longo, S., Guarino, F., and Cellura, M. (2021). Energy and environmental assessment of residual biowaste management strategies. *Journal of Cleaner Production* 285: 124815. https://doi.org/10.1016/j.jclepro.2020.124815.
15. Wu, C., Shu, M., Liu, X. et al. (2020). Characterization of the volatile compounds emitted from municipal solid waste and identification of the key volatile pollutants. *Waste Management* 103: 314–322. https://doi.org/10.1016/j.wasman.2019.12.043.
16. Tumusiime, E., Kirabira, J.B., and Musinguzi, W.B. (2019). *Energy Reports* 5: 579–583. https://doi.org/10.1016/j.egyr.2019.05.002.
17. Borek, K. and Romaniuk, W. (2020). Biogas installations for harvesting energy and utilization of natural fertilisers. *Agricultural Engineering* 24: 1–14. https://doi.org/10.1515/agriceng-2020-0001.
18. Marks, S., Dach, J., Fernandez Morales, F.J. et al. (2020). New trends in substrates and biogas systems in Poland. *Journal of Ecological Engineering* 21 (4): 19–25.
19. Keerthana Devi, M., Manikandan, S., Oviyapriya, M. et al. (2022). Recent advances in biogas production using agro-industrial waste: a comprehensive review outlook of techno-economic analysis. *Bioresource Technology* 363: 127871. https://doi.org/10.1016/j.biortech.2022.127871.
20. Czekała, W. (2018). Agricultural biogas plants as a chance for the development of the Agri-food sector. *Journal of Ecological Engineering* 19 (2): 179–183. https://doi.org/10.12911/22998993/83563.
21. Piwowar, A. (2020). Agricultural biogas-an important element in the circular and low-carbon development in Poland. *Energies* 13 (7): 1733. https://doi.org/10.3390/en13071733.
22. Mebarkia, M. (2022). Energy recovery - waste heat recovery. In: *Thermodynamics of Heat Engines*. Wiley. https://onlinelibrary.wiley.com/doi/book/10.1002/9781394188192.
23. Abanades, S., Abbaspour, H., Ahmadi, A. et al. (2022). A conceptual review of sustainable electrical power generation from biogas. *Energy Science & Engineering* 10 (2): 630–655. https://doi.org/10.1002/ese3.1030.
24. Weinand, J.M., McKenna, R., Karner, K. et al. (2019). Assessing the potential contribution of excess heat from biogas plants toward decarbonizing residential heating. *Journal of Cleaner Production* 238: 117756. https://doi.org/10.1016/j.jclepro.2019.117756.
25. Nguyen, L.N., Kumar, J., Vu, M.T. et al. (2021). Biomethane production from anaerobic codigestion at wastewater treatment plants: a critical review on development and innovations in biogas upgrading techniques. *Science of the Total Environment* 765: 142753. https://doi.org/10.1016/j.scitotenv.2020.142753.
26. Negash, D., Abegaz, A., and Smith, J.U. (2021). Environmental and financial benefits of improved cookstove technologies in the central highlands of Ethiopia. *Biomass and Bioenergy* 150: 106089. https://doi.org/10.1016/j.biombioe.2021.106089.
27. Karanja, A. and Gasparatos, A. (2019). Adoption and impacts of clean bioenergy cookstoves in Kenya. *Renewable and Sustainable Energy Reviews* 102: 285–306. https://doi.org/10.1016/j.rser.2018.12.006.
28. Garbowski, T., Bar-Michalczyk, D., Charazinska, S. et al. (2023). An overview of natural soil amendments in agriculture. *Soil and Tillage Research* 225: 105462. https://doi.org/10.1016/j.still.2022.105462.
29. Czekała, W. (2022). Digestate as a source of nutrients: nitrogen and its fractions. *Water* 14 (24): 4067. https://doi.org/10.3390/w14244067.
30. Czekała, W., Lewicki, A., Pochwatka, P. et al. (2020). Digestate management in polish farms as an element of the nutrient cycle. *Journal of Cleaner Production* 242 (118454): 1–9. https://doi.org/10.1016/j.jclepro.2019.118454.
31. Kozlowski, K., Dach, J., Lewicki, A. et al. (2018). Laboratory simulation of an agricultural biogas plant start-up. *Chemical Engineering and Technology* 41 (4): 711–716. https://doi.org/10.1002/ceat.201700390.
32. Czekała, W. (2021). Solid fraction of digestate from biogas plant as a material for pellets production. *Energies* 14: 5034. https://doi.org/10.3390/en14165034.

33. Czekała, W., Jasiński, T., Grzelak, M. et al. (2022). Biogas plant operation: digestate as the valuable product. *Energies* 15: 8275. https://doi.org/10.3390/en15218275.
34. Keck, M., Mager, K., Weber, K. et al. (2018). Odor impact from farms with animal husbandry and biogas facilities. *Science of the Total Environment* 645: 1432–1443. https://doi.org/10.1016/j.scitotenv.2018.07.182.

15

Practices in Biogas Plant Operation: A Case Study from Poland

Tomasz Jasiński[1†], Jan Jasiński[2], and Wojciech Czekała[2]

[1] *Tomasz Jasiński Biogas Consulting, Nowe, Poland*

[2] *Department of Biosystems Engineering, Poznań University of Life Sciences, ul. Wojska Polskiego 50, 60-627 Poznań, Poland*

15.1 Introduction

Electricity is needed every day, both in industry and municipal activities. In addition to electricity, thermal energy is used in many aspects of life and industry. Currently, when humanity struggles with the climate and energy crisis, an increased emphasis on renewable energy sources is being placed. Biogas plants are one of the installations where electricity and heat can be produced [1]. These installations can have a positive impact on both society and the environment. The owners of these installations can count on profits from the sale of electricity and heat. An additional income is also the sale of digestate residues from the anaerobic digestion process [2]. The revenues from biodegradable waste management services, primarily from the agri-food industry, should not be forgotten. Biogas plants can process waste from plants dealing with breeding, processing, and food production [3]. It is one of the most critical aspects of these installations.

Running an installation based on living organisms places much responsibility on employees. It involves having knowledge and experience in the biology and biochemistry

† Deceased

Biogas Plants: Waste Management, Energy Production and Carbon Footprint Reduction,
First Edition. Edited by Wojciech Czekała

of the process [4]. One should be aware that the expected effects depend on a sequence of events that are closely related to each other. It includes maintaining proper conditions for the life of bacteria involved in the entire anaerobic digestion process. The observance of primary factors such as temperature, pH, or dry matter content should be considered [5]. Therefore, the issue of feeding the installation is critical – the regularity of deliveries and their appropriate composition [6]. In addition to the typical biological and chemical aspects, it is crucial to ensure the necessary procedural requirements for the waste used. It is related to the catalog of waste approved for use by agricultural biogas plants. It is important because the installation may lose its license to conduct business activity in this area due to using impermissible groups of waste.

As seen from the aforementioned information, only completing all administrative, technical, and technological aspects gives a chance to succeed. Failure to follow the requirements of the process often results in a significant reduction in the project's profitability and, in extreme cases, a negative impact on the methanogenesis process or loss of authorization to run based on the existing law on the operation of agricultural biogas plants. The work aimed to discuss the operation principles of a biogas plant in Poland, which processes problematic waste into energy.

15.2 Legal Aspects Related to Running a Business in the Field of Biogas Production and Waste Management

If economically and environmentally beneficial, every form of obtaining energy is needed and should be developed. A suitable example in the group of renewable energy sources is biogas plants [7]. These are primarily installations that process biomass with the status of waste for energy production, especially waste that are difficult to use in other processes. Like any enterprise, an agricultural biogas plant must work according to applicable legal standards. Due to their specificity, biogas plants cover many legal areas, ranging from regulations related to energy, waste management, and fertilizers through many others related to Poland's membership in the European Union, along with national and local legal aspects. Among all legal acts, the most important is Ref. [8].

In addition to the legal regulations indicated by the European Union, national law is a significant aspect regulating the rules and functioning. The main legal act regulating the activity in the field of agricultural biogas and electricity production from agricultural biogas is the Act of 10 April 1997 – Energy law [9] and the Act of 20 February 2015 on renewable energy sources [10]. According to the latter of the cited legal acts, renewable energy sources are defined as "renewable, nonfossil energy sources, including wind energy, energy radiation of the sun, solar, aerothermal, geothermal, hydrothermal energy, hydropower, waves, currents error and tidal energy obtained from biomass, biogas, agricultural biogas, and bioliquid."

According to the latter legal act, agricultural biogas is defined as "gas obtained in the process of anaerobic digestion of agricultural raw materials, agricultural byproducts, liquid or solid animal manure, byproducts, waste or residues from the processing of agricultural products or forest biomass, or plant biomass collected from areas other than those recorded as agricultural or forestry, excluding biogas obtained from raw materials from landfills, as well as sewage treatment plants, including on-site sewage treatment plants for agricultural and food processing, where industrial sewage is not separated from other types of sludge and sewage."

A manufacturer conducting business activity in agricultural biogas must use only the substrates listed in the definition of agricultural biogas. In case of a breach of the obligation aforementioned, the Chief Director of the National Support Center for Agriculture is entitled to issue a decision prohibiting the manufacturer from conducting economic activity in agricultural biogas. Local law, which regulates the operation of such projects at the voivodeship, county, and commune levels, should also be added to all national legal conditions. They are essential because such installations run in the local environment. It includes aspects such as the proximity of human settlements, other enterprises, or conditions related to legally protected areas. Due to the activity's specificity and access to roads, the possibility of connecting to energy networks or heat distribution should be considered if a given entity plans to sell or share heat energy. Due to the type of activity, it is good if the installation is localized near agri-food processing plants. It results in significant savings in the availability and transport costs of the substrate used to produce biogas. It is also of immense importance in reducing CO_2 emissions related to transport. After obtaining "environmental decision," the investor can start designing the plant. The knowledge and experience of the designers are critical so that buildings and structures properly fulfill their tasks, starting from the infrastructure related to the course of the biogas production process, through the location of cogeneration modules, to the unloading places and office rooms.

15.2.1 Integrated Permit or Waste Processing Permit

At the stage of designing a biogas plant, a vision of how the installation will function is necessary. The basic assumptions should be based on the types of raw materials used for biogas production. Currently, for economic reasons, agricultural biogas plants are the best solution. Legal conditions allow for support for this group in the form of property rights granted to the electricity produced. An essential element consists of all the restrictions and requirements imposed on agricultural biogas plants. At this point, whether the installation is to work using only agricultural substrates, such as maize silage, slurry, manure, or other energy crops grown for this purpose, should be considered. It can also be a biogas plant that supports or is entirely based on agri-food industry waste.

As aforementioned, such premises should influence decisions at the design stage because an installation must be equipped with, among others, places for receiving the substrates planned for use or devices necessary for their preparation, e.g. shredders. However, the essential legal requirement is a waste treatment permit. Therefore, at the very beginning of the project, the investor should apply for "environmental decision." These offices, considering local conditions, issue a document based on relevant research and argumentation about the rightness of the investment. It is on their basis that the activity's type and manner of functioning are determined, including among others, the types of substrates used and the possibilities of electricity production.

The fundamental element of the installation operation is obtaining a waste processing documents. Based on the infrastructure and technologies of the anaerobic digestion process, it should be proven that the waste directed to it can be effectively processed in the biogas plant. Criteria are defined for the waste codes that can be disposed of at the facility. They must only apply to substrates that refer to biodegradable waste. However, it should be remembered that detailed legal regulations of supervisory institutions may prohibit individual waste, e.g. in agricultural biogas plants. Selected groups of waste that can be used in a biogas plant are presented as follows:

- from agriculture,
- from horticulture,
- from food processing,
- from preparation and processing of food products of animal origin,
- from preparation and processing of food products of plant origin,
- from sugar industry,
- from dairy industry,
- from baking and confectionery industry,
- from the production of alcoholic and nonalcoholic beverages.

From the beginning of the device and structure design, a biogas installation must assume that the batch will be weighed, i.e. it must be equipped with a certified scale. Another element is the place of unloading – part of the plant dedicated to unloading the arriving waste mass. At that point, storing systematically collected waste should not be assumed. It is related to problems connected to perishable items, but in contrast, because of changes related to the Act of 14 December 2012 on Waste [11], specific restrictions are on waste storage requirements. Ideally, the waste should be used directly in the process; alternatively, it should be directly fed to the so-called buffer tanks, where it is prepared (properly mixed and homogenized) before being fed to the fermentation tanks.

Technical aspects, such as the possibility of accepting waste, carrying out the anaerobic digestion process, technical possibilities related to the storage and sale of digestate but also legal aspects, i.e. the possibility of confirming Waste Transfer Notes, keeping records of suppliers, the amount of substrates received, and the amount of digestate pulp produced, should be included in the integrated permit.

15.2.2 Approval of the Plant by Veterinary Services for the Disposal of Waste of Animal Origin

Investors who intend to use waste of animal origin as input for biogas plants must be aware of the organizational and legal consequences. In the discussed case, the biogas plant is considered a plant for the use of animal byproducts. Based on the guidelines of the Act of 29 January 2004 on Veterinary Inspection [12] and acting in consultation with the County Veterinary Officer, the procedure should be defined. The most stringent restrictions apply to animal waste for biogas production. Its processing involves the need to adapt the biogas plant to veterinary requirements. Veterinary services in most plants producing this type of waste classify it in terms of the risk degree and, thus, how it must be managed. Waste animal tissue that can be used in a biogas installation is mainly in the second and third veterinary categories Act of 29 January 2004 on Veterinary Inspection [12]. In this case, it is essential to equip the biogas plant with devices to properly prepare the raw material before it is fed to the fermentation process. Regulation (EC) No 1069/2009 [13] define these conditions. In this case, category 3 waste must be pasteurized, while category 2 waste must be subjected to pressure sterilization, meeting the appropriate pressure, temperature, and time conditions. It is a significant expense for the operator of a biogas installation, but it is also necessary to dispose of this type of waste following the law and in an environmentally safe manner.

Such solutions, however, allow for a significant increase in waste, which can significantly affect both the yield of biogas and the economic effect of the installation operation. It should be mentioned that substrates based on waste animal tissue show a high degree of methane recovery.

Waste from meat production is an efficient substrate for biogas production. Its use, however, involves aspects that absolutely must be considered. It is essential to realize that the bacteria living in the fermenter should be prepared to process the substrate in question. Appropriate cultures of bacteria that can cope with this type of biomass must reproduce. For biogas production itself, this type of feedstock is very desirable. It is possible to obtain 200–250 m^3 of biogas with a methane content of approximately 55% from one ton of maize silage. However, from waste animal tissue – depending on the type – this value increases to 450–500 m^3. It is also vital that the methane content in biogas increases, usually reaching 70–75%.

Determining the type of animal byproducts and their categories allows for planning and building a line for processing the input material. If category 2 waste reaches the plant, according to the requirements of the Act of 29 January 2004 on Veterinary Inspection [12], it should be subjected to thermal treatment, i.e. so-called "sterilization." It is a process aimed at destroying all pathogens that could enter the process and thus the digestate. This method subjects the processed substrate to a temperature of 135 °C and a pressure of 3.5 bar. This process must take 20 minutes. After this time, the substrate can be added to the process. The described activities take place in sealed, pressurized tanks.

The biogas plant can also process category 3 waste. This group needs different treatments before being fed into the process. Waste must be thermally processed. In this case, the accepted mass must be heated to 70 °C and kept in such conditions for 30 minutes. After this time, it can be cooled down and fed to proper biogas production. It should be emphasized that when processing category 2 and 3 materials, the parameters of the process should be archived. It is necessary when conducting control activities by veterinary or other authorized services. Furthermore, it is about supporting the conditions set out in the Veterinary Act and the manner of managing this type of biological material.

15.2.3 Permit to Place Digestate on the Market

In functioning the anaerobic digestion process and obtaining the necessary permits and approvals, agricultural biogas plants must follow a logical sequence related to the gradual emergence of later production elements. At the stage of commissioning the installation, the biogas plant has a permit to accept and use waste and generate electricity from renewable energy sources before starting operation.

The digestate is produced due to the biogas production [14]. It is possible to obtain this product in a biogas plant after the process start-up and stabilization. Only after stabilizing the parameters, the biogas plant can continue obtaining a license to market digestate. Frequently, during this time, it uses the possibility of handling the digestate based on waste management because digestate can have such a status. This status is granted by the body issuing the waste processing permit. However, this is an embarrassing situation, as it limits the use of digestate on strictly defined land while keeping all the elements supported in such a permit.

A marketing permit is issued by the Ministry of Agriculture and Rural Development. As aforementioned, if the production process is stabilized, the procedure for obtaining such approval can be started. It is related to a strictly defined protocol, as installation operators, notifying their willingness to obtain a permit, must agree to collect representative digestate samples by qualified employees. The collected material is sent to independent state institutions, which, based on physicochemical and biological tests, issue opinions on the possibility of marketing and recognizing a given digestate as a fertilizer. In Poland, these are the Institute of Fertilization and Soil Science in Puławy, the National Veterinary Institute in Puławy, the Witold Chodźka Rural Medicine Institute in Lublin, the Institute of Institute of Environmental Protection – National Research Institute, and the Institute of Technology and Life Sciences – National Research Institute in Falenty. After receiving positive opinions on the tested digestate, an application for a marketing permit can be sent to the Ministry of Agriculture and Rural Development.

The approval process is lengthy because digestate and its properties are different at each installation. It largely depends on the substrates used for biogas production. Nevertheless, it is an activity that contributes to a substantial extent to the broader use of digestate on arable land. It is considered a full-fledged fertilizer, and its method and scope of application are determined by the regulations obtained during the approval process. Factors such as dosage, possible period of use, and type of agricultural crop on which digestate can be used are determined. There is no need to strictly record the number of plots on which this fertilizer is used. One more crucial advantage is that it positively affects the public's opinion because it is inevitable that independent state institutions have thoroughly evaluated such fertilizer, and its effects on plants and the environment have been thoroughly documented.

15.2.4 Permit to Introduce to the Electricity Distribution Network

The main product of the anaerobic digestion process is methane. The nature of the biogas installation determines its use because the infrastructure that uses the energy potential of biogas must be adapted to it. In the vast majority, it is used as fuel for engines, which drive generators producing electricity from methane combustion. At the design stage, the size of the devices should be considered in such a way as to maximize the potential of the installation. Each biogas plant uses electricity and heat for its needs, i.e. from driving mixers and pumps, through all installation components, to lighting. For this reason, a few percent of the energy produced is usually used in biogas plants.

Biogas plants are a desirable energy source because, with their proper operation and reliability of devices, they are a stable source of energy, positively affecting the stabilization of the distribution network. At the installation design stage, the fundamental element is obtaining a permit for connection to the electricity distribution network. It is a document that defines the potential possibilities of the biogas plant in terms of electricity produced. This document indicates the technical possibilities of connecting a biogas plant to the distribution network of a given operator. The place of connection, cable length, protection, and other aspects are strictly defined. Before the biogas plant is fully operational, electricity quality tests are conducted at the start-up stage. They are the basis for the final permit to feed electricity into the network. All these aspects aim to support the fullest possible protection of both the producer and the recipient of renewable electricity sources.

15.3 Biogas Plant Components: A Case Study from Poland

A biogas plant, like any installation, consists of individual components that form a properly functioning whole. Starting with the collection of substrates for biogas production, the biogas plant must have the proper infrastructure that allows for proper storage and preparation for feeding a given mass. It is worth mentioning the proper silos, hardened yards, halls, and tanks that are used for storing and homogenizing the substrate. Legal aspects are also essential in this case because when using waste from the agri-food industry, the guidelines of the Act of 14 December 2012 on Waste [11], veterinary regulations, and many other aspects required by law must be followed. Significant elements are supplied for treating waste animal tissue, i.e. a sterilization hall with devices, a receiving trough, and a cooling tank. The situation is similar in the case of the production process, i.e. fermentation tanks, power generators, and at the very end of the process, tanks for storing and distributing digestate. Appropriate interaction and reliability of each element prove the biogas plant's correct and effective operation.

15.3.1 Hall for Receiving and Processing Slaughterhouse Waste

Slaughterhouse waste is processed in the receiving and processing hall. It is a hall with an area of $560\,m^2$, of which approximately $425\,m^2$ is the sterilization area. The building is made of prefabricated walls, without a basement, with a flat roof. There is a reception point in the hall, a line for categories 2 and 3 slaughterhouse waste processing and sterilizing, and a vehicle wash.

15.3.2 Substrate Storage Yard

This facility is a reinforced concrete structure closed on three sides with walls designed to receive and temporarily store solid substrates from the outside. Substrates are delivered from outside and stored as a prism in the yard. The biogas plant has three silo tanks with dimensions of $77\,m \times 60\,m$. Substrates from the yard are transported to the mixing tank. Waterproof concrete is used to construct the retaining walls and the bottom plate (no leakage of leachates into the ground), and it is protected against the aggressive action of substrates. The surface of the bottom asphalt slab of the silos is additionally covered with geotextile foil. The bottom slab is made with slopes of 1.5–2% toward the sewage grates, through which leachates are discharged to the technological network of the biogas plant. The substrates are stored in a tight prism, reducing the environmental impact and protecting against weather conditions.

15.3.3 Solid Substrate Dispenser

Substrates are loaded into the solid's dispenser by the loader. The dosing container is equipped with a transport system. The material contained in it is moved to the hopper and then to the device, forcing the substrate directly into the buffer-mixing tank. The transport system is made as a moving floor, while the dosing system has the equipment necessary for proper operation, such as a screw conveyor, shut-off, and check valves.

15.3.4 Receiving Buffer Tank for Liquid Substrates

For the preliminary storage of liquid substrates, two buffer tanks for liquid substrates are built, with a capacity of approximately 200 m^3 each. The liquid phase, e.g. pig slurry and distillery decoction, is dosed from a buffer tank for liquid substrates of the combo type, i.e. tank in a tank. It enables the proper preparation of substrates by equalizing the temperatures: the descent of the hot distillery decoction and heating the cold liquid manure. The tank is insulated with a layer of the thermal coat with trapezoidal sheet metal construction protecting the insulation against the harmful effects of external factors. The tank is equipped with connection pipes for receiving material from the water tanker. Inspection hatches are installed on the tank's roof, ensuring the possibility of conducting service works.

15.3.5 Solid Substrate Buffer Tank

Due to the possibility of using water for the fermentation process, two receiving buffer tanks of liquid substrates are connected to the water supply network. The tank has connectors that enable material transfer from the water tanker. Each of the tanks has a capacity of approximately 200 m^3. Liquid substrates are received directly from slurry tankers or tank trucks through tight spigot connections, eliminating the possibility of odors escaping.

15.3.6 Mixing Buffer Tank

The mixing buffer tank is made of reinforced concrete and placed on a concrete bottom plate. The tank, with a capacity of approximately 280 m^3, is insulated with a thermal insulation layer with sheet metal construction. The tank has the necessary equipment: agitators, material connection pipes from the solid phase feeder, liquid substrate storage tank, and fermenters. The tank has a hatch that allows direct solid waste charging when the substrate feeder is not working.

15.3.7 Buffer and Mixing Tank

In buffer and mixing tanks with a capacity of approximately 380 m^3, streams of various solid and liquid substrates are thoroughly mixed while sterilized slaughterhouse waste is diluted. It enables the proper preparation of substrates for the fermentation process and ensures the best conditions for the decomposition of organic matter and biogas production. The tank is placed on a reinforced concrete bottom slab and thermally insulated with a layer of insulation with trapezoidal sheet metal covering protecting the insulation layer against the harmful effects of external factors. The tank has agitators, connectors for receiving material from the solid phase feeder, a liquid substrate storage tank, a sterilization line, and fermenters.

15.3.8 Technological Steam Generator

Technological steam is used as a heat stream in the sterilization process. The steam generator is placed outside the sterilization room in a separate container. The steam transfer installation allows transportation to the steam collector inside the sterilization hall. From the collector, steam is transferred to individual receiving devices. In receiving heat from technological steam, the receiving devices produce condensate, which is transported through the pipeline system to the condensate tank and reused in the steam generator. The heat exchange area is 104 m^2, the maximum operating temperature is 550 °C, and the maximum operating pressure is 0.5 bar.

15.3.9 Main Pumping Station

The mixed input material from the buffer-mixing tanks and the dilution of slaughterhouse waste is transported to the technical building, where it is pumped through piping and distributed to two fermentation tanks. Instrumentation for fermentation tanks is installed in the technical building. There are transfer installations of heat exchangers inside the fermenter connected to the circulation system, the role of which is to maintain the proper temperature of the fermentation process in the fermentation tank. The size of the designed heating system with heat exchangers ensures the correct temperature of the fermenting mass. The pumping station uses a system of pumps to pump the initial substrates to the fermentation tanks. In the building of the pumping station, there is a liquid substrate feeding system, distribution systems for transmission installations, and transmission installations for the heating and cooling installation of fermenters.

15.3.10 First-stage Fermentation Tanks

The dosing of substrates to the fermentation tank takes place through a hydraulic dosing system connected in a technological line with a solid phase feeder, a buffer tank for liquid substrates, mixing tanks, dilution tanks, and a heating system in the form of heat exchangers. A cylindrical tank made of reinforced concrete with a suitable strength class protects the structure against the corrosive influence of the fermentation environment. The fermenter is placed on a reinforced concrete bottom slab with a foundation. The tank's roof and the fermenter's gas part are protected against the harmful effects of biogas and other compounds produced in the fermentation process. The roof of the tanks is made of a rigid reinforced concrete structure. There are wall-mounted connection ports for gas transport, material transport, tank emptying, connection ports, and sampling stations with manual ball valves and safety valves. The fermenter is also equipped with other elements necessary for proper functioning, such as pressure, temperature, flow, and liquid-level sensors, as well as nonreturn and manual shut-off valves. On the tank roof or the wall, inspection hatches are mounted, supplying the possibility of performing service work, and a sight glass enables viewing of the processes inside the fermenter. The tank is thermally insulated with a layer of thermal insulation with trapezoidal sheet metal construction protecting the insulation layer against the harmful effects of external factors. To avoid sedimentation of the substrate, mixers are installed in the chamber: one vertical and two submerged. The mixing speed can be adjusted using an inverter. The capacity of each of the two tanks is approximately 3600 m^3.

Closed fermentation tanks in a biogas production installation have two main functions. In contrast, the anaerobic digestion process takes place in them, and in contrast, the tanks, thanks to covering them with a gas-tight roof, serve as buffer tanks for the resulting biogas.

In the tank, the fermentation substrate decomposes at temperatures between 38 and 42 °C. The fermenter works continuously, and the fermentation substrate is mixed to produce a uniform mixture. Material is supplied through a feeder that ends in the fermentation tank above the liquid level.

A mixture of prepared substrates is added to the fermentation tanks. There are agitators in each tank to mix the substrate, ensuring a uniform mix during fermentation. Agitators are only used in submerged mode. The complete control system in the technical building remotely switches on and off these mixers. The fermentation tank is equipped with a heating system to compensate for heat losses and to heat the freshly added raw materials. Substrates are heated using a central heating system installed inside the fermentation tank. The substrate level is controlled by a level sensor mounted in the tank and connected to a computer system in the technical building.

The tanks are equipped with solutions that eliminate the effects of system leaks or unsealing:

- operation of overpressure and underpressure protection systems and triggering an alarm when limit values are exceeded,
- protection against overfilling of tanks, which, when the maximum level of filling is exceeded, turns off the supply pumps and triggers an alarm,
- preventing the formation of foam in tanks,
- possibility of pumping the contents of the tank to any of the fermentation tanks and digestate tanks/lagoons.

15.3.11 Second-stage Fermentation Tank (3900 m^3) with Biogas Tank (1800 m^3)

The fermentation process takes place in two stages in sealed, closed tanks. The first stage occurs in two fermentation tanks, where a fresh batch of feed displaces the predigested and decomposed organic substance into one second-stage fermentation tank. Dosing material from the first-stage fermentation tanks to the second-stage fermenter takes place through a hydraulic dosing system connected in a technological line with a heating system in the form of heat exchangers or by gravity. A tank with cylindrical characteristics made of reinforced concrete with a proper strength class protects the structure against the corrosive effects of the fermentation environment. The gas part of the fermenter is protected against the harmful effects of biogas and other compounds produced in the fermentation process. The tank's roof comprises a two-layer plastic membrane – biogas storage. The membrane is connected tightly to the cylinder of the second-stage digester. The internal gas-tight membrane made of a special polyester PVC film collects biogas above the surface of the fermented liquid. This membrane is made especially for biogas storage. The second external, UV-resistant coating, reinforced with PVC foil fabric, is climatic protection that prevents the adverse effects of weather conditions. During gas filling, the membrane goes up; during gas withdrawal, the membrane goes down, according to the gas filling state. A device for measuring the filling level of the gas tank is supplied to indicate the filling level of the gas tank. It is used for remote signs and controlling the blower for gas transport and collection. Connection pipes for gas transport, material transport from first-stage fermenters, emptying the tank and connection pipe, and sampling stations with manual ball valves and safety valves are mounted on the walls. The fermenter is also equipped with other elements necessary for proper functioning, such as pressure, temperature, flow, and liquid level sensors, as well as check and manual shut-off valves. An inspection hatch is mounted on the wall of the tank, ensuring the possibility of performing service work, and a sight glass enables viewing of the processes inside the fermenter.

15.3.12 Condensing Circuit

The biogas produced is warm and humid, so it must be cooled and dried, as moisture could damage the internal combustion engine. Drying of biogas for combined heat and power systems (CHP) is done through a gravel filter for biogas coarse filtration, preventing foam from entering the biogas discharge lines and separating the condensate. The use of such solutions is an essential requirement of the CHP manufacturer. Leachates from the drying process are collected in drainage wells, returned to the substrate circulation system, or pumped to tanks included in the installation.

15.3.13 Biogas Refining System

Biogas has trace amounts of hydrogen sulfide, which must be removed before the gas can be used. The installation is equipped with a biological desulfurization system. In the desulfurization process, hydrogen sulfide turns into sulfate and elemental sulfur. The desulfurization process takes place in the first- and second-stage fermentation tanks by adding oxygen thanks to the membrane pump dispensing a maximum of 10% of the air volume. The flow of biogas through the desulfurization installation is approximately $1000\,m^3\,h^{-1}$.

15.3.14 Cogeneration Modules

Biogas, after purification, passes to cogenerators. These devices include a gas engine and an asynchronous power generator connected to it by a drive shaft. As a result of the combustion process in the engine chambers, mechanical energy is transferred to the drive shaft, and the cooling circuit receives thermal energy. The electricity is sent to the transformer station via a cable connection to the main power supply point indicated by the energy consumer. The generated electricity is supplied to the power network and sold. Thermal energy produced by the cogeneration engine is mainly used for the biogas plant's needs, e.g. for heating fermentation tanks and the production of technological steam used in the sterilization process. Any surpluses of unused heat are released into the environment. A gas sucker-blower is also provided in the piping system to ensure the proper flow rate and pressure of the gas supplied to the cogenerator. Independent cogeneration units with capacities of 600 and 1200 kW enable the production of electricity and heat from biogas combustion. The gas engines are mounted in a container structure, which allows for reducing the level of emitted noise, and in an exhaust silencer that reduces the noise of emitted exhaust gases.

15.3.15 Digestate Storage Reservoirs

Digestate should be stored in the period when fertilizing the soil is not possible. A maximum of $38{,}800\,m^3$ of digestate can be stored in the biogas plant, which allows for its storage for approximately 120 days. In the case of using a digestate separator and solid fraction management, this period may be extended. Two lagoons are made of membrane layers founded on the hardened ground and an earth embankment: lagoons equipped with a leachate buffer well and emergency and control installation. Through the connection in the anchor sleeve, the upper and lower membranes ensure tight separation of the digestate pulp from the external environment. The membranes are connected through double thermoair welds. A draw-off well is made for each of the lagoons. The wells enable the filling and pumping out of the digestate pulp and are connected to the lagoon through PVC pipes and a prefabricated reinforced concrete spigot.

15.3.16 Biogas Torch

The torch is used in emergencies and during engine servicing when the storage capacity of the gas tank determines the need to burn biogas. The assumed capacity of the torch is up to 1000 m^3 h^{-1}. The gas installation connection to the torch is made before and after the desulfurization system. Excess gas, which in such a situation has no place to be stored, is burned to avoid its emission to the atmosphere and to reduce the risk of explosion.

15.3.17 Biofilter

A biological filter is used to purify the exhaust air and eliminate odors from the slaughterhouse waste reception hall – a biofilter is adapted to automatic operation. This device makes it possible to use the natural abilities of microorganisms to convert odor nuisance and environmentally harmful substances contained in the exhaust air into environmentally neutral products. The device consists of a container with a technical room, a fan, a sprinkler column, and a control unit. The fan sucks in the odorous exhaust air, then it is pumped to the sprinkler column and, after optimal humidification, fed to the biofilter module. Contaminated air is microbiologically cleaned as it passes through the filter material. The stream of purified air is discharged directly into the atmosphere. The active area is adapted to the capacity of the exhaust ventilation system, not less than 3000 m^3 h^{-1}.

15.4 Functioning of a Biogas Plant Processing Problematic Waste: A Case Study from Poland

The installation in question has an electric power of 1.8 MW. The description presented in the chapter concerns specific constructional, technological, and organizational solutions implemented on the facility premises.

In the area of the described plant, there are facilities and installations used for the following:

- storage, preparation, and dosing of substrates,
- conducting the anaerobic digestion process,
- storage of the produced biogas,
- digestate storage,
- converting biogas into electricity and heat (Figure 15.1).

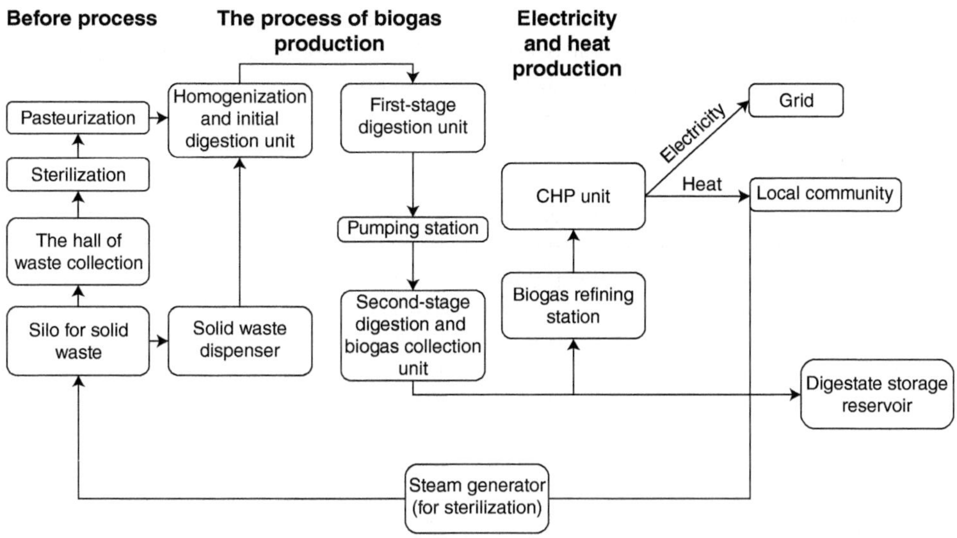

Figure 15.1 Functioning of an agricultural biogas plant (own study).

15.4.1 Searching and Obtaining Substrates

An agricultural biogas plant is mainly associated with corn silage, slurry, and purpose-grown plants for biogas production. Different assumptions must guide an installation based on agricultural and food industry waste. First, the situation regarding the raw material type, composition, and biogas efficiency should be named. It is the basis for considering the delivery method, frequency, and amount of a given type of waste. Such a mass cannot have chemical additives that may adversely affect the process of anaerobic digestion. In terms of biogas yield, slaughterhouse waste is a perfect substrate. Its remarkably high methane efficiency means that the amount needed to produce a certain amount of energy is lower than that needed for plant substrates. Biogas plants often use liquid manure, which results from cooperation with animal farms and, thus, slaughterhouses and processing plants. It turns out that even small slaughterhouses or processing plants produce waste that is successfully used in a biogas plant after its prior preparation (pasteurization or sterilization). To produce biogas, waste such as food and animal tissue such as blood, viscera, and other tissues are unsuitable for further processing.

The search for suitable substrates requires extensive knowledge of the provisions of the Act of 14 December 2012 on Waste [11] and the Act of 29 January 2004 on Veterinary Inspection [12]. The situation in the market is changing dynamically, and deliveries cannot be precisely planned in the long term. One of the essential elements is the proper identification of the plant in terms of the possibility of generating waste. Second, quantities and transport possibilities should be named. Another element is negotiating the financial conditions for waste delivery. Often, waste producers claim they do not have to bear the costs of waste disposal in a biogas plant. Charging for disposal is an additional source of income for the biogas plant, which significantly increases the business's profitability. Contractors using waste can be sure that their goods go to an installation with all the necessary permits and can be managed following the law. It should also be noted that the methanogenesis process often uses waste mass that is no longer suitable for other processing. Without a biogas plant, finding another profitable solution is not easy.

15.4.2 Receiving, Storage, and Processing of the Substrate, Feeding of Raw Materials

The waste obtained for agricultural biogas production must come from agriculture and the agri-food industry. Institutions supervising the work of agricultural biogas plants impose an obligation to properly select waste codes. In contrast, the methanogenesis process requires the proper choice of their type, chemical composition, consistency, and processing method. The employees of the biogas plant are obliged to obtain waste, which ensures the conditions specified in the national law but mainly in the relevant permits, such as a permit for waste processing or an integrated permit issued for a given installation. Waste arriving at the facility must be identified regarding its unloading process. It should be borne in mind that connecting proper devices for pumping out the organic matter and unloading it to proper troughs or feeders is possible. The quantities of supplied substrates must follow the processing capabilities, i.e. whether a given amount can be processed and fed to the process.

Organic waste are characterized by the emission of unpleasant odors, and its consistency often makes it impossible to store. Therefore, installations of this type do not have a storage and warehousing permit. They must be fed directly to the process in specially adapted halls. It prevents the escape of odors and access by people and animals. Halls equipped with biofilters for air purification are under the strict supervision of the biogas plant and veterinary services. They are also subject to meticulous records by finding the weight, supplier, code, and vehicle registration number.

After recognizing the amount of waste and potential methanogenic potential, the waste should be verified in terms of the given veterinary category. A veterinarian, assessing the risk to human and animal health of a given waste, assigns a category from 1 to 3. In biogas installations equipped with a line for sterilization and pasteurization, the law allows the use of animal byproducts of categories 2 and 3. Category 1 is provided for combustion in Polish conditions. Categorization also shows how a given batch of waste is managed. It is transported by specialized equipment to the biogas plant. Such transport and equipment are also needed to obtain proper permits issued by authorities such as the County Office and the County Veterinary Officer. Identified and properly marked waste enters the biogas installation in specially designated reception zones, where it is taken to specially designed troughs and tanks, from which it is further processed. A principal element is also the fragmentation of recycled material. It is essential in terms of technology for a given biogas plant and the method of feeding the substrate to the fermentation tanks. In units where the dosing of the substrate is based on liquid substrate and feeding it through pumps – this is especially significant. Choosing suitable pumps is crucial in this situation to prevent clogging of the system.

After shredding, the waste passes through a metal detection and separation system. The shredded material, devoid of metal elements, is transferred to a pasteurizer powered by a stream of hot water at a temperature of 90/70 °C, which comes from cogeneration. In the pasteurizer, the slaughterhouse waste is heated to a temperature not exceeding 75 °C. The

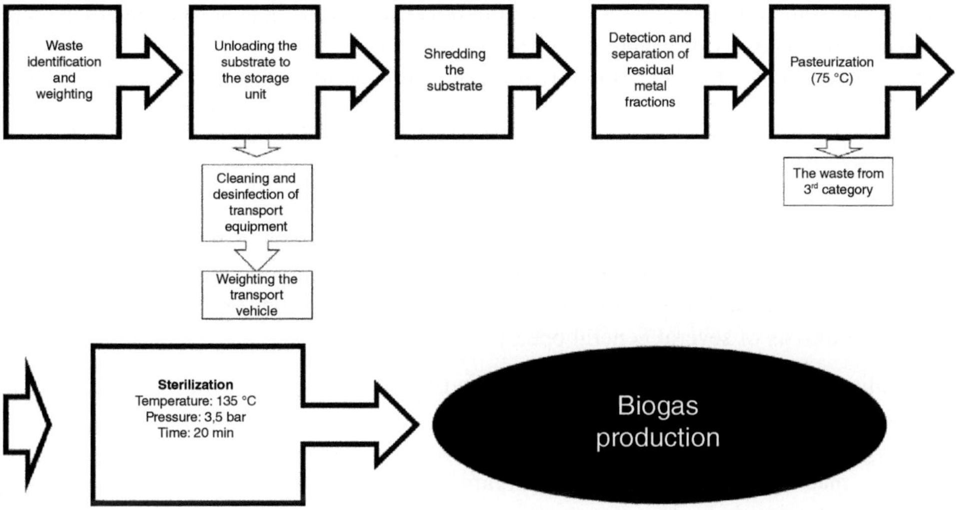

Figure 15.2 Scheme of handling waste requiring hygienization process (own study).

preheated waste is transported from the pasteurizer through screw conveyors. It enters the sterilizer with a working capacity of 8 m^3. Two sterilizers are used in the line to enable continuous intake of technological steam needed to sterilize waste, thanks to which it is possible to obtain favorable and even heat reception. Steam in the sterilization process supplied to sterilizers has a pressure of approximately 5 bar, corresponding to a temperature of 151.5 °C. Such a temperature on the sterilizer's jacket allows for a pressure of 3 bar inside the sterilizer (in the working tank) and a temperature of 133 °C for 20 minutes. Each sterilizer has a stirrer that allows the waste to be heated evenly during the sterilization process and the sterilized substrate to be unloaded. After the waste sterilization process, the unloading process takes place, and the sterilized slaughterhouse waste goes to the cooling tank and is then pumped to the biogas installation.

When filling the tank with material, it is heated. A stirrer is used in the destructor to mix the raw material and to collect heat from the heating surface of the destructor evenly. After filling the destructor, the waste is further heated until it reaches a temperature of 133 °C and a pressure of 3 bar. From this moment, the 20-minute sterilization period begins. Throughout this period, technological steam is supplied to the heating surface of the destructor (the destructor is heated). Saturated steam is generated in the destructor. The outlet valves are opened during cooling to make possible "dirty steam" to get to the cooler, where it condenses.

Figure 15.2 shows the method of dealing with categories 2 and 3 waste – the method of proceeding guarantees compliance with the requirements of the Act of 14 December 2012 on Waste [11] and the Act of 29 January 2004 on Veterinary Inspection [12].

15.4.3 Energy Production and Biogas Management

The biogas plant's task is to produce biogas and generate electricity and heat. Plant and animal waste, as well as other biodegradable waste, is an energy resource. The retention time and loading of the tank with substrates in the fermentation tank are adapted to the type of feed to guarantee its complete decomposition.

A biogas plant can be divided into two components based on technological processes:

- the biotechnological part, where the process of high-efficiency biogas production and the production of fertilizer from organic, renewable raw materials is conducted,
- the energy part, where the process of biogas conversion into electricity takes place.

The generated electrical power is received via a 15-kV cable line and a transformer station. Mesophilic fermentation is applied, which takes place at temperatures between 38 and 42 °C. An installation for high-efficiency biogas production from organic, renewable raw materials consists of several general operating units:

- facilities for storage, preparation, and feeding (dosing) of substrates for the biogas production process,
- integrated, closed fermentation tanks and gas tanks in which the fermentation process takes place, i.e. the conversion of organic matter into biogas (biomethane) and its storage, final tanks/lagoons for storing fermented substrates, the so-called fermentation residue, which is then used as a soil improver,
- facilities for the treatment and feeding of biogas for further highly effective processing,
- control and measurement equipment and automation used for qualitative and quantitative supervision of biochemical and mechanical processes of biogas production.

15.4.4 Digestate Management

The issue of fertilization is critical because it often determines the economic success of agricultural production. It becomes even more significant because the method of fertilization determines the quality and health value of cultivated plants for humans and animals. Depending on the size of the biogas installation, the digestate produced is, in contrast, a challenge for the installation operators because its amount depends on the amount of processed substrates. The annual amount of digestate produced in this installation is approximately 80,000 tons. The digestate mass enters the tanks, from which it is distributed to agricultural field farmers. The digestate produced from waste has been approved for trading by the Ministry of Agriculture and Rural Development. Considering the recommendations in the instructions for the use issued by the Institute of Fertilization and Soil Science, the annual production of digestate is enough to fertilize approximately 2000 hectares of arable land.

Digestate is used in agrotechnical periods when it can be mixed with soil on arable land. It is usually done before sowing cereals, rapeseed, sugar beet, or maize. Before removal, the digestate batch is thoroughly mixed, pumped into tankers, and transported to the fields. It is mainly transported by road transport, applied to the soil in the fields, and thoroughly mixed with it. In Poland, since there are very few biogas installations, the use of digestate in agriculture is not common. Interest increases only after applying and convincing farmers of their positive effects. In the era of the energy crisis, when mineral fertilizer prices reach exceedingly high levels, digestate has become an excellent alternative in fertilization. While it was a big problem for such installations in distribution, it has recently become in request. It should be noted that the beneficial effect of digestate on soil microflora is positive. It favors the reconstruction of beneficial microorganisms and stimulates the growth and development of plants. It also contributes to the faster decomposition of organic matter to a considerable extent.

15.4.5 Management of an Agricultural Biogas Plant

Considering biogas installations, the very essence of the process that takes place in the installation should be considered. For the process to occur, several conditions necessary for creating biogas must be met. It refers to anaerobic conditions that must prevail in fermentation tanks, proper temperature, pH, etc. All this is to create suitable conditions for the life of bacteria because they manage the proper course of the process, i.e. the demineralization of organic matter, leading to methane production. The purpose of the installation is to conduct the process so that methane production is as efficient as possible, not only in terms of the amount of biogas produced but also its chemical composition. It is desirable to have the highest percentage of methane in the biogas, but at the same time to reduce the production of undesirable compounds for the biogas, i.e. hydrogen sulfide, volatile organic compounds, or siloxanes. These components are a significant obstacle in properly operating devices such as cogeneration units. Biogas plants, the charge of which is based on waste organic matter of both plant and animal origin, are especially forced to control the relevant process parameters. High variability, both quantitative and qualitative, of the used charge forces constant monitoring of the proper balance in terms of the main elements, such as carbon and nitrogen, and the pH of the fermentation mass [15]. It determines the condition of biogas production.

Running a biogas plant involves work on many aspects. Each day begins with checking the basic parameters of the biogas production process. Foremost, the amount of biogas in the storage tank is checked, which proves the condition of the process, and at the same time, the power of the cogeneration engines is checked. Other analyzed parameters include the temperature in the fermentation tanks and biogas pressure. Basic tests of the fermentation mass, such as pH, fos/tac, dry matter content, and the amount of volatile fatty acids, are also performed daily. Monitoring these essential elements is critical because the waste constantly enters production; it is dosed at once into the process. Using too much waste can cause disturbances in biogas production, while a comparable situation occurs in the case of too small quantities. Care should be taken if the waste comes from numerous suppliers, and obtaining a homogeneous, stabilized mass is not easy. The threat of harmful chemical compounds in the waste is also worth mentioning, which may negatively affect the process.

The subsequent step in the installation operation is to check and monitor all mechanical elements, both in the electricity production process and partly responsible for dosing substrates, pumping, mixing, and other process elements. Reacting rapidly to all faults and failures is crucial because even the slightest failure can lead to further irregularities.

A vital element of the functioning of a biogas installation of this type is the constant search for waste that can be used. Since the installations do not have permits for storing organic waste, it is necessary to organize the frequency and quantity of deliveries in a way that allows for proper operation. Given the high supplier turnover, this task is complicated to implement.

15.5 Summary

Renewable energy sources are becoming increasingly popular in Poland and worldwide. Social awareness of environmental protection or sustainable development issues is increasing [16]. The use of technical and technological progress certainly makes it possible to meet the needs while significantly reducing the negative environmental impact. However, for this to happen, some steps must be taken. At the beginning of the agricultural biogas sector development in Poland, biogas production was based on maize silage and liquid manure [17]. Natural fertilizers such as pig or cattle slurry and manure are often used as additives. Biogas installations fulfill the correct function assumptions for using liquid and solid manure. In this way, they reduce the odor of the substrates used at the storage stage and in the field. As a result of anaerobic digestion, odor nuisance components, including hydrogen sulfide, the main factor influencing odor nuisance, are eliminated. After methanogenesis, these materials are better absorbed by plants, and undesirable processes related to the decomposition of organic matter, which contribute to carbon dioxide emissions, must no longer occur in the soil.

The choice of substrates is a critical issue in agricultural biogas production [18]. Popular biodegradable substrates include maize silage, liquid manure, and waste from the agricultural and food industries [19]. It should be emphasized that raw materials from crops and waste can be valuable substrates for agricultural biogas production. It is because organic matter, including waste, consists of the same elements as traditional substrates. It includes proteins, fats, sugars, starch, fiber, cellulose, and other organic compounds from which living organisms are built. These compounds are food for the microorganisms involved in the anaerobic digestion process, including methanogenic bacteria. Considering the aforementioned information, biogas installations should be entirely unusually perceived. This installation should be considered a place for waste management as part of the recovery process and not as power plants that take such desirable products from agriculture as plants or their parts. Biodegradable organic waste is an embarrassing issue for everyone. Left in landfills or improperly managed because of rot decomposition, it generates vast amounts of leachate, which ends up in both groundwater and surface waters, as well as odors. In sealed tanks, the organic matter contained in the waste undergoes decomposition and demineralization. As a result of a series of complex processes, biogas or biomethane is obtained, which is an ideal fuel for producing electricity and heat [20]. Recently, there have been increased possibilities for its use.

The society is often unaware of how much biodegradable wastes are generated daily in the agri-food industry. Such installations work perfectly near plants producing food and everything related to it [21]. Additionally, municipal waste related to human existence can be successfully used to produce energy in a broad sense. This situation has an ecological dimension, i.e. waste is disposed of in adequately controlled conditions and impacts the installation's economic conditions. There is no need to buy maize or other substrates. Recycling can typically be combined with charging a fee for disposal, which significantly affects the profitability of the biogas plant.

Such a system can be referred to as a renewable energy source that significantly contributes to the circulation of organic matter and nutrients in nature. As a result of the methanogenesis process, only carbon in the form of methane is extracted from the processed material. All the rest goes to the fields in the form of digestate.

References

1. Kowalczyk-Juśko, A., Pochwatka, P., Zaborowicz, M. et al. (2020). Energy value estimation of silages for substrate in biogas plants using an artificial neural network. *Energy* 202: 117729. https://doi.org/10.1016/j.energy.2020.117729.
2. Czekała, W. (2022). Biogas as a sustainable and renewable energy source. In: *Clean Fuels for Mobility* (ed. G. Di Blasio, A.K. Agarwal, G. Belgiorno, and P.C. Shukla), 201–214. Singapore: Springer https://doi.org/10.1007/978-981-16-8747-1_10.
3. Emmanuel, J.K., Nganyira, P.D., and Shao, G.N. (2022). Evaluating the potential applications of brewers' spent grain in biogas generation, food and biotechnology industry: a review. *Heliyon* 8 (10): e11140. https://doi.org/10.1016/j.heliyon.2022.e11140.
4. Wirth, R., Kovács, E., Maróti, G. et al. (2022). Characterization of a biogas-producing microbial community by short-read next generation DNA sequencing. Characterization of a biogas-producing microbial community by short-read next generation DNA sequencing. *Biotechnology for Biofuels* 5: 41. https://doi.org/10.1186/1754-6834-5-41.
5. Nsair, A., Onen Cinar, S., Alassali, A. et al. (2020). Operational parameters of biogas plants: a review and evaluation study. *Energies* 13: 3761. https://doi.org/10.3390/en13153761.
6. Eccleston, R. and Bongards, M. (2020). Determining conditions of intermittently fed digesters from biogas production rate data. *Chemical Engineering and Technology* 43 (1): https://doi.org/10.1002/ceat.201900354.
7. Ounsaneha, W., Rattanapan, C., Suksaroj, T.T. et al. (2021). Biogas production by co-digestion of municipal wastewater and food waste: performance in semi-continuous and continuous operation. *Water Environment Research* 93 (2): https://doi.org/10.1002/wer.1413.
8. Directive (EU) 2018/2001 of the European Parliament and of the Council of 11 December 2018 on the promotion of the use of energy from renewable sources (recast).
9. Act of 10 April 1997 – Energy law (Journal of Laws 1997 No. 54 item 348).
10. Act of 20 February 2015 on Renewable Energy Sources (Journal of Laws of 2015, item 478).
11. Act of 14 December 2012 on Waste (Journal of Laws 2013 item 21).
12. Act of 29 January 2004 on Veterinary Inspection (Journal of Laws 2004 No. 33 item 287).
13. Regulation (EC) No 1069/2009 of The European Parliament and of The Council of 21 October 2009 laying down health rules as regards animal by-products and derived products not intended for human consumption and repealing Regulation (EC) No 1774/2002 (Animal by-products Regulation).
14. Czekała, W. (2021). Solid fraction of digestate from biogas plant as a material for pellets production. *Energies* 14: 5034. https://doi.org/10.3390/en14165034.
15. Czekała, W., Jasiński, T., Grzelak, M. et al. (2022). Biogas plant operation: digestate as the valuable product. *Energies* 15: 8275. https://doi.org/10.3390/en15218275.
16. Czekała, W., Tarkowski, F., and Pochwatka, P. (2021). Social aspects of energy production from renewable sources, Społeczne aspekty produkcji energii ze źródeł odnawialnych. *Problemy Ekorozwoju* 16 (1): www.doi.org/10.35784/pe.2021.1.07.
17. Koryś, K.A., Latawiec, A.E., Grotkiewicz, K., and Kuboń, M. (2019). The review of biomass potential for agricultural biogas production in Poland. *Sustainability* 11: 6515. https://doi.org/10.3390/su11226515.
18. Seick, I., Vergara-Araya, M., and Wiese, J. (2022). Model-based analysis to increase the substrate efficiency of a biogas plant. *Chemical Engineering and Technology* 45: 00. https://doi.org/10.1002/ceat.202100370.
19. Czekała, W. (2018). Agricultural biogas plants as a chance for the development of the agri-food sector. *Journal of Ecological Engineering* 19 (2): https://doi.org/10.12911/22998993/83563.

20. Daniel-Gromke, J., Rensberg, N., Denysenko, V. et al. (2018). Current developments in production and utilization of biogas and biomethane in Germany. *Chemie Ingenieur Technik* 90(1-2). https://doi.org/10.1002/cite.201700077.
21. Tamburini, E., Gaglio, M., Castaldelli, G., and Fano, E.A. (2020). Biogas from agri-food and agricultural waste can appreciate agro-ecosystem services: the case study of emilia romagna region. *Sustainability* 12: 8392. https://doi.org/10.3390/su12208392.

Index

Biogas Plants: Waste Management, Energy Production and Carbon Footprint Reduction, First Edition. Edited by Wojciech Czekała